PLENTIFUL ENERGY

The Story of the Integral Fast Reactor

The complex history of a
simple reactor technology,
with emphasis on its
scientific basis for non-specialists

CHARLES E. TILL and YOON IL CHANG

ISBN: 978-1466384606

Library of Congress Control Number: 2011918518

Cover design and printing by CreateSpace, a subsidiary of Amazon

DEDICATION

To Argonne National Laboratory and the team of the many hundreds of scientists and engineers and other disciplines and trades who dedicated a substantial portion of their working lives to making real the many benefits arising from and inherent in the technology of the Integral Fast Reactor.

CONTENTS

FOREWORD

On a breezy December day in 1903 at Kitty Hawk, N.C., a great leap forward in the history of technology was achieved. The Wright brothers had at last overcome the troubling problems of 'inherent instability' and 'wing warping' to achieve the first powered and controlled heavier-than-air flight in human history. The *Flyer* was not complicated by today's standards—little more than a flimsy glider—yet its success proved to be a landmark achievement that led to the exponential surge of innovation, development and deployment in military and commercial aviation over the 20th century and beyond.

Nonetheless, the *Flyer* did not suddenly and miraculously assemble from the theoretical or speculative genius of Orville and Wilbur Wright. Quite the contrary—it was built on the back of many decades of physical, engineering and even biological science, hard-won experience with balloons, gliders and models, plenty of real-world trial-and-error, and a lot of blind alleys. Bear in mind that every single serious attempt at powered flight prior to 1903 had failed. Getting it right was tough!

Yet just over a decade after the triumphant 1903 demonstration, fighter aces were circling high above the battlefields of Europe in superbly maneuverable aerial machines, and in another decade, passengers from many nations were making long-haul international journeys in days, rather than months.

What has this got to do with the topic of advanced nuclear power systems, I hear you say? Plenty. The subtitle of Till and Chang's book "*Plentiful Energy*" is "*The complex history of a simple reactor technology, with emphasis on its scientific bases for non-specialists.*" The key here is that, akin to powered flight, the technology for fully and safely recycling nuclear fuel turns out to be rather simple and elegant, in hindsight, but it was hard to establish this fact—hence the complex history. Like with aviation, there have been many prototype 'fast reactors' of various flavors, and all have had problems.

Stretching the analogy a little further, relatively inefficient balloons, airships and gliders were in use for many decades before powered flight became possible, even though people could see that better ways of flying really did exist (they only had to look up in the sky, at the birds). Powered aircraft allow people to travel hundreds of times faster, and more safely, than lighter-than-air devices. Similarly, the type of nuclear reactors we have used commercially for decades, although far superior to other methods of generating electricity, have harnessed but a tiny fraction of the potential locked away in uranium. To get at that, you need a very different approach—a nuclear fission *Flyer*. Enter the integral fast reactor (IFR).

This wonderful book by fast reactor pioneers Charles Till and Yoon Chang, two of the foundational developers of the IFR during the fabulously productive years of research and development at the Argonne National Laboratory from the 1980s to early 1990s, explains in lucid terms the historical, philosophical and technical basis for truly sustainable nuclear energy. It's quite a story.

Imagine a reactor that passively responds to critical stressors of the kind that befell Three Mile Island, Chernobyl and Fukushima by shutting down without human operators even needing to intervene. Or one that includes a secure recycling and remote fabrication system that, almost Midas like, is able to turn uranium or even old 'nuclear waste' from contemporary reactors into an inexhaustible (and zero-carbon) fuel, as well as simultaneously solving the socio-political problem of long-term disposal.

Once you've read this book, you'll understand how this technological wizardry is performed and why other options—those alternatives to the *Flyer*—never quite worked out. Moreover, you'll have a much deeper appreciation of the true *potential* of fission energy in a low-carbon and energy-hungry world—and an insight into what has stopped it reaching its potential, to date. There is something here for the non-specialist scientist and engineer, but also for the historian, social scientist, and media commenter. It is wrapped up in a grand narrative and an inspiring vision that will appeal to people from all walks of life—indeed anyone who cares about humanity's future and wants to leave a bright legacy for future generations that is not darkened by the manifold problems associated with extracting and burning ever dwindling and environmentally damaging forms of fossil carbon, like coal, oil and gas.

For the sake of averting crises of energy scarcity, mitigating the ever mounting global problem of anthropogenic climate change, as well as drastically reducing the pressure on society to make huge swathes of productive landscapes hostage to biofuels and other diffuse forms of energy collection, we need to continue the historical impetus towards ever more energy-dense fuels. It's time for the *Integral Fast Reactor 'Flyer'* to take flight, because, as Till and Chang explain, the sky is the limit…

Barry W. Brook, Ph.D.
Sir Hubert Wilkins Professor of Climate Change
The University of Adelaide, Australia

CHAPTER 1

ARGONNE NATIONAL LABORATORY AND THE INTEGRAL FAST REACTOR

The Integral Fast Reactor (IFR) is a fast reactor system developed at Argonne National Laboratory in the decade 1984 to 1994. The IFR project developed the technology for a complete system; the reactor, the entire fuel cycle and the waste management technologies were all included in the development program. The reactor concept had important features and characteristics that were completely new and fuel cycle and waste management technologies that were entirely new developments. The reactor is a "fast" reactor—that is, the chain reaction is maintained by "fast" neutrons with high energy—which produces its own fuel. The IFR reactor and associated fuel cycle is a closed system. Electrical power is generated, new fissile fuel is produced to replace the fuel burned, its used fuel is processed for recycling by pyroprocessing—a new development—and waste is put in its final form for disposal. All this is done on one self-sufficient site.

The scale and duration of the project and its funding made it the largest nuclear energy R&D program of its day. Its purpose was the development of a massive new long-term energy source, capable of meeting the nation's electrical energy needs in any amount, and for as long as it is needed—forever, if necessary. Safety, non-proliferation and waste toxicity properties were improved as well, these three being the characteristics most commonly cited in arguments opposing nuclear power.

Development proceeded from success to success. Most of the development had been done when the program was abruptly cancelled by the newly elected Clinton Administration. In his 1993 State of the Union address the president stated that "We are eliminating programs that are no longer needed, such as nuclear power research and development." This included the IFR program.

This book gives the real story of the IFR, written by the two nuclear scientists who were most deeply involved in its conception, the development of its R&D program, and its management.

Between the scientific and engineering papers and reports, and books on the IFR, and the non-technical and often impassioned dialogue that continues to this day on fast reactor technology, we felt there is room for a volume that, while accurate technically, is written in a manner accessible to the non-specialist and even to the non-technical reader who simply wants to know what this technology is.

1.1 Introduction

Our principal purpose is to describe the technical basis of the IFR in adequate detail in a manner that is accessible to the non-specialist. The what, the why, and the how of the Integral Fast Reactor technology is what we try to convey. With a little willingness by the interested reader to accept approximate understandings and go on, a very adequate understanding of the technology should be possible without undue effort.

We have not limited our subject matter to the technical, as the IFR concept embodies a history that goes all the way back to the beginnings of nuclear power. In its background is much fascinating interplay between the indisputable hard facts of the scientifically possible, the changing beliefs of the possible in the politics, and the forces that made the politics what they were—the ephemeral perceptions of energy realities and indeed of nuclear power itself, at different times in the past sixty-plus years. The IFR's history is embedded in the history of nuclear power in this country—in its ups and downs, and in the plusses and minuses of nuclear technology itself. Its story starts sixty years ago with the first reactor that ever produced useful electrical power. It continues at a low level in studies and programs of the Department of Energy, and in programs around the world today.

Its development took place in two eras, separate and distinct: 1946 to 1964 and 1984 to 1994. The early period defined the fast reactor: choices of coolant and fuel were made, reactor configuration was selected, and two experimental fast reactors were built, EBR-I and -II. EBR-II, a complete power plant, did not cease operation when development along the EBR-I/EBR-II line stopped. It operated through the entire period from 1964 to 1994. The line of development it represented though was gone, terminated in 1964 by a combination of as-yet unsolved technical difficulties and by the politics of the time. And there the matter rested.

IFR development began in 1984, much of its technical basis coming from the earlier era. Solutions to the earlier problems were proposed and successfully tested, and a range of powerful new characteristics became evident, and were specifically identified, developed, and proven. This "advanced reactor development program," as it was called, was carried out for a decade at Argonne National laboratory; its result was the IFR. In 1994, although nearly complete, it was cancelled. In the State of the Union address that year the president stated that "unnecessary programs in advanced reactor development would be cancelled." The IFR was the nation's only such program.

Why then does the IFR have any importance today? A glance at today's energy realities will tell you. It is only a little simplification to say that the present world runs on fossil energy. Huge amounts are required. The strain required to maintain present production is increasingly obvious. The resource is finite, and depletion even now is straining the limits of the possible. Production declines are inevitable. Constant new discoveries are required simply to maintain production, and discoveries have lagged below depletion for decades now. This is the situation for the energy supplies of nations, the lifeblood of civilizations. The IFR deals at this level—energy supply for entire nations, truly inexhaustible energy for the future. Energy in massive amounts, in any amount desired, forever. Incredible? No. That is the promise it offers.

Magnitude is what is important. The magnitude of energy produced is what matters always. Surprisingly, this isn't always recognized as immediately and as specifically as it should be. When told about some new energy source, always ask how much it can produce. How important is it? Can it power civilized societies when fossil fuel production can no longer be sustained? The amounts needed to sustain our civilization are huge. What can replace them? The IFR meets the issue head on. That is its importance.

The U.S. has an electrical generating capacity of about one million megawatts. The capacity factor—the percentage of time of generation at full power—is about 45%. In 2009 the full-power generation, equivalent to 100% full power, was 457,000 MWe. The amount of electricity per capita used in the U.S. has increased by a factor of four since 1960 and continues to increase. These are the kinds of magnitudes that proposed energy sources must come to grips with, not units of 2 MWe, or 20 MWe, just to keep up with the combination of increased demand per person and the steady growth in population. Already increased use of electricity for transport is contemplated and transport needs are huge as well. Is electricity growth likely to decrease? It seems unlikely, very unlikely indeed. The IFR will be needed.

In this book, therefore, we lay out in simple terms the "whys" of the Integral Fast Reactor—why the characteristics are what they are, why we made the basic choices of materials we did, why we chose the design we did, and why those choices are important (and justified). It is not always sufficiently recognized that such choices lead to fundamental differences in the most important characteristics between the different variants of the fast reactor. One way of looking at the possible characteristics is whether one decision is truly better than another. Discriminating choices in the materials and the choices in the design matter. They matter a very great deal.

As we go along, we hope it will become evident why we thought it important to push its development as we did. The IFR is technology that the authors had the principal role in drawing together, defining and developing. If we can provide the reader with anything of value, it is surely on this subject. In writing of our experience, we hope to make the future developers of nuclear power aware of its possibilities. We also hope to make it all as easy a read as we can. We want the answer to be yes to the question, "Is there any book that, in simple language, tells what the IFR is and why it works the way it does?"

The most complete of the highly technical works is a special issue of *Progress in Nuclear Energy*, entitled "The Technology of the Integral Fast Reactor and its Associated Fuel Cycle." [1] Edited by W. H. Hannum and authored by a fairly complete cast of the people who developed the IFR, this volume, written for the specialist, gives excellent in-depth overviews of the principal technical areas of the IFR. At the other end of the spectrum are two very good books by non-specialists, Tom Blees [2] and Joe Shuster [3], who don't seek to provide the technical bases underlying the IFR technology, but who do an excellent job of summarizing the capabilities of the IFR and why it matters.

The IFR story is an Argonne National Laboratory story. It's completely an Argonne story; no other laboratory or company was important to its development. In a very real way its story is the story of the history of Argonne National Laboratory itself. In the period during which the IFR was an on-going program, it was only a part of the programs of the laboratory—a big part certainly, but still only a part. But really it was more than that; the IFR was the culmination of all the years of fast reactor development at the laboratory. And fast reactor development was at the center of Argonne's own development as a great national laboratory. The history goes all the way back to the beginnings of the national laboratory system—indeed, to the beginnings of nuclear fission itself.

The authors came to Argonne National Laboratory from very different backgrounds, and became the closest of scientific associates and friends. We have been both now for fully thirty-five years. Till was the instigator of the IFR program. He was the Associate Laboratory Director for Engineering Research with responsibility for all nuclear reactor research at Argonne. Chang was the General Manager of the IFR program from its beginnings, with the responsibility for seeing it done properly. He arrived at Argonne about ten years after Till, and worked with him almost from the beginning.

In these first two chapters we'll describe something of what made a great national laboratory what it was, and what it was like to work there. We'll have something to say about what it takes to make a laboratory great—what it takes to allow it to accomplish important things. Our observations, of course, are our observations only. They depend on where we were in the hierarchy of the laboratory at various times through its history. However, we can say that we have seen the workings of a great national laboratory as scientists working each day, and then at every supervisory level, and in every political environment the laboratory experienced over a time nearing fifty years. We start with the laboratory in early times, the most productive in its history, and we'll see how the laboratory changed with events and time. Till saw the tail end of the early years, they were gone by the time Chang arrived.

What was it like to live through the years at the laboratory and eventually to start and to almost finish a task whose importance still can't really be gauged, but which could eventually supply the energy for the entire world? In Chapter 3 we'll try to bring those years back and make them come alive as best we can. In this chapter we'll talk about this task, almost finished, but not quite, and we'll go into the whys of that as well.

We will be describing events and circumstances that substantially affected the laboratory. Our view is certainly a view from an Argonne perspective. We won't even say it is from *the* Argonne perspective, as undoubtedly there are many such views. Our description of the views, actions and reactions of the reactor program at Argonne to events through the years, and our observations and particularly our conclusions as to their significance are just that—our own. And as in other fields, there is competition between laboratories and organizations, and people in other organizations undoubtedly saw many of the same events quite differently than we did. Our goal is to describe as accurately as we can events as they were seen, and interpreted, from within Argonne at the time they took place.

We have no doubt that the other nuclear national laboratories had very similar histories in striving for excellence in their programs, and in their experience in striving to create and maintain the conditions necessary for excellence to be fostered. The laboratories all felt the same forces over the years. We expect Argonne is typical in the larger scheme of things. In any event, it is Argonne we know, and it was at Argonne that fast reactor development, culminating in the IFR, took place over all those years.

In the IFR, we undertook to develop a complete nuclear reactor technology whose fuel resources would be limitless, and at the same time to try at least to minimize concerns about nuclear power. There was nothing new about limitless fuel supplies being put foremost in development. That went back to the very beginnings of nuclear power at Argonne. But what now had to be faced was that concerns were, and are, honestly felt by the public, though certainly amplified (when not invented whole) beyond any recognition of basic fact by organized anti-nuclear politics. But by the early eighties, their amplitude had increased to the point that nuclear power in the U.S. had not only been influenced by the organized opposition, its further growth was at a standstill. We set for ourselves the task of advancing as far as we could in eliminating, or at least ameliorating, the list of concerns that were always pointed to, and to do it solely through technology itself. The technology was that of the Integral Fast Reactor, the IFR.

We will trace a little of the history of nuclear development, in particular for electricity production. These were the "civilian nuclear" programs that were undertaken around the world following the proof of the chain reaction and the eventual drama of the explosions of the two atomic bombs that ended WWII. We want the reader to experience a little of what the civilian nuclear enterprise was like in the early years and what research in nuclear at Argonne was like and how it changed in almost every decade. What kind of people were at Argonne, and how did they arrive there? The scientists had different backgrounds but all had pretty much the same education and a similar outlook in technological matters too.

But how was business conducted? Remember, the whole field of nuclear energy was new. Everyone felt the newness of it; there was a huge amount that was still unknown. The first nuclear reactor in the world had gone critical barely twenty years before Till joined the laboratory in the spring of 1963. A few nuclear engineering schools, the source of much of the laboratory staff later, had begun at a few major universities, but they were not yet a major source of laboratory staff. The scientists at Argonne tended to be young, with no particular specialized nuclear engineering training when they came to the laboratory, but they came out of schools with good scientific and engineering credentials and they learned as they went along.

In these introductory chapters I (Till) will describe the Argonne of those early days, and trace a little of my own technical history as a more or less typical example of the backgrounds of people coming into the laboratory at that time. Argonne was acknowledged everywhere in the world to be the leader in reactor development—or again, "civilian reactor development," as it was called then. I was familiar with the respect Argonne generated worldwide. Before I came to Argonne in 1963 I'd seen it firsthand.

To be part of it, that was the thing. Not everyone has such an opportunity. The projects, the people, the approximations we made in calculation, the short cuts in analysis and experiments we used in those early days, are all gone now. Several years ago Richard Rhodes, the Pulitzer Prize winning author of the "The Making of The Atomic Bomb" and other distinguished historical books, who at the time had written a small volume on contemporary nuclear power [4], kindly offered help in a history "of the second stage of nuclear reactor development." "You were," he said, "at the center of it. People decades from now would see it through your eyes."

But I knew a lot had gone on in those years that I knew nothing about at all. Reactor development had advanced quickly. It was true that I had been very privileged to be at Argonne through all those years. I saw it as a young man starting out, wanting to accomplish something, something important for my time. And through the years as the laboratory changed, and as I became more deeply a part of it, I saw how a national laboratory fights to maintain its goals and ideals—its integrity really.

It may strike the reader that "integrity" is a strong word—too strong—and perhaps overly dramatic. But remember, scientific integrity is natural, the norm in national laboratories like Argonne, who in the main hire the most highly qualified of scientists and engineers, Ph.D. level people—often brilliant, always competent. The laboratories attract such people because of the scientific freedom they offer to them. Advanced degrees are expected, generally, and many come out of the best and most prestigious colleges and universities. People who have pursued scientific knowledge this far have scientific goals themselves; they have scientific ideals and they have scientific integrity. They do not regard their careers, their life in science, as "just a job." A laboratory must fight to retain these qualities the scientists have naturally, if it is to give the nation the kind of science and technology that the nation deserves and that, after all, the taxpayer pays for. Integrity is foremost. The science produced must stand up under the most detailed and careful of examinations. Failure to demand these kinds of standards is deadly. It can result in the kind of crises of skepticism currently faced in climate investigations. Once lost, reputations based on scientific integrity are hard to regain.

To understand Argonne well, some history will help. The first thing to understand is this: At any time in its history it could have been said quite truly that Argonne faced a difficult situation. Through its history Argonne was always in trouble, always in one battle or another—indeed, sometimes several at the same time. Always the issue was the same; the Laboratory fought to establish and maintain itself as a laboratory doing first-rate work *for the nation*, with all that that implies. Always there was tension, sometimes over control of research with its federal sponsor—in the first decades the Atomic Energy Commission; later its successor agency, the Energy Research and Development Administration; and still later, its successor agency, the Department of Energy. For decades, the lab was at odds periodically with the various Argonne-related associations of Midwestern universities, which changed over the years, but whose interests quite naturally lay in more control of the lab than the laboratory was willing to relinquish. And sometimes the laboratory had to defend itself in political infighting between the parties in Congress, acting for their own purposes. But in spite of all that, it was always a place where things could get done, initiatives taken, discoveries made. A fine place to work, to accomplish something, where "good work"—new knowledge—was given the highest respect and honors; it was a life to be proud of. By the time I retired in 1998 I had been part of Argonne for two thirds of its history.

1.2 Argonne National Laboratory

Argonne National Laboratory spreads over a pleasantly pastoral site in a still fairly rural area of DuPage County about thirty-five miles southwest of Chicago. (Figure 1-1) It occupies a few square miles of open grassland, with largish forested areas and winding roads. Its buildings and layout are vaguely university-like, but it is protected by a high security fence, has a uniformed security force, and everyday access to it is limited. (In its early days after the war, the syndicated radio personality, Paul Harvey, attempting to demonstrate lax security, climbed partway up the fence. An Argonne guard had him in the crosshairs, but the guard, for once, showed good judgment and let him climb down.) Experimental nuclear reactors were built there in the days after WWII. The site was made large enough to keep the reactors a mile or more away from homes or roads in general use. The buildings are mostly redbrick; each one as a rule is for a single scientific discipline—chemistry, say, or physics. The buildings have been there for more than fifty years. They are set apart from each other on curved and meandering forested roads and avenues. The silver domes and the temporary high buildings of the experimental reactors of the early days are gone. This is a federally funded scientific laboratory, in its early years focused on non-military nuclear research; now, while still doing some R&D in nuclear matters, it is a mature laboratory working in basic science, computations, and basic materials, with a common thread of work on energy-related fields in general.

Figure 1-1. Argonne National Laboratory main campus
in southwest suburb of Chicago

Nuclear development for peaceful purposes began here in the mid 1940s. It came to the laboratory directly, as a portion of the distinguished scientific group at the University of Chicago, who, under the direction of the great Italian physicist, Enrico Fermi, had built the world's first nuclear reactor. They had assembled it in a squash court under the west stands of Stagg Field (Figure 1-2) on the campus of that fine university.

Figure 1-2. West stands of the Stagg Field of the University of Chicago, the site of Chicago Pile-1

Chicago Pile-1 (CP-1) first "went critical," beginning operation on December 2, 1942. ("Pile," it was called, because it was assembled from uranium chunks in graphite blocks piled high enough to achieve criticality. This term for a nuclear reactor was used for some years, but the usage has long since died out.) (Figures 1-3 and -4) More reactors were to be built, different reactor types for a variety of purposes, and the University was situated in a densely populated area on Chicago's South Side. Given that the war was on, a more remote site had actually been originally chosen for CP-1, but it was not complete when the scientists were ready to go ahead. So, a squash court just south of downtown Chicago became the site of the world's first nuclear reactor.

Figure 1-3. World's first reactor Chicago Pile-1

Figure 1-4. Cutaway view of ChicagoPile-1 showing the graphite blocks

The next reactors would be assembled in a more remote site, out of the city in the Argonne Forest southwest of Chicago. A Cook County forest preserve, the name Argonne came from the final battles of World War I where in the Argonne forest in France American troops in strength went into action for the first time. The site chosen for the new reactors, "Site A" as it was called, in the Argonne forest, did not become the site of the new Argonne National Laboratory after the war as might have been expected. (Figure 1-5) The forest preserve had granted occupancy for the duration of the war plus six months only, so the site of Argonne National Laboratory was established a few miles away, out of the forest preserve. But the name Argonne stuck when, on July 1, 1946, Argonne National Laboratory was officially created; the first of the great national laboratories charged with investigating the scientific mysteries of the atom.

Until recently Argonne has had a satellite laboratory in the Idaho desert, thirty miles west of the small city of Idaho Falls. (Figure 1-6) There, the Argonne site occupied a small part of the National Reactor Testing Station (NRTS), an expansive desert site tens of miles across in every direction. Established in 1949, the NRTS was to be the site for the first experimental reactors; Argonne would build them and conduct experiments on them. Argonne would later build more over time, and other contractors would do the same, until eventually over fifty experimental reactors were built on this site. But on these two Argonne sites, the main Argonne campus in Illinois and its satellite in Idaho, much of the history of world reactor development would come to be written. For Argonne in those early years, the late forties, and the early fifties, was the laboratory charged with the principal role in the nation for development of nuclear power for civilian purposes. It was here, in the main, that the U.S. did its early nuclear reactor development.

Figure 1-5. Argonne at Site A, 1943 to 1946

Figure 1-6. Argonne-West site in Idaho (now merged into Idaho National Laboratory)

1.3 National Nuclear Power Development

The likely importance of nuclear power was recognized early. Nuclear electricity can be generated in any amount *if it is allowed to do so.* There need be no limitation from fuel supplies, an almost incredible promise. In the late forties and early fifties nuclear development programs began in many countries. In part, this was the early realization of nuclear energy's promise of unlimited energy; partly it was national striving for "energy independence"; and partly it was simply that this was *the* new, the exciting technology. Nations did not want to be left out. It could even be said that it was fashionable to be a part of it. The U.S. itself had made a huge nuclear development effort during WWII, and afterward, it began a very broad program of civilian reactor development. In every nuclear field, in fact, the U.S. led the way. It is hard, really, to credit today just how dominant a role the U.S. played in the early decades of nuclear development. With very few exceptions, other nations simply followed the U.S.'s lead, from the beginning of their programs or soon after; so advanced was U.S. technology, so wide its range and so rapid its development.

Perhaps it seems quixotic to say this today, accustomed as we are to the ceaseless attacks on nuclear power that started in the late sixties and early seventies and have continued to this day, but the early development of nuclear power was actually driven by non-proliferation considerations. When first introduced in the 1950s and 1960s nuclear energy was certainly not a must. In no nation was it necessary to meet its energy needs of that time. Populations were smaller, per capita energy use was smaller, fossil fuels were plentiful and inexpensive, and the local pollution impact of coal was not high on the public agenda. Nuclear power development began as an exploration of the possible. It was viewed principally as a possible prudent hedge against an energy shortage far off in the future. But very soon nuclear power got a huge boost from U.S. national policies that evolved to deal with a threat entirely different from possible limitations in energy supply.

The stimulus for early introduction of civilian nuclear power came from a now-familiar imperative: the need to find some way of dealing with the looming threat of proliferation of nuclear weapons. It was a principal concern of President Eisenhower on taking office, one that led him to a determined attempt to reduce the danger to the world of what seemed to have become inevitable—that there would be growing numbers of nuclear weapons, in the nations with nuclear development programs. The effect was to accelerate the introduction of practical civilian nuclear power.

An international bargain was to be offered by the U.S. Civilian nuclear power, proven nuclear power, would be the bargaining chip. The growing number of nations capable of developing such weapons was thought to be at least a dozen by the mid-fifties. With his preferred course blocked when his initial two-party proposals to the Soviet Union for weapons limitations were rebuffed, and his options then limited, Eisenhower took a principal role in drafting a landmark speech. In it, he outlined his Atoms for Peace proposals to the United Nations. [5] It offered this trade: U.S. nuclear knowledge, technology, and even U.S. nuclear fuel for civilian purposes, in exchange for an undertaking by recipient nations themselves to halt indigenous weapons development. The U.S. was in a commanding position internationally in all things nuclear, but this new policy depended on U.S. ability to maintain this position; U.S. demonstration of the technical and economic feasibility of civilian nuclear power would make the bargain desirable.

The bargain was accepted. From these proposals flowed the entire international non-proliferation structure, the setting up of the International Atomic Energy Agency, and the definition of its role as monitor of such activities. In place now for fifty years, with a few notable exceptions, the arrangements have worked. Nuclear power programs around the world were materially aided by the U.S. release of mountains of hitherto guarded nuclear information at two giant nuclear conferences in Geneva in 1955 and 1958. [6-7] (Figure 1-7) Civil nuclear power therefore began as implementation of U.S. foreign policy, not as a need for a new energy source.

Figure 1-7. A scale model of the Experimental Boiling Water Reactor (EBWR) was exhibited in the Argonne booth during the Second Geneva Conference of 1958

The most promising lines of reactor development took some time to settle. In the 1950s, where other nations concentrated effort on one or two reactor types, the U.S. undertook a very broad program of development, one aimed at settling the question of the best reactor technologies for the future. Many combinations of fuel, coolant and moderator were tried, and four quite different reactor types were constructed in the Power Reactor Demonstration of those years. All were "thermal reactors," where neutrons slowed by "moderators" provided the principal nuclear reactions. (Capture by a uranium atom and its resulting fission is greatly increased for neutrons of low speeds; the slowest are at 'thermal" energies. Moderators are light elements, hydrogen in water or carbon as graphite, whose atoms slow the neutrons in collisions like those of billiard balls. Properly arranged arrays of moderator and fuel can slow the neutrons right down to thermal energies, where their effectiveness in fission is at its maximum.) "Fast reactors" or "fast breeder reactors," shortened to "breeder reactors" in this book are different in kind from thermal reactors. Fast reactors keep the neutron energies as high as possible, because fast neutrons allow significantly more breeding than any design of a thermal reactor.

The impetus for early development of thermal reactors was mostly their simplicity. Simplicity in understanding their behavior, simplicity in their low enrichment (natural uranium could be used if the deuterium isotope of water, or sufficiently pure graphite, was used as moderator), simplicity in their construction; all offered promise of early economic success. Of the various possibilities, the U.S. selected the light water reactor (LWR), a uranium oxide fueled reactor, moderated and cooled by ordinary water, in two variants: the pressurized water reactor (PWR), the choice of Rickover for submarine propulsion, and of Westinghouse and others for commercialization, and the boiling water reactor (BWR), the choice of GE for commercialization. In the former high pressure stops boiling in-reactor, in the latter the coolant is allowed to boil in-reactor, but both operate under very high pressure to allow the water temperatures of hundreds of degrees Celsius necessary for an efficient steam cycle. Both came to be built in quantity in the next decades, and these two reactor types, along with the independently developed Canadian D_2O moderated system, have become the common choices around the world.

Success in development led to a brief period of rapid construction. Technical feasibility was established very early, almost from the beginning, but economic feasibility, which meant competing with the cheap fossil fuels of the time, was more difficult, and it became the central focus of development from the late fifties onward. By the late 1960s success had been achieved. Nuclear power plants were not only technically feasible, but they were economic, so much so that turnkey contracts began to be offered to the utilities by commercial reactor suppliers. (For a fixed price, the supplier would turn over a finished reactor plant, "ready for the key to turn it on.") A boom in orders and construction began, and in the few years between the late 1960s and 1974 slightly over one hundred large nuclear plants eventually were built in the U.S. The orders then slowed and stopped; ultimately, no plants were constructed from orders after 1974.

Organized opposition had begun, arguing environmentalism initially, and then joined by proliferation-related attacks. In the last year or two of the sixties the attacks had begun and with growing influence, by the mid-seventies the anti-nuclear groups had had their way. Their strategy focused on driving up the cost of nuclear power plant construction, so far up that plants would be uneconomic, if possible. To do so they attacked every issue that could be used to insert the legal system into interference with construction decisions, blocking construction progress by any means possible. In so doing they introduced very lengthy construction delays. Success in delaying nuclear construction while interest on the borrowed construction funding kept increasing and increasing eventually made their argument self-fulfilling. They had made their assertion a reality: nuclear construction was now expensive. Every possible facet of the legal system was used. Plant after plant with financing in place for billions of dollars, and interest charges running up, had construction held up month after month, year after year, by one legal challenge after another, as a rule related in some way to environmental permits. Nuclear opponents could congratulate themselves; they had destroyed an industry. Their strategy had been a brilliant success. To what purpose, though, may one ask? It stopped orderly progression of nuclear power development and implementation by the U.S., and, indeed, led to similarly destructive movements in other countries too. The world then went back to fossil energy and hundreds, more probably thousands, of new fossil fuel plants have gone into operation in the years since then.

Attacks on civil nuclear, specifically those citing proliferation concerns, accelerated in the 1970s as their effectiveness in denigrating nuclear power became apparent. In the early 1970s, articles began to appear in scholarly journals such as "Foreign Affairs," critical of the existing international non-proliferation regime (see, for example, Quester [8]). The institutional arrangements flowing from the Eisenhower administration initiative fifteen or twenty years before were led by a newly created International Atomic Energy Agency (IAEA) placed in Vienna, staffed by personnel from the countries with nuclear energy programs, and with powers of inspection. An elaborate structure, it had worked pretty much as had been hoped in that many nations discontinued their weapons programs. But there were significant exceptions, and the compliance of nations with IAEA measures was strictly voluntary.

Like the organized environmental activists, the non-proliferation people were politically active, and were often in positions of considerable political influence. Many became part of the Carter administration when it came to power in 1977. Their position, which became the position of the new administration, was that the IAEA regime was insufficient; that the very existence of nuclear power represented an unacceptable threat, civilian or no; and that the U.S. must lead by example in constraining nuclear development. Reprocessing of spent fuel was forbidden, breeder development was seriously curtailed, and the growth of existing reactor capacity was stopped. (The "breeder" reactor is a reactor type designed to use up none of its fissile fuel material; it "breeds" replacement fuel by converting otherwise useless "fertile" (non-fissile) uranium to fissile plutonium fuel as it goes along.) Over the next decade this had the chilling effect on U.S. nuclear development they desired. While damaging the U.S. program badly, with the concomitant decrease in U.S. leadership once taken for granted, in the main this had little effect on other nations. With one or two exceptions, it had no effect on reprocessing in other nations, a principal goal of the new U.S. non-proliferation policies, nor did it have much effect on the progress of successful nuclear programs of other nations. The U.S. took the blow, in the main. By its lingering leadership it did damage to some degree the civilian programs of its allies, but not nearly to the extent that it damaged its own.

Significant nuclear electrical capacity did come on line in the U.S. from the 1980s onward. Although the rush of orders stopped in 1974, the plethora of previous orders resulted in dozens of new nuclear plants that began operation through the late seventies and eighties, and continued to improve their operational efficiency throughout the 1990s and 2000s. Today those plants, roughly one hundred in number, produce a fifth of the electrical power in the U.S. But it is also true that to the present time, with seemingly plentiful oil for transportation, plentiful coal for electricity production, and plentiful natural gas for home heating as well as electricity production, nuclear power is a source that could still be regarded as optional. Its need is not absolute, although even a cursory glance at fossil fuel resource estimates today makes it apparent that this is becoming less true, day by day. It's important that the outlines of the fossil fuel situation be understood, because they underlie everything in the discussions of nuclear power. We will explore today's apparent realities in Chapter 4.

1.4 Beginnings: The Early History of Argonne and Argonne Reactor Development

In 1984, when Argonne began the IFR development it was with the full knowledge that it was going to be very controversial. The views of the organized anti-nuclear groups dominated the media; by and large, the public had been convinced that nuclear power was both dangerous and unnecessary. Our initiative was going to require extraordinary efforts across a whole range of political and technical fronts and it was going to require resources that would have to be gathered. Most of all, it was going to involve very real risks to the Laboratory. It meant going against interests that had power to damage the Laboratory, and going against a lot of the trends of the time. But in a sense Argonne was used to that; it had been in almost constant conflict right from its beginnings. In one way or another, its early history had involved freedom to pursue the R&D directions the laboratory found the most promising, freedom for its R&D staff to innovate, freedom to "call its own shots." Sometimes it won its battles, sometimes it didn't. But Argonne always took initiatives; Argonne tried. And some pretty amazing things resulted. We'll look now at a little of the early history, because that history formed the Laboratory's attitudes and practices through all the years that followed.

Argonne National Laboratory was born in controversy. In describing this early history, and portions of the history that follows, we have been fortunate in having Jack Holl's careful and detailed account of Argonne at hand, published in 1994 [9], and we have drawn on it liberally in following the flow of events over the years. When the Laboratory was created on July 1, 1946, the newly appointed director faced a nightmarish combination of national and local politics, academic rivalries with Midwestern universities, intergovernmental disagreements, staffing difficulties, and on top of this, the technical problems that had to be solved. A new Argonne site had to be established, one that was out of the Argonne forest, isolated from population centers, but accessible for commuting and possible evacuations. Only after bitter quarrels, protests from some two hundred land owners, and political scores that had to be settled, were the 3667 acres assembled for the new laboratory. It took a strong and demanding director, and in Walter Zinn the Laboratory had the right man, for Zinn was a person to be reckoned with. [9]

Walter Zinn was a Canadian émigré, naturalized a U.S. citizen in 1938. He had come to the University of Chicago from Columbia with Enrico Fermi in 1942. He and Herb Anderson, another of Fermi's protégés, had then used the squash court under the west stands to build exponential piles—pieces of a reactor, really—that give accurate information on important reactor characteristics, and in particular the characteristics that the CP-1 would have. He had been Fermi's deputy in directing the construction of the next reactor, CP-3 on Site A. (CP-2 was CP-1 reassembled on Site A.) Alvin Weinberg, Director of Oak Ridge National Laboratory, noted later that Zinn was close to Fermi, was intelligent and was tough, and commented, "Zinn was a model of what a director of the then-emerging national laboratories should be: sensitive to the aspirations of both contractor (University of Chicago) and fund provider (Atomic Energy Commission), but confident enough to prevail (over both) when this was necessary." [10] Attentive to the need for science programs and to the needs of the scientific personnel, Zinn strove to make Argonne the laboratory it became.

The AEC regarded Zinn as the nation's leading expert on reactors and the work at Argonne as the most promising, in addition to the work being the first historically. In 1947 "Zinn's Breeder" was approved for construction, and going one step further, the AEC decided all reactor development would be concentrated at Argonne; Zinn was not consulted. The more reactor-focused the lab would be, the more the AEC would lean toward universities for its basic science, Zinn thought, and this very decidedly did not please him. Zinn accepted the exclusivity of the reactor role only on the promise that basic research at Argonne would be helped by it. Alvin Weinberg wasn't in agreement with the whole idea either. Weinberg and Zinn circumvented the centralization as early as the spring of 1948, by sharing the responsibility for the design of the high flux reactor. This reactor, conceived and guided in design by Eugene Wigner, was one of the three selected by the newly established Atomic Energy Commission for immediate construction. It became the Materials Test Reactor on the Idaho site, the workhorse for studying the effects of radiation on all kinds of materials for decades thereafter. With all this going on, it was 1948 before Zinn got any buildings, Quonset huts to house the engineering division actually doing the reactor design work. And an uneasy arrangement was set up that brought Midwestern universities in as a board of governors.

The new centralization meant that work on the pressurized water reactor for the nuclear submarine was also moved from Oak Ridge to Argonne. In the spring of 1946 Hyman Rickover, with a small group of officers, had gone to Oak Ridge to learn nuclear physics and engineering. In a year or so they had a working knowledge of their subject and Rickover persuaded Harold Etherington to apply his pressurized water technology to submarine propulsion. The Submarine Thermal Reactor (STR) was essentially a pressurized version of the high flux reactor. With Etherington and his division moved to Argonne, Argonne expected complete responsibility for the reactor. Rickover objected strenuously, maintaining that it was "his reactor." Zinn would not have it; a reactor should not have two bosses. Zinn held his ground, Rickover was not yet in a position to have his way, and Zinn prevailed. The attendant ill will lasted for decades afterward.

At the same time, while work with the Midwestern universities was to be an important element of the laboratory's responsibilities, and there were now dozens of university researchers associated with Argonne, liaison had become more and more difficult. Increasing tension with the Soviet Union had increased the security necessary at this nuclear laboratory. Argonne's reactor work was largely classified and it accounted for fully half of Argonne's total budget. The basic science side of the Laboratory and the Board of Governors that had been assembled from the Midwestern universities felt that basic science was being sacrificed to reactor development. The AEC had by now formed a reactor development division in Washington. Their position held that no regional board would get between themselves and Walter Zinn. With no genuine management responsibility then apparent the Board dissolved itself. Relations between the universities and the laboratory were in no way helped by this.

More trouble arose when Zinn's breeder project needed a site more remote than the new Illinois site. When in 1949 the Idaho site was selected, the AEC wanted its new Idaho field office to run the project. Zinn objected vigorously. His objections led to a compromise wherein the Illinois architect-engineering firm he had selected was chosen, and in practice worked directly with Argonne.

Only in 1950 were the first permanent Argonne buildings completed. With EBR-I (CP-4 for those counting) critical in December of 1951, the Materials Test Reactor in Idaho critical in May of 1952, and the STR Mark-1 submarine prototype on March 30, 1953, Argonne began to construct CP-5 on the Illinois site. A follow-on from CP-3, this was a five MW research reactor for use by the basic sciences, which remained in operation until 1979. Since it was a heavy water moderated reactor like the DuPont tritium production reactors in prospect at Savannah River in Georgia, Argonne accepted a group of DuPont professionals to train, in what turned out to be an excellent relationship. And after some negotiation, in May 1955 the Experimental Boiling Water reactor (EBWR) began construction as well on the Illinois site. It went on line just two years later. (GE went on to commercialize the concept in the next decade.) In still another controversy with the Midwestern universities—this time over the appropriate site for a large accelerator—and with the accumulation of other aggravations, Zinn resigned in 1955. Staying on for a grace period, in 1956 he left the directorship and the lab and established his own firm, the General Nuclear Engineering Company (GNEC). (It was later acquired by Combustion Engineering Company, with Zinn as the head of its growing Nuclear Division.) With him went a number of key laboratory people.

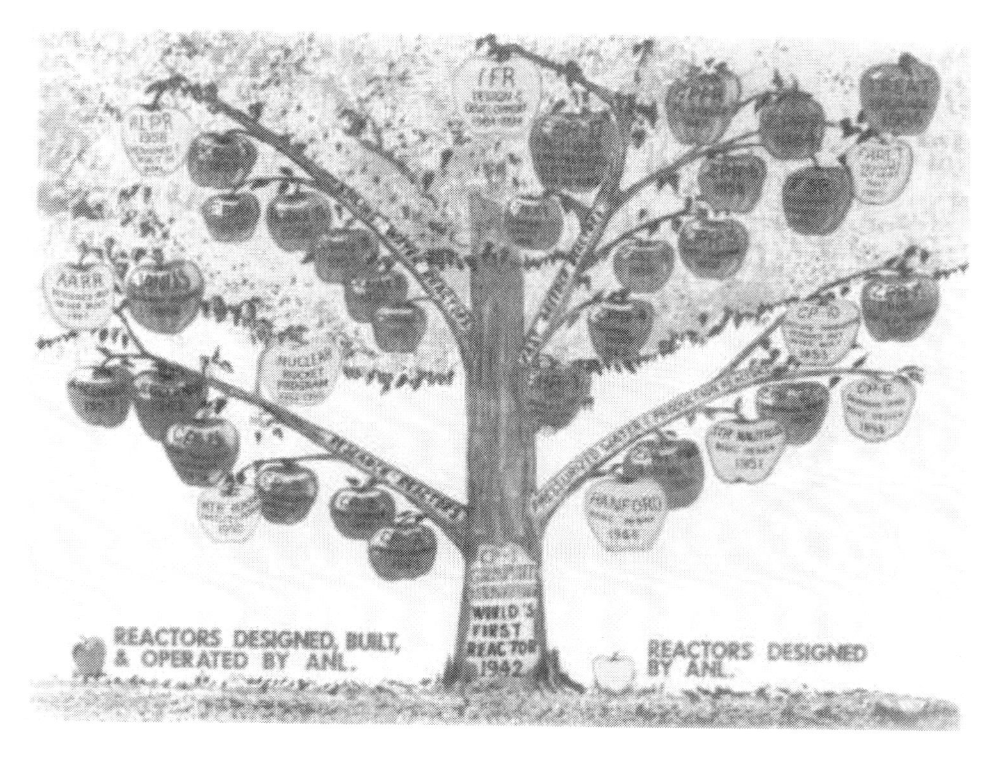

Figure 1-8. Reactors developed by Argonne

He had had a commanding effect on U.S. reactor development. In the choices to be made among the dozens of possible power reactor concepts, Zinn's view was decisive. The paths followed were those suggested by Zinn: there were to be two basically different paths. Both came to be accepted worldwide, the PWR and BWR on one hand, and the fast breeder on the other. And Zinn was largely responsible for the establishment of the NRTS in Idaho and influential in the formation of the American Nuclear Society as well as being its first president.

Zinn's ten year experience as Argonne director had been a constant battle to establish a laboratory meeting his expectations, and at the same time to rapidly develop the most promising reactor type for each important application of the day. His deputy, Norm Hilberry, succeeded him and continued Zinn's policies. The laboratory ran according to principles that Hilberry described in this way:

"A first rate R&D laboratory staff can be led only by indirection; the leadership must lead, not order. External factors like education and training and extensive experience cannot modify genius much. Management must have the respect of creative scientists, and to have this, management must have risen through the ranks. Further, management must provide a buffer between the creative scientists and bureaucratic non-essentials. Security and financial freedom must be provided to allow freedom to think, free from coercion, distraction or fear, so the staff can judge and plan purely on scientific grounds. Recognition of creative stature will always be the chief motivator. It is the people who are important, not the facilities." [9]

These were policies that established Argonne National Laboratory as the laboratory its management intended it to be. Controversy and confrontation were continuous. But the policies for managing the laboratory's staff outlined by Hilberry in the passage above were just what I found them to be when I arrived in 1963. I don't think any of my colleagues had any notion of events at the top of the laboratory, in the confrontations to establish and maintain laboratory programs. The description of my early days at Argonne in Chapter 3 will make this even clearer. Remarkably enough, through all this the laboratory was advancing reactor technology very rapidly. The management policies came to be tested later, and were tested particularly severely in the decade from 1965 on. But we'll come to that in due time, so let's now catch up with Argonne's fast breeder reactor development program.

1.5 Fast Breeder Reactor Technology at Argonne: The Early Years, 1946 to 1964

The first thing that needs to be said is that the IFR came from a distinguished past. It was based on ideas, concepts, discoveries, developments, and technical approaches that reached back to Argonne's earliest days. Argonne's first reactor was almost the personal product of Argonne's first laboratory director, and a fast breeder reactor as well. It began it all. The Experimental Breeder Reactor Number 1 (EBR-I) was to start the world along the path to develop a commercial breeder technology, and to do it in the earliest years of nuclear development. But the path ended, suddenly, in the mid-sixties, uncompleted, its technology no longer pursued, no longer in fashion. The technology existed, of course, in the minds of those who took part in it, and, if you knew where to look, in papers and proceedings of the conferences of earlier years.

It existed most tangibly and undeniably in the existence of EBR-II. In May of 1954 Zinn had proposed a new reactor, an Experimental Breeder Reactor-II, which was big enough to test full-scale components and applications. EBR-II eventually became an Argonne triumph. It was a small but complete fast reactor power plant. It was based on EBR-I experience, but it had a number of innovations exactly right for fast reactors of the future. But the line of development was never followed up. Fast reactor development changed radically in the late 1960s; the directions of fast reactor technology development changed, more than anyone probably really realized at the time; and the Argonne directions of those early years were abandoned. EBR-II at this time was only in its earliest years of operation.

But its reason for being there, its purpose, was gone. It was no longer on the development path. The U.S. was now to pursue a quite different fast reactor type, and. the Argonne technology receded into the past. The papers, the conferences, the very basis of its line of development dimmed in memory. A new generation came into Argonne who had no connection to that past. By the 1980s traces could be seen, if you looked—a paper here, a reference there. Relics really, turning up from time to time, like pieces of the wreck of a civilization now gone, half-buried, unconnected, and unremembered.

But the entire panoply of development of that reactor concept lay just under this surface at Argonne. As a line for further development, it had been abandoned. The people who did the work remembered of course, and many of them were left at the Lab. But they no longer worked on that technology, buried as it was by the onrush of the new. Their battles had been fought and lost, and they moved on.

What had happened in the mid-1960s? Ironically, the first thing was that the U.S. breeder development program was given the highest priority for development of a commercial reactor system. But the U.S. program then turned away from the Argonne line of development, adopting both a different fuel and a different reactor configuration. These changes were to have very significant effects.

A new organization for directing the breeder program was put in place in the Atomic Energy Commission. It would make huge changes in how breeder development would be done. A sea change in the directions of U.S. reactor development, and a sea change in the organizational framework for carrying it out, had taken place. The new AEC organization in DC would direct the development in detail. Importantly, and vitally important to Argonne, the breeder center, was that much of this new program would be done in places other than Argonne. Argonne would have assigned tasks in reactor physics and reactor safety (where its expertise was unmatched), in mechanical components development, and in the operation of EBR-II.

Things move fast in the early stages of technological development; there's little time and no wish at all to look back. It's in the nature of technology development to look ahead. Scientists want to discover, to make important contributions to new knowledge, to make their own reputations. They don't look back. The early development had been supplanted. It had been a living line of development, but that meant little twenty years later in the Argonne of the mid-1980s, when the first stirrings of what became the IFR program took form. The early development hadn't been carefully secreted away at Argonne, ready to spring out at the first opportunity. It had been buried and forgotten.

This early work had been very well thought out and very well done. It hadn't all worked, and it wasn't complete, most certainly, but there was more technical sense in it than in any of the development that displaced it. The ideas, the fundamentals, were right.

Walter Zinn had personally laid out the basic materials choices and an approximate design for a breeder reactor. One of the relics, an unpublished Argonne paper, turned up decades later. As I remember the Argonne central library may have been pruning its files. In any event, a forgotten paper by Walter Zinn appeared—written in the manner of a man reasoning his way along, a little tentative, in uncharted waters, not always perfectly grammatical, obviously written swiftly and in simple terms. But there it was: the basis for the design of the Experimental Breeder Reactor-I.

It was dated 1946, the year Argonne was formed. With uranium thought to be a scarce resource, and the fissionable isotope of uranium even scarcer (only comprising 0.71% of the uranium found in nature), it couldn't be used carelessly. There might not be much more. Creating more fuel with a reactor that "breeds" its own fuel might be essential if nuclear power was to amount to anything. And the physics of the fast neutron in a reactor assembled of materials to maintain neutron speeds, although still largely guessed at, did suggest it might be possible. The wartime work had demonstrated that plutonium could be bred—created—from the 99.3% of uranium that otherwise was useless as fuel. And plutonium is very good reactor fuel.

"Zinn's Breeder" was another of the first three reactor concepts given priority for development by the Atomic Energy Commission. In that 1946 paper he had selected sodium as the coolant (a sodium-potassium mix, liquid at room temperature, was actually used in EBR-I), chosen metallic uranium as the fuel, estimated the likely breeding, and suggested a layout for how the reactor might look. Under Zinn's supervision, and less directly, with Fermi's suggestions, Argonne began building the reactor on the newly established site in the Idaho desert. The Experimental Breeder Reactor Number One, EBR-I, came into operation in December of 1951. EBR-I was a test reactor. It did what it was built to do: it proved the breeding concept. But it also produced the first nuclear-generated electricity ever in the world. And it established many other things, all of them firsts. (Figures 1-9 and -10)

Figure 1-9. Experimental Breeder Reactor-I,
designated as National Historic Landmark in 1966

Figure 1-10. EBR-I produced the first electricity from nuclear, supplying power to the reactor control system as well as the building and a machine shop

In what was meant to be one of its last experiments, in 1955, with its coolant flow intentionally shut off, the reactor was given a sharp increase in reactivity to rapidly increase its power. This would rapidly increase its temperature and allow effects of temperature alone on reactivity to be measured. This was judged a risky experiment, but worth it in the knowledge it would give. Unfortunately, an operator mistake let the power increase too far, and a portion of the football-sized core melted. Radioactivity was detected in the control room. However, there was no explosion, little damage outside the core, and no injuries. The reactor was cleaned up, and in 1962, it became the first reactor ever to operate with a plutonium-fueled core. It had served its purpose, however; EBR-II was coming into operation, and EBR-I was officially shut down in December of 1963. It was designated a National Historic Landmark by President Johnson in 1966. It can be visited today, and is something of a tourist attraction. It is well worth a visit by anyone travelling through eastern Idaho, perhaps on the way to Yellowstone or to Sun Valley.

EBR-I had been designed and built by a remarkably small number of engineers, all young—in their twenties and thirties—and numbering about a dozen in all. Several went on to lead the design and construction of EBR-II. But EBR-II was not only the logical next step in the scale up of the breeder reactor, it was a trail-blazing concept in itself. By the time it was shut down, for purely political reasons that we'll go into later, it had forged a proud history, and it too had a large number of firsts to its credit. It had incorporated many sound insights. One of the EBR-I young engineers, Len Koch, became the Director of the EBR-II Project. In its early days EBR-II was referred to as "The Koch Machine," which gives a hint as to the importance of Koch to the project (as well as the pronunciation of the director's name).

EBR-II was a scale up in power by a factor of sixty or so from EBR-I, but unlike EBR-I, it was no test reactor. This was a power plant, and a complete one, producing 20 MW of electricity, supplying power to the site and power for sale to the appropriate utility. In a major first, it had an on-site facility for processing its own spent fuel and fabricating new fuel with it. Processing was done without delaying to allow the spent fuel to "cool"; the fuel was returned to the reactor "hot," highly radioactive. (Figure 1-11)

In another major first, EBR-II was given a "pool configuration". In a pool, all the radioactive components—the core itself, of course, but also the associated piping and equipment that circulates sodium coolant through the core, and whose sodium is radioactive as a result—are immersed in a large pool of sodium coolant. (Figure 1-12) This is easy to do with liquid sodium coolant because it needs no pressurization (unlike in water-moderated reactors). It is liquid at low temperatures (97°C, about the boiling point of water) at room pressure, and stays liquid without pressurization at temperatures far above operating temperatures of a reactor. No pressure containment is needed for the primary tank; it can be made any size desired. No need for the thick walls of the pressure vessels needed to hold water in a liquid state at reactor operating temperatures in a water-cooled reactor. This feature has many advantages, not least of which is that any leaks of sodium over the reactor lifetime would not involve radioactive sodium. We'll go into that, and the other advantages, in a later chapter.

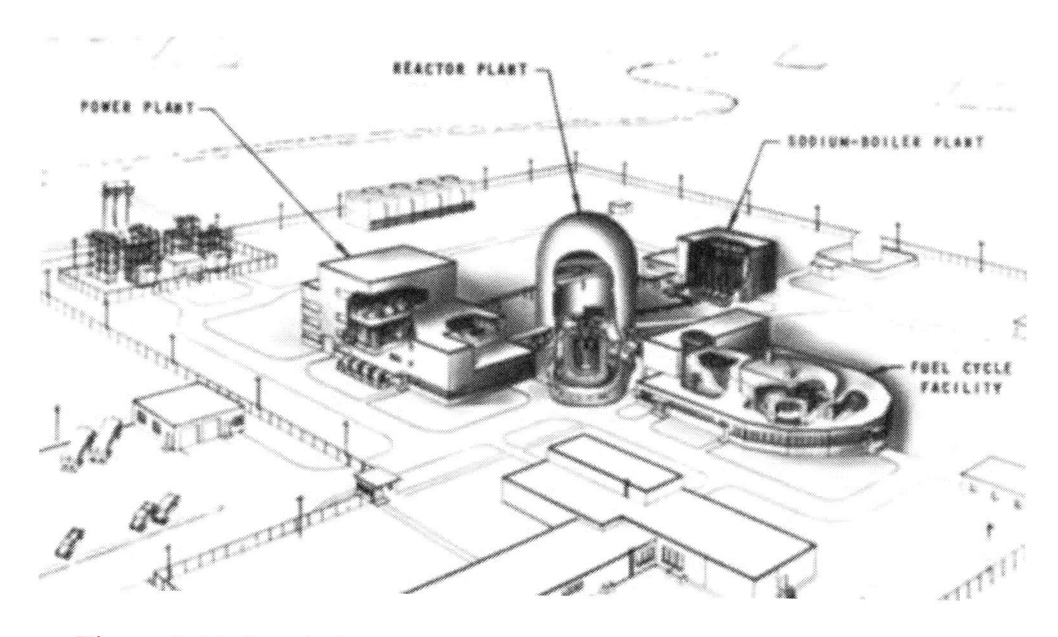

Figure 1-11. Rendering of the EBR-II Plant and its Fuel Cycle Facility

The downside was the fact that the metal fuel used by EBR-II was not at all satisfactory. It would not withstand even reasonably lengthy irradiation in the reactor. A uranium alloy, it swelled substantially under irradiation, and it would burst its steel cladding if left in the reactor more than a few months. In fact, it was just this problem that was the pressing reason for providing the reactor with an onsite processing facility at all. The fact that fuel would have a short life was accepted and provided for. After its short irradiation time, it was taken out of the reactor and simply melted. Melting itself partly purified the new fuel. It released the fission product gases and a chemical reaction with the melt-vessel extracted more of the fission products, though many were left. It was a crude beginning for processing, but it did work. Importantly, it introduced the thought that onsite processing might be simple, feasible and desirable.

Nonetheless, a much longer irradiation life was going to be necessary if there was to be a successful scale-up to commercial size. This would have been a feature of EBR-III, the reactor that never was. In the sweeping political changes of the mid 1960s, the most important immediate technical change was to discard metal in favor of oxide fuel. When problems with metallic fuel burnup became evident in the early 1960s, General Electric led the way in advocating a move to oxide fuel for fast reactors. The experience with oxide in the thermal reactors that were then starting to commercialize had been favorable, and it seemed natural to move in that direction. It was known that oxide would withstand irradiations of considerable length, and, as a further incentive, the very high melting point was thought to be favorable for safety.

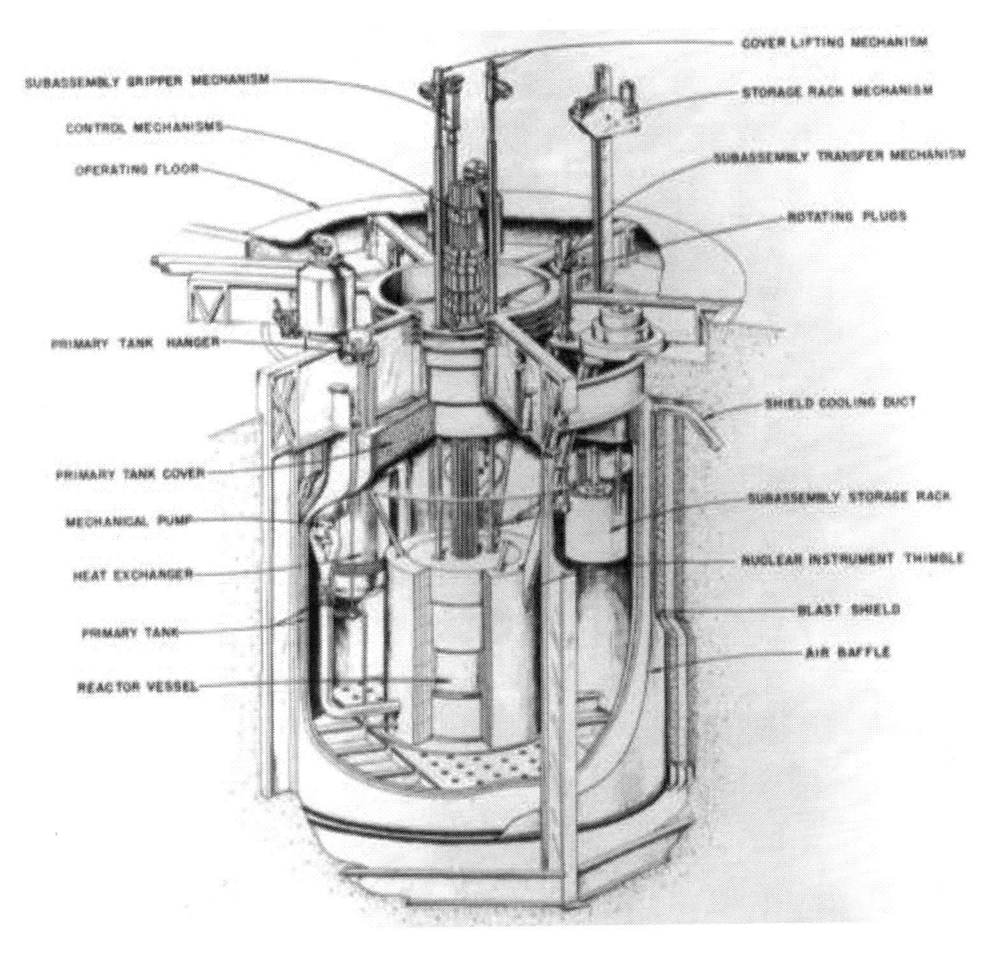

Figure 1-12. Cutaway View of EBR-II Pool-Type Primary System

Oxide fuel had been successfully demonstrated in non-breeder reactor prototypes, in particular for the Light Water Reactor (LWR). Oxide is hard and intractable; it is a ceramic like pottery. It was certainly capable of much longer irradiation times than the EBR-II metal fuel of the time. (A British fast reactor prototype at Dounreay in Scotland, and the Fermi-1 fast reactor prototype built by Detroit Edison, both had metal fuel too, and both suffered similarly from short fuel lifetimes.) Before any EBR-III could realistically be contemplated, oxide, not metal, had become the fuel of choice. This was to be so for all subsequent reactors, fast reactors or the thermal LWRs and Canadian heavy water reactors (CANDUs) as well.

It can be noted in passing that this choice serves as a prime example of the unintended effects of a quick choice of a most important design variable, a choice made on what seemed to be perfectly obvious grounds. Oxide fuel had unrivaled burnup capability. But it brought irritating problems later, in breeding—in inherent safety, in processing, and in almost every other important area of fast reactor performance. But the decision had been made and it was final. EBR-II became the last of its line.

The priority to development of the fast breeder reactor had begun with an influential report to the president in 1962. The AEC then intended to use EBR-II, the Fermi-1 reactor built by APDA for Detroit Edison, and the SEFOR reactor built in Arkansas by GE, to establish reactor fuels and materials, safety properties, and operating characteristics of a first generation of fast reactors. Argonne was authorized to build a Fast Reactor Test Facility (FARET) on its Idaho site, as an advanced facility for testing components.

EBR-II began operation in 1964 and was still operating in the mid-eighties when IFR development was initiated. Its purpose and mission then had little to do with its purpose when it was built. It was to be the pilot plant for a whole new technology. The "oxide revolution" had overtaken it. Not only the metal fuel, but the pool configuration, and the on-site processing—all were abandoned. The national direction had turned away from these Argonne selections. The new direction was to be oxide fuel, "loop" configuration (only the core inside the reactor vessel, the coolant piped to heat exchangers outside the vessel), and centralized spent fuel processing in a large standalone plant located somewhere else, not on-site. A complete reversal of direction, really, for in addition to its technical rationale, the dramatic change was driven by unusually significant changes in the personnel of the Atomic Energy Commission and in the accompanying congressional support. These developments had impacts on Argonne that shook the laboratory to its very core.

Impressed by the rapid development of the submarine reactor under Rickover's single-minded direction, powerful elements of the Joint Committee on Atomic Energy of both houses of Congress felt that progress on the breeder had been too slow. Research, they said, had been preferred over a single-minded emphasis on going ahead and building fast reactors. Construction was what was required. Led by Milt Shaw from Rickover's staff in Washington, in late 1964 ex-Naval Reactors personnel were put in charge of fast reactor development at the AEC.

Shaw modeled fast breeder development on Rickover's success with the Shippingport reactor. [11] Shippingport was a 60 MWe PWR prototype for civil power generation and for aircraft carrier propulsion. Its construction was directed by the government. It began operation in 1957 and operated for about twenty-five years. Applying this experience to the breeder development program meant managing the breeder program as a construction project whose technology was a settled issue. Detailed management from the AEC in Washington meant drastic changes in management structure and national policies for breeder development. One basic design variant of the breeder reactor was selected, and that decision was frozen in place. There was to be no more explorative development. There would be proof by testing of the components of the selected reactor type. It was made plain to the Laboratory that that was all that was going to be necessary, and certainly all that would be supported.

Oxide fuel and the "loop" configuration of the reactor cooling system similar to the submarine reactor were now to be the designated design of the fast reactor. At the time, these did not seem unreasonable choices. The possible advantages of metallic fuel hadn't yet been identified and its disadvantages were well known and thought to be serious. On the other hand, the disadvantages of oxide fuel in a sodium-cooled reactor had not been recognized, and if they had, they probably would not have been thought important. And the advantages of the pool, particularly its possibilities for greater protection against serious accidents, were not seen at all.

In this way the "oxide-loop" variant of the fast reactor became the international choice. U.S. influence was still strong at this time. Argonne, for almost twenty years the R&D center for fast reactor development, was not convinced that the new directions were sufficiently well researched. Argonne was not opposed to new directions, but thought the whole field of fast reactor development still too new to decide on a single direction and put all eggs in that one basket.

The decision to implement Rickover's methods was fateful. Everything followed from it. It wasn't recognized at all that Rickover's mission was far different from the appropriate mission of a breeder program at the stage it was, and that that mattered. For submarine propulsion, a perfected version of the pressurized water technology, and that only, robust in coping with movement, and with a very long-lived fuel, along with haste arising from defense needs to get on with its introduction, were the main requirements. Perfection was vital in the hostile environment of a submarine, perfection right from the start. Cost was secondary. None of this applied to the breeder. The breeder program needed more research to establish its best directions, needed to be developed in the way Argonne had done, solving problems as they arose, and using, preferably, the experience and developed skills of the nation's only proven fast reactor capability, by far its most complete capability in any case, at Argonne. The unstated assumption in adopting a version of the Rickover methodology was that the right breeder technology was a settled issue, and it was only necessary to strictly enforce Rickover's methodology to get it built perfectly. The need for haste, in the breeder case, was greatly overstated; the very much expanded program resulted in costs for breeder development that even now seem exorbitant. The constantly stated need for haste was belied by the painfully slow progress of the program itself. There is little evidence that these obvious facts were ever recognized.

EBR-II continued to run—and run—and run. In concentrating on the oxide-based fast reactor technology, the new people in the AEC-HQ breeder program offices overlooked the increasingly clear advantages demonstrated by EBR-II. After its shakedown in early years it just performed so well. But it did point in directions quite different from the accepted breeder development path. Its efficient operation was not trumpeted in any way, as such successes normally are. In fact, it became something of a national secret. By the eighties, EBR-II had been long settled into its assigned role; a patient, very efficiently run irradiation facility, to be used for testing the oxide fuel concepts of others. But this was no relic. This was tangible and this was real. The reactor operated superbly.

EBR-II technology was the beginning of specific tailoring to the use of sodium as coolant to give the best features of the fast breeder reactor. The AEC choice of oxide/loop took over the choices made for good and sound reasons for water moderated and cooled reactors. Implicit in this was the acceptance of any disadvantages, and of advantages foregone, in using oxide in the metallic sodium coolant.

In those years, as LWR commercialization spread, a very large government supported fast reactor R&D program spread over many organizations. The largest funding piece went to Hanford, which was being brought into breeder development in a very big way. The Fast Flux Test Facility (FFTF) was to be built there. This was not the Argonne breeder reactor design in any respect. Breeder development had begun in answer to a uranium resource thought to be rare; now, when large uranium discoveries changed this, breeding was still to be the goal, but the technology to be developed from this point would follow few of the Argonne ideas.

1.6 Argonne in the "Shaw Years" of Reactor Development, 1965-1973

The changes that came with Milt Shaw and his Reactor Development and Technology Division (RDT) in the AEC in Washington took a little while to be felt at the Lab. The new people had had little experience with fast reactors. They applied their naval light water reactor experience to the fast breeder, in all the major things: the reactor configuration, the choices of materials, the programs to pursue, and how they would be directed. Directing the laboratories as to how they would take part in the fast reactor development, there was to be no straying from the RDT line: the word of the time was "compliance" (with orders from AEC and later DOE headquarters). Other key words included "disciplined development," which meant a slow step-by-step march along a pre-determined route; "Quality Assurance" (QA), a detailed documentation of every decision, with many-person checking and re-checking, and oversight separate from the engineer or scientist actually doing the work; and "RDT Standards," a rigid following of specified procedures and requirements that must be developed and then enforced, taking much effort. Time eventually caught up with all of this, but nearly twenty years was to elapse in the meantime.

Argonne was critical of all of this. Jack Holl [9] quotes the laboratory director, by then Albert Crewe, at the American Power Conference in April of 1965. Crewe criticized in detail the Rickover methods now being applied to AEC programs. In brief, his point was that whereas health and safety were paramount in the submarine or in space, and therefore high levels of QA through exhaustive procedures and proof testing were justified, and similarly they were justified where health and safety issues had to do with the eventual reactor developed by the program, nevertheless such high levels of QA could not be justified in development itself. They would slow development, add to cost, and cause scarce resources to be misdirected. "If Rickover instead of Fermi had been in charge of CP-1," Crewe said, he "doubted whether the United States would have achieved the first nuclear reaction." He went on to say that existing nuclear technology rested on the experimental approach: project construction that began early, and problems solved as the project progressed. A "cult of perfection," obsessed with avoiding occasional or even imaginary embarrassments, would so retard the breeder reactor development that the U.S. eventually would end up buying breeder technology from Europe. Today, it is hard to argue against these prescient words. The U.S. did not buy breeders, but its development certainly was so slowed that time and political trends had caught up with the breeder before it could show what it could do.

Shaw cancelled FARET, dismissing it as "too small" and Argonne's program as "too unambitious." FARET was designed to test fast reactor components, its design was complete, its cost, $25 million, was reasonable, and it was ready for construction. Instead, a much larger and very much more expensive test reactor, the Fast Flux Test Facility (FFTF) would be built at Hanford to test fuels. Thus Argonne was stripped of its principal new breeder development project. A Liquid Metal Fast Breeder Reactor (LMFBR) "program office" was set up at Argonne that reported directly to Shaw's staff in Washington, effectively weakening Argonne's ability to plan and manage its breeder development. Diplomacy was dispensed with. Shaw simply informed the Laboratory that now Argonne would "serve as an extension" of Shaw's office. No independent initiatives would be allowed. Using Rickover practice, he had members of his own staff set up "watchdog" offices at Argonne, in Illinois, and in Idaho. When challenged by one of the Lab's University of Chicago review committees on the importance of independence of research, Shaw would have none of it; he simply charged that the Laboratory had "not been responsive" to his direction, and demanded further changes.

Turning on EBR-II, he stated he had no interest in supporting it as a research reactor. Because EBR-II had been designed as a prototype, Argonne had planned on proceeding with necessary experimentation to get information to be used later on the reactor itself to establish the characteristics of its type, and on its fuel cycle, fuel fabrication, processing, and recycling. Shakedown problems arising from minor equipment failures, procedures, fuel leaks, and concerns about small anomalies in reactivity were being worked on, and were straightened out in time. Not in any sense primarily a commercial power producer, Argonne viewed the fraction of time that EBR-II was at power as a useful gauge of performance but as secondary to the needs of the program of R&D on this prototype. Shaw very much did not. He demanded EBR-II be operated solely as a fuel irradiation device. Time at power mattered. He pointed to Argonne's low numbers for time at power as proof of incompetence, and demanded that changes be made in management. EBR-II already had detailed day-to-day involvement of RDT people; no action related to reactor startup or operation could be taken without the approval of Shaw's staff. (One on-site representative had a gauge installed in his office so he could monitor the power level without moving.) Humorous now, an RDT phrase of the time was "malicious compliance," a bureaucratic stifling applied when in some area or other the Laboratory had complied to the letter with RDT commands but the result had worked out badly.

The development work spread around the national laboratory system and among several private companies was driven by artificial deadlines. Arbitrary dates for breeder introduction were introduced as though there were some externally imposed imperative for construction, and then they were reset. Experience with and knowledge of fast reactors, of course, had been pretty much confined to Argonne among the national labs. But Shaw's stated purpose was to build a very large industrial and institutional base to build fast reactors in quantity. This would be done by instituting a very large and very costly program of proof testing of individual elements of the reactor, not only at the laboratories, but at commercial firms too, directed by Shaw's RDT in Washington. Ignored was the evidence, obvious from LWR development and deployment, that when U.S. commercial firms decide a technology is ready for deployment and there is demand, they go ahead and deploy it, without ceremony. The sole real question was whether the breeder technology Shaw selected was the right technology. It wasn't that some large industrial base had to be built at this stage of breeder development. LWR experience had shown that commercial incentive, implementing sound technology based on experimental work, and follow-on development where necessary by laboratories skilled at just that, would build an industrial base.

The switch to oxide fuel wasn't really any issue at Argonne. In several internal meetings related to this at that time, it was clear that Argonne technical leaders saw sense in this and accepted its likely usefulness for the fast reactor without any particular reservation. "Worth trying" seemed to be the view, but the rigidly enforced direction that was to come wasn't foreseen. In fact, the amused look on the face of the experienced leader of one meeting I attended at this time, when it was suggested that metal should continue to be the basic fuel, told the story. Oxide seemed to have proven itself as a reactor fuel, it was worth trying it in a sodium-cooled system, and the AEC wanted to pursue it. With an indulgent chuckle he went on with the oxide discussion.

But more and more procedures, documentation, inspections, and all the rest of the bureaucratic norms were gradually forced into place at Argonne. Argonne was not alone in this, of course. The other labs now in breeder development were in the same position. But Argonne had been the leading power in reactor development. It had been productive; EBR-I, EBR-II, and sound development work done in all the major fast reactor fields in every case led the world. The time-wasting effort was felt all the more keenly at Argonne.

No competitor to the oxide-loop technology directed by RDT was countenanced, from Argonne work, or indeed from any other fast reactor projects such as the Fermi-1 reactor of Detroit Edison. Argonne was to work only on the line of development selected by the AEC. EBR-II was allowed to continue to operate, and EBR-II fuel development to extend burnup was allowed to continue, because EBR-II was needed for tests on oxide fuel, at that time quite untried in a fast spectrum environment. Fueling EBR-II with longer burnup metal fuel, if it could be developed cheaply, would keep costs of oxide fuel tests down.

At this time also, the European fast reactor programs— the U.K., which had met with the usual experience of the time with metal fuel, Germany, and France all began designing prototype fast reactors of about 1,000 MWth, with electrical generation in the range of 300 MWe, and a little later Japan joined in. For these new prototypes, all chose oxide fuel like the U.S.; two chose the loop configuration and two the pool. The French Phenix reactor, a pool design, went into operation in the early 1970s, and was much the most successful of these prototypes. As in EBR-II, Phenix depended on simplicity, was inexpensive to build, and operated (with time out for one or two significant problems) into the twenty-first century.

In the meantime, another AEC decision turned out to have fateful ramifications for successful development. It was decided that the U.S. would not build a prototype, not yet anyway. A curious decision this, in a way, given the earlier stated need for haste in proceeding with construction. Rather, the new program would embark on a slow and expensive development program that would answer all questions before building a prototype. It would involve many laboratories and many industrial entities. No one of these would have overall responsibility for success development of the breeder reactor. Certainly not a laboratory—laboratories were seen as emphasizing research, as not "keeping their eye on the ball" and not focusing every effort on the directions decided by the AEC. Managing the program was the AEC itself. They would coordinate the work, manage the many organizational entities, keep them focused, and make the technical decisions, and in detail, on how and what work should be done.

Contradictory to the need for haste initially stated, this slow step-by-step proving of the technology at scale was undertaken. The replacement for the cancelled Argonne FARET advanced test reactor was a new fuel test reactor of 400 MWth without electrical generation, the Fast Flux Test Facility (FFTF), to be built at Hanford. Hanford had a distinguished history in wartime nuclear development, and had the advantage of having the two senators, Warren Magnusen, a very influential member of the Appropriations Committee, and Henry Jackson, chairman of the Energy Committee, representing the state of Washington. Beginning with a slowly evolving design, and ending with slow construction, well over ten years would elapse before FFTF would begin operation. No power-producing breeder demonstration plant prototype was ever finished. The less ambitious Argonne experimental power reactors had taken less than half this time; two or three of them, much less ambitious, had taken little more than two years in construction.

Eventually, in June of 1968, matters came to a head. Shaw charged that Argonne "could not provide the necessary technical leadership in the overall liquid-metal fast breeder program." Changes in the management of EBR-II were made, and then more changes were made, changes that went across what was left of Argonne's breeder programs. Eventually, in April of 1972, Robert Laney, a trusted veteran of Rickover's naval reactors program, was brought in as Deputy Laboratory Director to manage the Argonne breeder programs. This eased the pressure on Argonne, but by that time it was a quite different place than it had been a few years before. And Shaw himself had other worries. Dr. Dixy Lee Ray, a marine biologist, and very much a scientist, had been appointed a Commissioner on the Atomic Energy Commission itself (the five-person ruling body of the AEC). She was named chairman (as she referred to herself) of the Atomic Energy Commission on February 6, 1973.

Ray quickly gave priority to restoring confidence in laboratory and field management of the breeder program. FFTF construction under the Shaw management methods was costing huge sums, the costs had risen by at least a factor of ten over the initial estimates. The eminent Los Alamos physicist, Louis Rosen, had observed that LMFBR development was "spinning out of control" because Shaw was trying to run it as though it were a production program rather than an R&D program that required extensive and bold creativity and imagination. Hans Bethe, Cornell's distinguished Nobel laureate, went further to predict that when (not if) the breeder program failed, the University of Chicago would share the blame with the AEC for not insisting that fundamental research be done in materials, reliability and safety. [9] Something had to be done and done quickly.

Within a few months, Ray had engineered a reorganization that left Shaw with only a part of his original RDT organization, having assured herself of the necessary majority on the commission to make it stick. Shaw correctly saw this as a vote of no confidence, objected strenuously, and demanded even more direct control over the reactor people at Argonne. But the AEC itself was facing an uncertain future in the energy crisis of the time, the commission ignored his objections, and on June 8, 1973, Shaw resigned. On the June 14 the commission meeting noted Shaw's decision and moved on to other matters. [9]

1.7 The Decade of FFTF and CRBR

Shaw's departure was greeted with satisfaction at the Laboratory. He had damaged Argonne's reactor capabilities. He had driven good, experienced people, still young, out of positions where it was certain they could have done much for the nation. These men had decades of invaluable experience, with proven capability for discriminating judgment on breeder developments, large and small. Several of the very best left the laboratory entirely, pursuing careers outside fast reactor development.

Argonne was left with three breeder divisions in Illinois, in physics, safety, and components. The substantial chemical engineering programs on reprocessing and other reactor-related chemistry were gone, as were such programs in materials in the metallurgy division. There was no longer any organized reactor design capability in a reactor engineering division at the laboratory. In Idaho, EBR-II continued operation, and there were large facilities for experiments in other areas of development. A big hot cell, Hot Fuel Examination Facility (HFEF), provided remote handling services; a physics reactor core mockup facility, Zero Power Physics Reactor (ZPPR) was completed and put to use; and a reactor for carefully monitored accident tests, Transient Reactor Test Facility (TREAT), was kept in periodic operation. But there was no longer an Argonne breeder reactor development program. Argonne's assets were used for technical support to the FFTF, then under construction, and later for the Clinch River Breeder Reactor (CRBR)—the demonstration of the oxide-loop technology at scale that time caught up with and was never finished.

But by this time, the oxide-loop fast reactor design was beginning its last years. The AEC and its successor agencies had long put off building a demonstration plant. The 350 MWe CRBR was authorized in 1976, with construction to start in 1978 in Tennessee. But in 1977 the new president, President Carter, had announced cancellation of CRBR as part of his policy to slow civilian nuclear and terminate principal developmental directions. The Barnwell fuel reprocessing plant, soon to come into operation in South Carolina to reprocess civilian reactor fuel, went no further, and was terminated in a few months. The Carter policies would have significant ramifications for U.S. energy policy and for U.S. influence in nuclear matters around the world. But Congress had to agree to CRBR cancellation. It did not. Until 1983, under difficult conditions, the project continued. Finally, in November of 1983 the Clinch River Breeder Reactor (CRBR) was cancelled.

Its purpose was to demonstrate the breeder reactor technology selected in the mid-sixties. The fuel for the CRBR demonstration plant, when it was finally undertaken, was made identical to that for FFTF, because that had been proven in the exhaustive series of irradiations in EBR-II. This seems a small thing, but one questionable decision can lead to the next, which can lead to the next and so on. Passed over was the fact that fuel optimized for a test reactor has much different requirements, the pins considerably smaller in diameter, from fuel optimized for breeding. The small pins made the breeding poor enough that the goal for demonstrating breeding was difficult to meet with certainty. This led to the adoption of a French idea, wherein some uranium-only assemblies were substituted for fuel assemblies in the core. This brings the total amount of uranium in the core up, which gives an effect rather similar to larger pins. But it also has an effect on safety coefficients, possibly beneficial, but needing confirmation, and so on.

At Argonne the view was that the CRBR, as a joint project with Commonwealth Edison, had a good chance of success. The laboratory had worked with Commonwealth Edison in the past and had a high regard for their capabilities, and with the new president in 1980 it seemed we had a government that intended to really go ahead and build. But the success of the project would be measured by building it on schedule, and on budget, and knowledgably operating it when it was completed. Only then could it be a real success, demonstrating, once and for all, very important things: Sodium cooled reactors can be built economically, and with dispatch, and are very easy to operate. These are things EBR-II had shown, but now they'd be demonstrated at real scale. Argonne was not entranced with the oxide-loop technology, but if such practical things could be successfully demonstrated, that would be important.

The FFTF was different. It had been under construction for some years and was proceeding slowly. The Argonne budget had been cut periodically to make funds available for the FFTF construction, but the laboratory had survived, and occasionally the funds were borrowed—and actually returned. Still, the Argonne view was cool as to the justification for the reactor—time had passed the FFTF by. Year after year had passed since it was authorized in the mid 60s, and the need for such a facility had slowly eroded away. This was to be a facility that gave required information on oxide fuel in fast reactors. Irradiations in EBR-II had by then accumulated much detailed information on oxide fuel behavior in a fast neutron spectrum. About the only variable EBR-II couldn't investigate was the greater length of the fuel in the CRBR design. And the demonstration plants of other countries, France and the U.K., as Crewe's prescient speech to the American Power Conference had implied, were now in operation, and they had not found anything noteworthy in fuel length effects.

It could be seen now that this line of breeder development had been passed by. Time, events, and political decisions had weakened the case for both CRBR and

FFTF. The atmosphere for things nuclear had changed, and changed dramatically. Argonne's scaling up, based on the successes of its line of development, had been passed over a decade before. Whole new cadres had to be developed and trained at the other institutions that were brought into the program. The unwieldy nature of the program of many participants of varying expertise wasn't helpful. The result was delay and decisions made and remade.

It is possible that had Argonne been allowed to proceed with an EBR-III, with the expertise and experience it had accumulated, the fast reactor could have been successfully developed before the climate for nuclear changed so radically. The cost of the Argonne program would certainly have been a tiny fraction of the cost of the program that replaced it. Direction of a program in a single lab is easy enough; direction of a program of a dozen or more organizations, each with its own interests, isn't. This is so no matter how determined the hand of the directing entity. As events a decade or so later were to show, Argonne might have been able to bring the nation a new and more complete fast reactor technology, and a working plant based on it, in that same time period, had it been allowed to. All this is arguable, of course, what is certain is that the path taken expensively failed.

Congress continued to allocate civilian reactor development funds. The FFTF at Hanford continued construction and was completed and put into operation in 1982, operating for a little over a decade. The need for more data on oxide fuel became more and more remote. The FFTF irradiated metal fuel test assemblies fabricated at Argonne, which gave information on the lack of length effects on the metal fuel as well. Its last use was to be a loading of IFR fuel. The FFTF had been termed by the AEC the "flagship" for breeder development. But the very purpose of the big Fast Flux Test Facility had been swept away in the cancellation of CRBR. It was meant to lay the basis for this demonstration of the oxide breeder technology. The problem was that this wasn't what was needed; EBR-II, already in operation, could provide this. FFTF should have been a demonstration plant from the beginning, ten years before CRBR was contemplated.

The breeder program continued for some months after the cancellation of CRBR without any obvious change in direction, but it was equally obvious that a new direction was vital. It was not clear that a new direction was possible, but it was certainly clear that it was necessary.

1.8 Summary

Argonne National Laboratory came into being on July 1, 1946 as the first in the network of large national laboratories created in the wake of World War Two to investigate the atom and its implications in all aspect of nuclear energy. Argonne's experience illustrates very clearly the tension between the political and social trends of each period, which the management of the laboratory at each time had to deal with, and the need for scientific freedom for its highly qualified technical staff to do what they were there to do—to develop the knowledge needed for discriminating use of the power of the atom. In the main Argonne was successful in maintaining a balance between the two competing imperatives, more so at some times than others, but the need to protect scientific freedom and integrity was always put foremost. Argonne's institutional history is one of constant struggle, constant controversy, as the laboratory maneuvered to balance the competing requirements of the Atomic Energy Commission, later the Department of Energy, its sponsor; the University of Chicago, its contractor and its supporter; regional universities wishing to benefit from the presence of the Laboratory; and its own need to get on with its work. In the main, Argonne was successful; at the working level the Argonne scientists and engineers were almost completely buffered from the concerns of management at the top of the laboratory, and remarkable results were produced. All the principal reactor types now proven to be successful around the world were Argonne products; if not invented at Argonne, they were developed there.

Walter Zinn, the first laboratory director, the nation's foremost reactor designer in the very early years, quite truly invented the fast reactor. Under his guidance, breeder development began early at Argonne's Illinois site, and in 1949 construction of the Experimental Breeder Reactor Number One began. Operational in December of 1951, this reactor proved the breeding principle, and with it the concept of unlimited fuel supplies for nuclear energy; it generated the first electricity by nuclear means; it demonstrated the use of plutonium as nuclear reactor fuel; and it achieved many other "firsts in the world."

The Experimental Breeder Reactor Number Two followed, beginning operation on the Idaho site in 1964. This was no "proof of principle experiment," this was a complete power plant, with a number of very sound and very important innovations. Producing twenty megawatts of electricity, EBR-II demonstrated for the first time the pool concept, wherein all radioactive items, including the coolant, are kept inside the reactor vessel, along with the reactor core. It demonstrated simple fuel manufacture by remote means and simple reprocessing of spent nuclear fuel, and its return to the reactor. In a number of important areas it pointed the way.

Its technology was not perfect; more development was certainly needed to bring the Experimental Breeder concept to practical commercial completion, and more research was needed in some key areas. Solutions to some of the problems were by no means obvious. Before such work could be undertaken, and before the next step, an EBR-III, could be contemplated, a sea change took place in the support of the Atomic Energy Commission. The Experimental Breeder line of development was terminated. The nation was to take a different path to its fast breeder reactor. It would adapt light water reactor design choices to the fast breeder reactor. They fitted the breeder but not well. Twenty years passed; new people, a new generation in fact, had come into the laboratory, and the technology of the Argonne line was largely forgotten. Forgotten except for the very real presence of EBR-II, that is, but even EBR-II by this time was thought of as a sort of Argonne aberration, useful as an irradiation tool, but in no way representing a feasible line of further development.

However, the twenty years from the mid-sixties to 1984 had seen a number of advances and discoveries at Argonne, some relevant to the Experimental Breeder line of development. Some seemed to make possible characteristics that had not seemed possible before—indeed had not been recognized as important before, until time and experience had very much increased their importance. They have been unnoticed really, until in late 1983 when the line of development undertaken in the mid-sixties came to an abrupt end with the cancellation of CRBR. Suddenly these advances at Argonne seemed more important.

References

1. W. H. Hannum, Ed., "The Technology of the Integral Fast Reactor and its Associated Fuel Cycle," *Progress in Nuclear Energy*, 31, nos. 1/2, Special Issue 1997.
2. Tom Blees, *Prescription for the Planet: The Painless Remedy for our Energy and Environmental Crises*, 2008. http://www.prescriptionfortheplanet.com.
3. Joseph M. Shuster, *Beyond Fossil Fools: The Roadmap to Energy Independence by 2040*, 2008.
4. Richard Rhodes, *Nuclear Renewal: Common Sense About Energy*, Whittle Books, 1993.
5. Ira Chernus, *Eisenhower's Atoms for Peace*, TAMU Press, 2002.
6. *Proc. First U.N. International Conferences on Peaceful Uses of Atomic Energy*, Geneva, 1955.
7. *Proc. Second U.N. International Conferences on Peaceful Uses of Atomic Energy*, Geneva, 1958.
8. George Quester, "More Nuclear Nations: Can Proliferation Now Be Stopped," *Foreign Affairs*, Council on Foreign Relation, October 1974.
9. Jack M. Holl, *Argonne National Laboratory: 1946-96*, University of Chicago Press, 1997.
10. Alvin M. Weinberg, "Walter Henry Zinn: December 10, 1906 - February 14, 2000," Biographical Memoires, The National Academies Press. http://www.nap.edu/readingroom.php?book=biomems&page=wzinn.html
11. Shippingport Atomic Power Station: A Historic Mechanical Engineering Landmark. http://files.asme.org/ASMEORG/Communities/History/Landmarks/5643.pdf.

CHAPTER 2

THE INTEGRAL FAST REACTOR INITIATIVE

The concept of the Integral Fast Reactor is that of a complete system composed of a safer, more fool-proof reactor and a new process that allows recycle of its spent fuel and creates a waste product with a much reduced radiological lifetime. All this is on the same site, self-contained. The nuclear reactor development that formed it flows from the history we have described. By the mid-eighties there had been no major new reactor concept undertaken in decades, but we now believed we had an important contribution to make, and we thought we knew pretty much what technical steps would be needed. Yet it is one thing to know what to do, and quite another to know how to get it done.

2.1 Beginnings

It seems easy now to look back at the early beginnings, to look at EBR-I and EBR-II, and see what they represented—a magnificent beginning. It is also easy to see that the IFR followed from it. But in the years before the IFR's beginnings, it was not obvious at all.

The line of development was dead, abandoned twenty years before, and where could the start come in picking up the threads? The technological blocks were significant. Metal fuel had had very poor irradiation performance; moreover, it had melted partially in two fast reactors, EBR-I and the Fermi-1 reactor of Detroit Edison, and in another sodium-cooled reactor experiment, this one a thermal reactor, the Sodium Reactor Experiment (SRE) of Atomics International, a division of North American Aviation about thirty miles north of Los Angeles. And this metal fuel was uranium metal. A substantial amount of plutonium would be necessary in the fuel cycle, and plutonium was known to reduce the melting point of fuel still further.

Further, the very heart of the recycling process, the processing of spent fuel had been done in only the crudest way in EBR-II, and nothing more promising had been tried. Pyroprocessing was in fact an historical artifact. Next, the safety aspects of metal fuel were suspect given that melting had occurred in three different sodium-cooled reactors. And finally, in the absence of any better processing technology, there was no reason to think the waste product would have noteworthy properties either. So there was little incentive to go much further than noting the reasons why the original directions were abandoned.

The beginnings of what became the IFR program began in the work done by Argonne in 1977-79 for the International Nuclear Fuel Cycle Evaluation (INFCE), a thrust of the Carter administration toward non-proliferation, in an attempt to sharply constrain reprocessing and limit it to as few nations as possible. To form a sound technical basis for such an evaluation, Argonne was asked by the Department of Energy to examine all possible fuels and fuel cycles for all feasible reactor types.

The result turned up a number of very interesting things, useful to the purpose of the effort, but also very useful in stimulating thought on improving reactor technology generally. The effort is described in a very complete ANL document, ANL-80-40. [1] It described in detail what every practical fuel can do. It was this impetus that caused the first analytical work on metal fuel to be done, apart from the calculations necessary for routine EBR-II operation, in at least fifteen years.

Thinking began again on safety characteristics of metallic-fueled cores too, and on metal fuel alloys that would be used in a metal-fueled breeder, and on reprocessing possibilities. All this was part of the INFCE study. Interesting things turned up in the look at metallic fuelled reactors in the study. There had been advances of several kinds since the sixties, particularly in the techniques and accuracy of calculations. And this new investigation was being done with the benefit of this new knowledge.

Argonne had done a lot of work on analyzing accidents in CRBR, and in particular, the phenomena in oxide fuel under severe accident conditions. Metal fuel, it seemed, might have some surprising advantages over oxide in accidents whose probability was small, but whose consequences could be substantial. It was an important insight into the mechanisms of such accidents: with metal fuel, if they did occur, they would be easily contained—a very significant finding.

The work on the reactor core showed very favorable physics characteristics, the high breeding held up in uranium-plutonium versions of the lower density zirconium alloy fuel now being specified for EBR-II operations. Changes amounting to discoveries in the design of metal fuel pins had for several years been incorporated into the fuel for EBR-II. The short time in-reactor problem of metal fuel had been solved—for uranium fuel. Plutonium had not been tried. Plutonium lowers the melting point, and it mightn't make a feasible metal fuel. But data on alloys and some early experiments suggested that with the new design, metal fuel containing plutonium could work.

A line had been pursued that the reactivity feedback characteristics of EBR-II, which could automatically adjust reactor power to meet demand, might be important in safety behavior as well. This culminated later in full-scale demonstrations of the ability of EBR-II to shut itself down in the face of accidents that had hitherto not even been contemplated as possible to handle safely.

Each of these things was important in itself. But the assembly of characteristics began to take shape; only fuel cycling was missing.

2.2 The Integral Fast Reactor Initiative

In the fall of 1983, the CRBR project was cancelled. Good or bad, it was gone. There was no longer any need for the work at Argonne on oxide-fueled reactors in physics—in safety particularly—or in several other areas, specifically for CRBR. The days of laboratories proposing whole new reactor development programs were long gone. But with CRBR gone, perhaps a whole new reactor concept might be acceptable, and possible now based both on the old work and on the newer findings. If such an initiative was to be possible, the time for it was now.

The final technical piece also came from the INFCE work, the Argonne look at reprocessing in the study. It was concluded that electrorefining should be adaptable to the materials and flows from a fast reactor. And it could be done on a scale that did not require huge investments.

The decade of IFR development ran from 1984 to 1994. The IFR was to have characteristics not possessed by the current generation of nuclear reactors. These characteristics should amount to a revolutionary improvement in the prospects for nuclear power, particularly in the massive amounts needed in the future. Revolutionary improvement in literally all the important areas of nuclear power was possible in fuel efficiency, safety, waste, and non-proliferation characteristics.

The word "integral" denoted that every element of a complete nuclear power system was being developed simultaneously, and each was an integral part of the whole. (The word "fast," incidentally, simply denotes the energies of the neutrons in reactor operation, useful to know but not central to this discussion.) The reactor itself, the processes for treatment of the spent fuel as it is replaced by new fuel, the fabrication of the new fuel, and the treatment of the waste to put it in final form suitable for disposal—all were to be an integral part of the development and the product. Nothing was to be left hanging, unresolved, to raise problems later. (This had not been the case for the present generation of nuclear power, where reactor construction had moved ahead before acceptable means of dealing with its spent fuel had been worked out.)

The safety of the reactor was to be buttressed by bringing out "inherently safe," or "passively safe," characteristics made possible by the materials choices and the design. The reactor would respond to any event that could lead to an accident by "instinctively lowering" reactor power to safe levels, right up to complete shutdown if necessary. No need for any operator action, or indeed for any device at all. The reactor responded this way inherently, just due to materials used in its construction. "Passive," then, denotes the fact that no movement of control rods, or any other mechanical device, was needed when a potential accident situation arose, nor was any action by the operators: the reactor responded to trouble passively, simply taking it all in stride.

These safety characteristics were made possible by the development of the new fuel type for the IFR. A metallic fuel alloy, a liquid metal coolant, and a pool configuration provide the right nuclear and heat removal properties to accomplish this. In this way the reactor is made invulnerable to the most serious accidents that can befall a reactor. There are two kinds of these, and both types have actually happened, one at Three Mile Island-2 in 1979 when the means of transferring the heat the reactor was producing was lost and the reactor overheated and melted fuel (Loss of Heat Sink), and the other at Chernobyl in 1986 when the coolant flow through the reactor was lost, the fuel melted, and the graphite moderator caught fire (Loss of Flow).

The new fuel type also allowed new technology to be developed for processing the used fuel to cycle it back into the reactor. This gave huge benefits. It enabled the used fuel to be cleaned up and used again and again, potentially extending fuel supplies more than a hundred-fold, and, also extremely important, it very much shortened the lifetime of the radioactive waste. The process is small in size and cheap to implement. It didn't require a huge commitment for a commercial plant. (Present commercial plants are very large in size and very costly—many billions of dollars are needed.)

The final benefit from IFR fuel and fuel processing lay in the fuel product itself, as it came from the refining process. The methods of reprocessing commercial nuclear fuel in current use in several nations (but not in the U.S.) were actually developed originally to provide very pure plutonium for use in nuclear weapons. The commercial plants have that same capability, not a desirable situation from the point of view of weapons proliferation. The IFR process, on the other hand, provides a fuel form with many different elements in it, one useless as it stands for weapons purposes, but perfectly acceptable as a fuel material in a fast reactor. The process cannot purify plutonium from the IFR spent fuel; it is scientifically impossible for it to do so. The IFR technology should not contribute to weapons proliferation. On the contrary, by replacing present methods it should substantially reduce such risks.

The IFR refining process also produces a waste with less volume and a shortened radioactive life. The materials carried along in the fuel product that ruin its value for weapons are the very ones that give current nuclear "waste" (more accurately, used fuel) its long-lived radioactivity. But because they remain in the IFR fuel throughout the cycle, they are burned when recycled back into the reactor, and do not appear in the waste in significant amounts. The reduction in lifetime of significant radioactivity is dramatic, from tens or hundreds of thousands of years down to a few hundred at most. And the IFR program included the development and proof testing of very stable, inert waste forms for final disposal of this product.

The IFR technology was one in which all the pieces fitted together, dovetailing to make each part of the system complement the rest, and to make possible an entire system that could have had a truly revolutionary impact on nuclear power for the future. The implications of its termination on energy supplies for the future are plainly and painfully obvious. This was no marginal supplier of energy. It dealt with entire electrical energy needs of nations.

Argonne in 1984 still had broad reactor research and development capabilities. They weren't focused on any concrete goal at this time, and they were thin in some areas, but there were divisions of a hundred or more in reactor physics, reactor safety, metallurgy, engineering components, and chemical engineering at the main site in Illinois. By no means all R&D personnel were working on reactor technology, but they were there, and could be called upon. In Idaho, Argonne had the big reactor test facilities that had been painstakingly assembled over the years, which by then did include all the facilities needed for complete reactor system development. At the center was EBR-II, which might be made an IFR prototype, but there were also other large facilities: ZPPR for physics, where large reactors could be mocked up and operated at low power; TREAT for safety, where accidents could be simulated; the fuel manufacturing facility; and extensive remote handling hot cells, where highly radioactive materials could be handled as if in your own hands. These invaluable facilities had been carefully husbanded through the years. So Argonne still had much of the capability to do what was about to be proposed.

The Argonne discoveries of the past decade or so related to metal fuel hadn't seemed important at the time they were established. But now they didn't seem so unimportant. Most were results of experiments; some were improved understanding of relevant phenomena; some had been made possible by improved analytical techniques; and in sum, they suggested that pretty revolutionary improvements might be possible, some in ways that went all the way back to the earliest breeder development at Argonne.

The fundamental basis for the change in outlook for the EBR line of development was the radical change in the prospects for a long-lived metal fuel. This had come from tinkering with how best to design long-lived fuel for EBR-II as an irradiation facility. In this way Argonne had solved the fuel swelling problem for uranium metal by the early seventies, but in the atmosphere of the time, it went unnoticed. But in the IFR the fuel could not be just uranium; it would always contain substantial amounts of plutonium, and a substantial alloying element would be necessary as well to elevate its melting point. We needed a ternary alloy (three metals) that could behave the same way as uranium under irradiation, and then we would have our start.

The reactor would be a variant of the breeder, sodium-cooled, but with properties quite different from the oxide line of development. Safety characteristics would improve, and a better passive safety response would be possible, but just how much better we did not know when IFR development began. The reprocessing of spent fuel would be based on electrorefining, which would give a number of new advantages, not least of which would be that it could probably be developed cheaply. The waste product would have less volume, be less radioactive, and last a much shorter time. The basis for development would always remain the breeding characteristic, the potential for complete recycling and reuse of spent fuel. The IFR would have substantially better breeding characteristics than the oxide system.

2.3 Assembling the Pieces

An initiative like the IFR demands a number of things. A great national laboratory provides scope to assemble them. Largely, they are these: a self-evidently promising technology, enthusiastic support from the honored figures of the nuclear field, political support, and enough of the relevant R&D facilities and first-rate personnel to move forward rapidly initially. [2]

Argonne had superb analytical capabilities. But although calculations would help in understanding, they would not provide new basic information, nor test the new ideas. An experimental program would be needed, quite a big one. Not at first, but soon. How to get started? Step by step was the only possible way—no big program could realistically be expected to be funded right from the start. But step by key step, if each step succeeded a program could be built.

Most of the players who could affect our initiative were known. The organized anti-nuclear groups were against us, but as the start was small they didn't pay much attention to us, at least until the program was well underway. There was broad support from scientific leaders in the nuclear field, vital to credibility. Congressional support came, starting with the Idaho and Illinois delegations. Several members were on committees dealing with energy and with appropriations.

We also needed the support of the director of the laboratory, Al Schriesheim and the support of the University of Chicago as well. Schriesheim had been made director of Argonne a year or two before. He had been director of research at Exxon, and he came to the laboratory as a familiar figure in the nation's research establishment. He and his wife Beatrice, herself a chemist, worked as a team to upgrade the laboratory in many areas, from the buildings and fixtures on one hand to the quality and the national importance of the laboratory programs on the other. Specifically, Schriesheim wanted programs that had the potential for real importance to society, programs appropriate for a great laboratory. The IFR initiative fitted perfectly.

It is only the truth to say that the IFR could not even have begun, nor could it have grown into the major program it did without Al Schriesheim. In his good-humored, even-handed way he gave the IFR his active and constant support, including generous portions of the small amount of overall laboratory funding that he had at his discretion. He was unflagging in his belief in the importance of the IFR and in the importance of the laboratory's pursuing it to a successful conclusion.

In a few weeks, an IFR Review Committee was assembled. Their unimpeachable credentials, and the eminence and objectivity of every one of these founding members were obvious. Hans Bethe, the Nobel Laureate long known for his fair-minded expert testimony on all kinds of nuclear issues; Manson Benedict, founding chairman of the Nuclear Engineering Department at MIT; Max Carbon, ex-head of the nation's Advisory Committee on Reactor Safety; Lombard Squires, ex-head of the huge reprocessing installation operated by DuPont at Savannah River and long time member of the Advisory Committee as well; Spencer Bush of Pacific Northwest National Laboratory, expert on nuclear fuels and on the effects of radiation on materials, whose experience in nuclear matters, interestingly, reached all the way back to Trinity test in 1945, the world's first "nuclear device"; Richard Wilson, Mallinckrodt Professor of Physics at Harvard, at Harvard since 1955, and a distinguished scientist of wide ranging interests; Wallace Behnke, Vice Chairman of Commonwealth Edison and head of their considerable responsibilities for the CRBR project, David Okrent, Professor of Mechanical and Aerospace Engineering at UCLA, expert on reactor safety; and from time to time one or two others - Joe Hendrie, Brookhaven National Laboratory, and ex-Chairman of the Nuclear Regulatory Commission; John West, ex-Vice President in charge of nuclear power at Combustion Engineering; John Taylor, Electric Power Research Institute Nuclear VP; and Mel Coops of Lawrence Livermore National Laboratory, expert in electrochemistry. But the core group, largely intact, stayed with the program all the way through. This committee reviewed our program at the start and our progress every year thereafter, and provided the University with an insightful written report, useful in guidance and useful in persuasion as well.

The actual start came in a meeting with the President's Science Advisor specifically on the IFR proposals, after which his deputy arranged with the DOE to provide Argonne with two million dollars specifically to investigate the IFR. In large technology R&D, this isn't a large amount; the key thing was that it provided charter to begin to the IFR investigations. It provided legitimacy.

A plutonium fuel fabrication capability was set up in Idaho. The most important go/no go condition was the behavior of plutonium-bearing metallic fuel under reactor operating conditions. Such fuel needed to be fabricated and started under irradiation in EBR-II. The fuel fabrication facility was in operation and fuel in the reactor in only a few months. We were on our way.

The work expanded rapidly after that first year. With all the resources in one lab, work can move along quickly. The specialists know what they have to do, and they know the priorities. Lab-scale electrorefiners were built in Illinois and put into operation there. They worked for uranium fuel; they worked less well for plutonium-alloyed fuel but they did work. The separations, it turned out, were crude. There were three products: most of the fission products, as one product, to go to waste; a fairly pure uranium product, to be the bulk of the new fuel as the second product; and an uninviting mixture of uranium, plutonium, and the minor actinide isotopes, such as neptunium, americium, and curium, to enrich the new fuel to the fissile level needed for reactor operation. This product isn't useful for thermal reactors, nor, as we hoped, was it useful as it stood for weapons. It was admirable fuel, though, for a fast reactor, as all the higher actinides fission well in a fast neutron spectrum.

These two elements of the system, the fuel and the fuel cycle, were the mandatory first steps. They had to work satisfactorily. It would take a while before the necessary data could come in. But by this time (April 1986) we had prepared for many months for the series of EBR-II safety tests that would prototype what could be expected in an IFR. These were accident simulation tests, where the most serious kinds of accidents were initiated with the control and safety systems disabled. No operator action, no operation of the safety systems would be needed for the reactor to just ride through the two worst accident-initiating events that can befall a reactor. Both the Loss of Heat Sink, as happened at TMI-2, and Loss of Flow, which had not happened at that time anywhere in the world, were initiated at full power, one after the other. Both gave the expected result: The reactor, unaided by anyone or anything, reduced its power and shut down without harm.

Although there was a considerable international audience present to observe these tests, the press release put out by the DOE stirred no obvious interest. Then in an amazing coincidence, in that same month of April 1986, the Chernobyl accident happened. The contrast between the violent explosion there and the quiet shutdown of the reactor in Idaho, when a similar initiating event, this one on purpose, had happened to both, brought a blaze of publicity, and with it, the interest of congressional committees. After the Chernobyl accident occurred, an alert science reporter for the Wall Street Journal, Jerry Bishop, a man with a long history of reporting on nuclear power development, had the Idaho demonstration press release at hand and recognized the importance of the Idaho demonstration. His article in the Journal caused a sudden increase in support for the IFR, and enabled us to accelerate the pace and to considerably widen the scope of IFR development with the increases in funding that came in specifically for IFR development.

The new fuel passed 10 percent burnup, our goal for a commercial IFR, and went on to approach 20 percent before being taken out of the reactor for the final time. There had been no fuel pin failures at all. The fuel was a complete success. The fuel cycle work went on; the size of fuel batches was increased; waste forms for final disposal were developed; improved fabrication was being looked at; and the key thing, the full-scale work with plutonium fuel electrorefining was being readied in a new, big electrorefiner.

But by 1994 a new administration had settled into place.

2.4 Termination of the IFR Program

It was in President Clinton's second State of the Union address in early 1993 that the bad news came. Development of the reactor that consumed much of its own waste, was largely proof against major accident, and was so efficient that existing fuel supplies would be inexhaustible, was to be terminated immediately.

The Clinton administration had brought back into power many of the best-known anti-nuclear advocates. The implications of this were obvious. Ten years of development work were behind us. From tiny beginnings midway through the Reagan administration, success after success in the development work had allowed a broad and comprehensive program to be put in place. Every element and every detail needed was being worked on. With the momentum existing then, another two years or so should bring successful completion of the principal elements.

In 1994, Democrats were in the majority in both houses of Congress. Anti-nuclear advocates were now in key positions in the Department of Energy, the department that controlled IFR funding. Other anti-nuclear people were now in place in the office of the president's science advisor, in policy positions elsewhere in the administration, and in the White House itself. The IFR had survived the first year of the new administration on its unquestioned technical merits, but only after some debate within the administration.

But now the president's words in his 1993 State of the Union address were chilling: "We are eliminating programs that are no longer needed, such as nuclear power research and development." With only one such program in the nation, there could be no doubt as to who this was aimed at.

The president's budget, submitted to Congress, contained no funding for the IFR. There was no funding source to tide over a program at a National Laboratory when funding was cut off. The program was dead and that was that. Democratic majorities in the House of Representatives were nothing new; in themselves they were not especially alarming to us. During the previous ten years the votes on IFR funding in the House had always been close, and although a majority of the Democrats were always opposed, enough supported us that IFR development squeaked through each year. The Senate votes on the IFR, sometimes with Republican majorities, sometimes without, as a rule went easier. But this was a very different year: the administration had gone from weak support of the IFR program to active opposition.

Congressional staff, some of whom later moved to staff the White House, began coordinating the opposition to the IFR, in support of the administration's decision to terminate its funding. Argonne had its say in the usual Congressional hearings that followed; testimony pro and con was offered, and in the end the House of Representatives upheld the president's position. In the Senate everyone knew the vote was going to be close. The key to the Senate position was Bennett Johnston, a Democrat from Louisiana and chair of the Energy and Water Subcommittee of the Senate Appropriations Committee. This committee oversaw IFR funding.

At conclusion of the hearing, Johnston announced he had decided to fight for continued IFR development. That set the stage for a full-scale Senate floor fight, which took place over a period of several hours. The pro-IFR forces were led by Johnston himself. He had like-minded colleagues in both parties speak in support, and he himself summarized the need for continued development of the IFR. Johnston had been involved in energy matters for decades, knew his subject, and matter-of-factly put the case for the IFR. He stressed the likely need in the light of the vastness of future energy needs.

The anti-IFR forces were led by John Kerry. He was the principal speaker and the floor manager of the anti-forces. He spoke at length, with visual aids. He went through the litany of anti-nuclear assertions, articulately and confidently.

After both sides had their say the vote came, and the pro-IFR forces prevailed. But now the funding bill had to "go to conference." When this happens, a compromise committee of both houses is assembled from the relevant committees of both; its job is to consolidate different versions passed by the two houses into one bill to be sent to the president for signature into law. There was brief hope that IFR development could continue even in the face of the powerful opposition.

But the conference committee, behind the closed doors normal to such meetings, upheld the House position. There was to be no IFR funding. The IFR was dead.

A few weeks later, the mid-term elections brought Republicans into power in Congress. The IFR votes had always been politicized. With some significant exceptions, such as Bennett Johnston and the Illinois Democrats and some from other states—in fact, just enough each year to fund the IFR—the vote had generally been along party lines. Had the IFR been able to hang on for a few more weeks its development almost certainly would have gone on to completion. Instead, it became the path not taken.

2.5 Accomplishments and Status of the Integral Fast Reactor Initiative

Ten years of development grew a full scale development program from tiny beginnings—ten tumultuous years. Ten years of accomplishment for the technical staff. A lot had been accomplished—things were now known that had not been known before, important things that could truly affect what would be possible for reactors of the future. Some things were surprising, some we had guessed at and established as fact. New technologies were still possible in nuclear power, technologies that could improve every part of a complete reactor system.

But the 1994 termination of development stopped scaled up plutonium-containing fuel processing work before this vital element of the program could be finished. The equipment was being readied; then the work was stopped. Later, in connection with the EBR-II spent fuel waste disposal program that followed the termination of EBR-II operation along with rest of the IFR program, four such runs were done. Three of these have been published and will be discussed in detail in the appropriate chapter of this book. More development is needed.

Important specific things had been established:

1. Very high burnups are possible with uranium-plutonium fuel. Burnups neared 20 percent of the total uranium content in the fuel in the long burnup tests, which means that fully one fifth of the bulk fuel can be used in a single pass. There were no pin failures. The fuel could actually have gone on further had the program termination not stopped further tests. The implication of these high burnups is important: the flows to spent fuel processing are lessened, with correspondingly less equipment, less personnel, and lower fuel cycle cost.
2. Such fuel can be fabricated remotely, simply and easily.
3. The fuel can be reprocessed electrochemically. In fact, EBR-II metallic uranium fuel is being processed at pilot plant scale today as a waste disposal method.
4. The processed fuel remains highly radioactive. It requires remote handling and from a diversion viewpoint it is self-protecting. This is inherent in the

process. Quite simply, a lot of radioactivity stays in the product, harmless to reactor operation, requiring remote handling.

5. Waste can be largely stripped of the long-lived actinide elements; their amount is decreased by two orders of magnitude, and without additional steps or cost. They stay in the recycled fuel where they are burned. It is these elements that gave the waste its "forever" reputation. The lifetime of radio toxicity affecting the environment above permissible standards decreases from hundreds of thousands of years to a few hundred years.
6. The waste forms for permanent disposal have been developed and they are simple too. The waste is of two kinds. One is a metal waste form of steel from the fuel assembly hardware, cladding from the fuel elements, and zirconium from the fuel alloy. It is cast as an impervious metal alloy. The second waste form is comprised of fission products immobilized in a ceramic.
7. The excellent heat transfer characteristics of the metal fuel and coolant efficiently lower the power in proportion to need in the face of serious accident initiators like those at Chernobyl and Three Mile Island. These accidents would not have happened in an IFR.
8. The size, the scale of these things, is on a human scale. Huge installations are unnecessary.

In total, accomplishments like these and more represent a largely proven reactor technology with remarkable characteristics. Lessons can be drawn from the IFR experience, some obvious, some not. Principal among them are, first, that exciting new technologies can be developed still at this nation's great national labs. Second, that R&D success on large complex projects takes money; but not only money, commitment is required. The development of a completely new reactor system requires national political will that must last between administrations. The early research stage can be done at the nation's national laboratories, if the necessary talent and will to undertake the task happen to be present. But even for this, the national will to fund the research, and to continue to fund it, is necessary. There is no savings account at a national laboratory. Funding each year, every year, is needed from Congress at the start of the fiscal year. If not, the program must collapse and the scientists and engineers working on it must be transferred or let go.

The bigger the program, the bigger is the risk to any laboratory's well-being. But the commitment of the best of the laboratories to accomplish important things for the nation should not be underestimated. If the program is scientifically sound and important, from forces generated within them, laboratories will take such risks. To bring them to fruition, long-lasting political will must be patiently cultivated.

2.6 Summary

In late 1983, the line of development undertaken in the mid sixties came to an abrupt end with the cancellation of CRBR. Suddenly technical advances at Argonne seemed more important. Once again Argonne launched a controversial effort to bring the nation a complete new reactor system, as soundly based as development based on the original Argonne line and the advances, discoveries, and innovations of past two decades could make it. The initiative was the technology of the Integral Fast Reactor.

An initiative like the IFR requires a self-evidently promising technology, enthusiastic support from honored figures in the nuclear field, political support, and sufficient relevant R&D facilities along with first rate personnel to move the project forward rapidly initially.

Ten years of development of a complete reactor system began from tiny beginnings in 1984. Technical successes followed one after the other until a broad comprehensive program was put in place. Every element and detail needed was being worked on.

The end came when in first year of the Clinton administration "all advanced reactor development" was terminated because "it was unnecessary." The IFR was the only advanced reactor development program in the nation, so it was clearly the target. A political struggle in Congress followed. In the end the administration prevailed and the IFR program terminated on September 30, 1994.

References

1. C. E. Till et al., "Fast Breeder Reactor Studies," ANL-80-40, Argonne National Laboratory, 1980.
2. Charles E. Till, "Reminiscences of Reactor Development at Argonne National Laboratory," W. B. Lewis Lecture at Canadian Nuclear Society Meeting, June 4, 2007. http://www.ecolo.org/documents/documents_in_english/IFR-integral-fast reactor07

CHAPTER 3

THE ARGONNE EXPERIENCE

The development of an Argonne staff scientist

The ways the authors, Till and Chang, came to the careers we did, and our paths to Argonne National Laboratory, were very different. But really, they were typical of the time and were in no way unusual. It is a commonplace to say that America is a great country and many who were born in other nations have been welcomed and made their contributions to America. Certainly Argonne had its share of accomplished scientists from elsewhere. Indeed its first laboratory director was Canadian by birth. But we think it important to make a point not made as often as it probably should be: Both of us (coming from Canada and Korea) are conscious of the opportunities we were given and grateful for the lives we led and the satisfaction that came with our careers here. Argonne did that for a lot of scientific and technical people, then and now. The Argonne experience molded us both scientifically and technically in very similar ways.

The whole field of nuclear energy was new, barely twenty five years old. Much of what we did then is now common knowledge, and the practices, techniques, and even attitudes, have receded into the past. It may be of some interest, perhaps historical only, to relate a little of what it was like then, and how Argonne formed the views, the scientific techniques and scientific procedures, the ambitions, and the accomplishments as well, of the staff scientists at this great laboratory.

3.1 The Argonne Experience (Till)

In the spring of 1963, when I arrived, Argonne National Laboratory was the place to be if you wanted to be at the center of nuclear reactor development. I was sure of it. Development of nuclear fission for peaceful purposes was the big field of the future, and working in it, you were treated by most people with quite flattering respect. To be called a "nuclear physicist" was to be made out to be exceptional, like a "rocket scientist" or a "brain surgeon." I was twenty-eight years old.

I had had some experience in research, even some in calculations and experiments on reactors, that I thought might be useful in what I supposed work at

Argonne would be. My Ph.D. at Imperial College, University of London, had come in 1960, after a couple of years there. A fellowship had paid all my expenses—lucky indeed that was—and I felt very grateful to the British fellowship committee; I knew my life would be changed by the experience, and indeed it was—opportunities came my way later that I could not have dreamed of then. My PhD thesis work dealt with graphite-moderated gas-cooled reactors, Britain's choice for nuclear electric power. The British were just bringing the first big commercial ones into operation. It was my first brush with reactor calculations. The techniques were fairly simple algebraic expressions whose constants had been derived from experiments on that specific reactor type. In the most basic calculation, that of reactor criticality, a four-factor formula, as it was called, was used. Basically, each of the four factors accounted for the change in neutron population given by a single phenomenon, and the four, multiplied together, gave the reactivity in the absence of neutron losses due to leakage. Such techniques were crude; they could be, and were, carried out by hand, and because they had been modified to fit experiment on precisely that reactor type they gave pretty accurate answers, but of course they had little generality They were similar in kind to those used in calculations for all thermal reactor types at that time. (Fast reactor calculations had to be done quite differently.)

Figure 3-1. The authors in 1998 when the baton of Associate Laboratory Director for Engineering Research was passed on from Till to Chang

I had had some experience in research, having spent a couple of years doing engineering research at the Canadian National Research Council, a federally funded laboratory complex which was a sort of amalgam of a national laboratory and of the laboratory of the National Bureau of Standards here. The years were from 1956 to 1958. My work had nothing whatever to do with nuclear. But I had learned the techniques of research: the library work, the trial and error, even the attitudes and the persistence necessary for anything to be accomplished. My boss was G. O. Handegord; he was my first professional mentor, a recognizable, admirable Canadian type; a fine engineer of unmistakable integrity, enthusiastic about knowledge, talented, disciplined, humorous, and gentlemanly. I tried to be like him.

Upon returning from London a fresh PhD in early 1961, I joined Canadian General Electric Atomic Power Division. CGE had the contract to construct the Nuclear Power Demonstration reactor, NPD-2, a 20 MWe pilot plant for Canada's home-grown and successful heavy-water-moderated CANDU reactors. (I have no idea what NPD-1 might have been.) My job was to start it up. I was to specify the start up procedures, direct the startup myself, and specify and carry out a program of measurements of the physics on it. It is a mark of expectations in those days that at this point, as a young man at the start of his career, I had never been in the control room of an operating reactor, but was given this kind of responsibility anyway. I had a lot to learn, and a little over a year to do it. This was real responsibility. And I knew how little I really knew.

My senior colleague in the small physics group at Canadian General Electric was Chic Whittier. Chic had been recruited from the first-rate reactor physics group at the Chalk River Laboratory of Atomic Energy of Canada Limited (AECL). Of the many things I learned from him, one thing was particularly valuable. Chic, you see, could *estimate.* He could tell you in a minute, or more likely in seconds, what the answer to a reactor problem was likely to be, how big the effect was, and he'd get it right, more or less. I hadn't seen this done before, and it fascinated me. I went over how he thought until, perhaps not as well, I could do it too. It was his gift and it lasted my whole career. (Basically it involved identifying the principal single factor, in your head estimating its size by analogy with something you did know, and then throwing in a guessed correction for other factors influencing the answer. There's nothing particularly exceptional in the concept, but to see it done, that was the thing.) It helped me later in assessing directions for development when I was given responsibility for such things.

I had had a number of "learning experiences" in my time in the British nuclear program. When I arrived there in the autumn of 1958, two very big international conferences in Geneva had taken place, outgrowths of the U.S. "Atoms for Peace" initiative. The primary purpose of the "Geneva Conferences" was for the U.S. to share with the world the depth and breadth of its reactor development. For reactor-related work, the conferences seemed to be dominated by Argonne papers. In 1958, Argonne even shipped, and in a few days in front of everyone had assembled a working small reactor there. (And then at the end, to an equal audience, they disassembled it.) The proceedings of the conferences were many-volume sets, and in Britain they were studied intensely. I'm sure they were everywhere. They were my first acquaintance with the preeminence of Argonne in civilian reactor development.

I had worked at the United Kingdom Atomic Energy Authority laboratory in the north of England, at Windscale in the Lake country, through the summer of 1959. The reactors producing the weapons plutonium for the British program were situated there. They were big "dual purpose" reactors—electricity production as well as plutonium production for the weapons program—and were close by at Calder Hall. They were the first nuclear reactors I had ever seen. In essence, electrical generating equipment had been added to reactors that were designed to harvest plutonium. But they generated the first substantial electricity by nuclear means in Britain.

At Windscale, I joined a group doing reactor physics calculations on a version of the Advanced Gas-cooled Reactor, an improvement on the Calder Hall type. Arguments were common. The best way, or even a right way, to do many of the calculations wasn't obvious, and often it most certainly wasn't agreed on. But it didn't take long for me to notice that every argument was settled by reference to a thick Argonne report, a "greenback" document, ANL-5800. They called it "The Bible."

Famously, in 1957, there had been a fire in the graphite of one of Windscale's plutonium production reactors. The Windscale production reactors were primitive reactors, cooled by air once through, with the cooling air passing through the reactor and releasing through a tall stack. My new colleagues told me that a couple of years before when driving into work one morning, they saw black smoke coming out of one of the stacks. "That can't be good," one of them said. And it wasn't. The reactor had overheated. Its fuel elements were glowing red hot and releasing radioactive elements. Radioactivity had spread over the countryside.

I knew something of this before I arrived there because the head of the nuclear engineering group at my college and my nominal thesis advisor, Professor J. M. Kay, had led the subsequent inquiry. I remember very little of the result, other than that the fire was caused by the uncontrolled release of heat energy from the reactor graphite caused by irradiation—Wigner energy release, it was called, after the distinguished scientist who had initially identified it. Once it was realized that this was the cause, afterward it was easily annealed out, and indeed, nothing like an uncontrolled release ever happened again. And the contamination was taken in stride by the public; there wasn't much concern expressed anywhere. This, after all, was a country that had been bombed pretty thoroughly not much more than a decade before. Principally, they simply made sure that milk from the pastoral countryside wasn't used until the contamination died away, a matter of some months.

But Imperial College, where I was, was a remarkable place in its way, and J.M. Kay was certainly a memorable character. A small man with a commanding bulldog presence, he oversaw the nuclear power program like a Victorian schoolmaster. The buildings of Imperial College were not the "dreamy spires" of Oxford. They were red brick and they all were wedged into a city block or two in the busy London district of South Kensington. But the faculty was impressive, with a number of Nobel laureates; the college had a fine reputation, and it was said that it had recently been described by the government as "Britain's MIT."

The British graduate student experience was very different from that in the U.S. or Canada. The British custom was that graduate students certainly had access to any number of applicable courses that were offered as a matter of course. But we were not obliged to attend any of them. You decided what knowledge you needed. You were assessed as you went along. If you were deemed "a good man" as the English said then, you had a lot of assistance, advice, and help. But if you were not so favored, it was a pretty lonely place. When I went to see Professor Kay about starting my PhD program, he looked at me evenly and said, "Ah yes, Till, and what do you propose to write on?" When I started to stammer a reply he added, "Well! I suggest you go to the library and read on your subject, and when you decide what you wish to write on, come back and see me."

After some time I did, and he said, "Yes, well, go and see Geoffrey James (one of the professors)." He then smiled his wintry smile and added, "And keep at him." He himself gave a class or two, which I was careful to attend. A couple of years later, when my actual advisor, Professor C. G. James, told him my thesis work was complete and I was ready to graduate, he said, "Have him give a colloquium." I did so, a day or two later, in front of an impressive army of professors. I had derived some mathematical expressions that were new and when I derived them on the blackboard for this august assembly, in their honor I skipped about every other step in the reasoning. He asked me afterward to come to his office, and when I did he told me immediately that he would arrange everything to have me graduate as soon as the University of London's two-year minimum requirement was up. He then said with a hint of a smile, "Well Till, I must say you have an *impressionistic* way with mathematics."

I talked to him again when he led my PhD examination. It took less than forty minutes. I owe Imperial College a great deal. I believe the experience reinforced an unfortunate stubborn tendency to proceed on my own, but may have helped in what came later.

By late 1962, NPD-2 had started up, without incident, thankfully, and I had completed the series of physics measurements on the operating reactor. The measurements bore out the accuracy of the design calculations quite remarkably well, except for criticality. That wasn't predicted as well as it should have been; generally it's about the easiest thing to get right. Later it was found that the engineers designing the fuel had installed small spacer springs of a nickel-chromium alloy and because the design change had seemed (to them) so trivial, hadn't bothered us with it. But the extra neutron absorption in the nickel-chromium had made a difference. It didn't make any difference in eventual reactor operation, but in startup more reactivity had to be added than should have been necessary, and that remained a puzzle until the reason appeared later, very much as an afterthought from the fuels folk. Last-minute design changes shouldn't be made without noting them—rather an obvious lesson, but one that occasionally has to be relearned, it seems.

We were done with NPD-2, the utility had taken it over, there wasn't another reactor on the horizon yet, and so I thought I'd move along to where the fields were greener. In a few weeks I had offers from nuclear firms in the U.S., but when Argonne made a generous offer, I jumped at it. I was going to the premier lab in the world in civilian nuclear power development. I was on my way.

I describe these things to give an idea of what the nuclear field was like in those early days, and so you can see pretty much what I was when I came to Argonne. In writing of the laboratory I can describe things only as I saw them, of course. I might not have always seen things clearly, and I want to be clear on what I knew then and on my own limitations. One thing: I didn't fool myself. I knew some reactor science. But my knowledge wasn't deep. I knew that. I expected to be overwhelmed by my new colleagues. I wasn't particularly apprehensive, really. I could learn. I'd done it before. I intended to stay a couple of years, learn all I could, and then move on to where I imagined the money was better. In the event, of course, my lifelong career turned out to be at Argonne.

3.1.1 Argonne Life in the Early Years

Argonne that spring was a laboratory of perhaps twenty scientific divisions, all with some connection to practical nuclear matters; the connection was direct for the reactor development divisions, but rather faint for divisions like high energy physics. Scientists and engineers ran the place. Scientists and engineers with nuclear knowledge and backgrounds were scarce still, and we were treated that way. The administrative divisions were there to help, and help they did, willingly and fully. But if they didn't give us the things we needed, when we needed them, well, the "program people" had little patience. Changes were made, and made rapidly. Meetings were called by afternoon, changes made by the following morning.

I was assigned initially to the Reactor Engineering Division. This was a very large division of some few hundred people; it contained all the reactor physics, reactor safety, reactor engineering, and the reactor-related construction projects at the Laboratory—in fact, all the central elements of reactor development and design. Most Argonne divisions had a hundred or so scientists, engineers, technicians, and support people. Reactor Engineering was a lot larger. It was broken into two separate divisions later that fall, and later again it split into three.

There was remarkable freedom in those years. Only the quality of your work mattered. Of course, the closer you were to the development side, as opposed to pure science, the more likely you were to be asked to investigate a particular area, generally part of a larger developmental area; a new reactor type, perhaps, or possibly a phenomenon that was not yet understood. In my case, I was asked to plan a thorough program of plutonium fueled Light Water Reactor experiments. What kind of conversion of non-fissile uranium to plutonium might be possible in reactor cores more tightly packed with fuel than normal had been the question; now the question was how the characteristics changed with plutonium fueling. The supervision ended right there. That was all the instruction you got and that was typical. You said "yes, I can do that," and you were on your own. You pursued your own path. Nobody supervised you. You defined what needed to be done; you requested, and got, whatever you needed—equipment, time on the reactor to do your experiments, anything. When you were ready to explain what you proposed to do, if it was a big project, and always if there were safety questions, you presented it to a meeting called for the purpose by the division director. All the leadership of the division was present. If your proposal wasn't sensible, it would be stopped at that point. But up to then you had had all the resources and time you needed to get it right.

If your program passed this test, and yours was a major initiative involving the laboratory's reactors, you went before the laboratory's Reactor Safety Review Committee (RSRC). This committee was composed of the best of people from divisions with specialties in all the major disciplines. Their responsibility was to protect the laboratory from mistakes with the laboratory reactors that could have really serious implications. I was privileged to appear in front of this committee a time or two. I think of my performance in front of the committee with some amusement today, but I assure you I didn't then. It was an active committee. Questions came one after another, aimed at every conceivable technical weak point, and the follow-up questions demanded detailed answers. Logical weaknesses were ruthlessly, and I imagined joyfully, exposed. It was exactly what such a committee should be.

I remember vividly driving home one night, after I'd been at the lab for a few months, with a sense of exhilaration unlike anything since. I was thinking exactly this: "I am a member of a minor aristocracy. Nobody tells me what I have to do. I am doing exactly what I know how to do. I am given all the equipment and all the help I need to do it. And I am very well paid. What a life this is."

In the division were the men whose names were on the papers of the Geneva Conferences that I had read in Britain. I hadn't studied them as much I should have.

But I had gone over and over the ones in my field. At the laboratory, the generation of Walter Zinn was mostly gone—men like Joe Dietrich, a man with unusual depth in his views of the nuclear enterprise generally; John West, the engineer who built the Experimental Boiling Water Reactor, later the GE commercial reactor system; Harold Lichtenberger, who had been principal in the building of EBR-I. These men had gone with Zinn when he left to form his own company. But just behind them were Bob Avery, Harry Hummel, Len Koch, Harry Monson, Al Smith, Walt Loewenstein, Dave Okrent, and others, very much also names I knew, men whose names were on the Geneva papers, were internationally known, and who were the leadership of the division.

I soon found that Argonne, as might be expected, was not composed entirely of first-rate scientists and engineers. There were more than a few. And they stood out. But the really talented, the truly exceptional were just that; they were exceptions. And it was on those exceptions that the reputation of the laboratory had been built. It's always that way, in every field, I imagine. At Argonne the good ones were very good indeed.

A prime example was Argonne's noted reactor physicist, Robert Avery. Bob Avery was especially respected, and for good reason. An exceptional talent, with unusual and enviable powers of logical thought, he was widely knowledgeable and effortlessly influential. He was wise, in a way that few are, not only in technical matters but in political and administrative matters too. He was about to be promoted, for at the time he was head of the Reactor Physics theoretical group, and he was about to become director of a newly created Reactor Physics Division later that year when the big division was divided.

Avery's influence in all reactor development, and in the affairs of the entire laboratory, only increased in the thirty years that followed. And very important to my decisions on undertaking the IFR later, he influenced my thinking a lot; Bob always represented the attitudes of the Lab as they had been in the first years I was there. All through the years, he took *initiatives*. He pushed and prodded and persuaded the DOE to underwrite programs and projects needed for successful development. This or that program would establish some important property of the fast breeder reactor. In later years, they were reactor safety programs, and without them, key elements of the IFR would still be in doubt.

In those early years he was reserved, dignified really, more than he chose to be later. A session with Bob was always a mental workout. He looked at problem from every angle. When Bob was finished there wasn't much uncertainty left to talk about. And he was a critical judge of technical talent. He had infinite patience in explaining his position to those with talent, and little with those who could not or would not understand. Particularly dreaded was an abrupt sentence I chose occasionally to use myself later, when, after perusing an unsatisfactory analysis that he had asked for, he said dismissively, "I could have written this myself."

Further, the very *attitudes* that Bob represented I came to accept as my own. It seemed to me it was up to us, at the laboratory, to identify, and to do everything possible to get accepted, the directions that fast reactor development must take. We had firsthand knowledge of what the R&D was saying. But there was more to it than that. It was one thing to come to believe you knew what direction development should take. It was quite another to believe you could actually make it happen. Bob showed us that you could, if you knew how to go about it. You could actually make big things happen. Perhaps in the later years attitudes like this remained more common than one would think, and sometimes in places where they might least have been expected.

Figure 3-2. Till, Avery, Schriesheim, and Lewis (*l to r*)
at the Avery's retirement reception

An anecdote: A few years into the IFR program, a Soviet delegation of reactor specialists, bureaucrats and politicians, visited the laboratory. The head was the equivalent of our Secretary of Energy, a man with some technical background and an unusually combative conversational style. After we had exchanged courtesies, he began by emphatically stating that the Soviet program was superior to ours. I responded in kind, saying that we had tried the directions they were pursuing and that they were not good enough. They would lead to failure, I said, and good luck to them. I then described the IFR program. He was silent for a while and then said, "Where did these ideas come from?" Knowing his equivalency to our Energy Secretary, I said they came from the program of the Department of Energy. "No," he said, "They came from your laboratory; the good ideas always come from the laboratories." Incidentally, later at dinner we found a good deal of common ground and passed a pleasant enough evening.

In my first years at Argonne, I found that it was perfectly acceptable to pick up the phone and ask technical advice of someone very eminent in his field. If it might help progress, it was expected. I took full advantage of this. Many times in those first years I was on the phone with men I'd never seen or spoken with previously, who willingly and enthusiastically shared their specialized knowledge with me. They didn't know me from Adam.

The collegial technical meetings were free-wheeling, no-holds-barred discussions, often argumentative, sometimes impassioned. The disputes came over the right interpretation of an experimental result, argument whether a calculation had been done correctly, dispute over the degree to which a phenomenon was being properly understood. To say this was free and full discussion doesn't quite catch the flavor. But things got settled, and mostly they were settled correctly. And at the end, people walked out of the room still colleagues, still friends, laughing and talking. This was just Argonne doing business.

When I was done with my first draft of the plutonium experiments, I went over to the division's theoretical group to ask some questions, and I was told that Bob Avery would know, and was ushered into his office. When he heard my first question he asked me the context, and then suggested I describe my work to him topic by topic. All through this description, which took probably an hour or so, he merely looked at me mildly, silent, and after I'd finished a section, and waited, he said, "Sounds about right." Then silence again until I'd finished the next section. Then a similar gentle comment came. He had no responsibility at that time for my work. He was making his time available to a very junior colleague he had not yet met. At the end he got up, came around the desk, and said, "That'll do allright, Chuck. Good to meet you." I was satisfied that what I'd done was okay.

Later I found that he had been responsible for the volume that the Windscale people had called "The Bible," ANL-5800. This Bible was modestly titled "Reactor Physics Constants." Avery had edited chapters contributed by a large number of people. Discussing it once, he said to me, "I just assumed there was a mistake on every page. It was up to me to find it." That's "editing."

Although Walter Zinn had left the Lab several years before, his influence was still strong, and could be felt in the number of Zinn stories told again and again. Usually they reflected no credit whatever on the story teller, and were savored for that reason. Zinn apparently had a very limited appetite for foolishness and none for delays and excuses. I met him many years later at a conference close to his home in Florida. He asked me about the temporary buildings the lab had started with in the late 1940s and whether they were still there. He said, "When they told me how long it would take to put up permanent buildings, I said just give me Quonset huts. *Everything that is worth doing will be done in ten years anyway.*" That was an attitude to be respected.

There were a few people left who had worked with Fermi, but by then not very many. There were occasional references to him in the early years I was there. One thing was the "salary curves" that the lab used to correlate pay with performance. The top one was labeled the "Fermi curve." And later, I worked with a very gentlemanly colleague, Bill Sturm, who was a graduate student of Fermi's, and also present when CP-1 went critical. Bill occasionally reminisced about him. Fermi had observed in a wondering way that the fear people had of radiation was interesting; after all, it was a perfectly natural phenomenon. But generally, it can be said there weren't that many references to Fermi. Zinn, though, stayed a presence for a very long time.

By early fall, my meeting with top division technical management had been arranged. The plutonium experiments were to be assessed. The plan was complete. It covered all the necessary experiments and measurements to establish the behavior of plutonium oxide fuel in water-moderated reactors, material I had gone over with Avery and more. There was an additional important point, a difficult point that I hadn't tackled when I talked to Avery. The safety of the proposed program of measurements had to be assured. The building provided no containment to speak of, the reactor was operated manually, had limited over-power capability, and the smaller amount of delayed neutrons from plutonium fission gave less margin for mistakes in reactivity.

I had made the only possible case. There wasn't going to be any further construction done on the building itself under any circumstances. The containment that the light building offered was all the containment there would be. I had therefore set bounds on the allowable experiments, so that with reasonable care, any possible accident would be mild. In such an accident the steel cladding on the fuel rods should not be damaged, so there could be no radioactive release under any circumstances that I had outlined. The analysis came with detailed calculations and I felt sure of the results.

But the reactor was on the Illinois site of Argonne, not in the remote desert of Idaho. After a full review, and a thorough discussion, with opinion from everyone, the decision of the division director, by now the same Bob Avery, came with quiet finality. "No such experiments will be undertaken." No further discussion, no further advice or approval was needed or sought. The "program people" had used their judgment. They were the best in the world, and they had decided this was inadvisable, and that was that. No fuss, no procedural delay, no waste of effort; this potential set of experiments had been looked at, understood, and terminated before their start for good reason. It was a lesson in the Argonne way.

Argonne by this time was one of several national laboratories. Each laboratory had its own specialty, and by intent there was considerable overlap in scientific fields between laboratories. Argonne's was civilian reactor development, of course. I was once told by someone very much in a position to know that Zinn, to his credit, not only did not think Argonne should have a monopoly on such work, but thought that having other laboratories with other ideas, and other reactors, would increase the pace of development, and that was what was important.

Financed by the federal government through the Atomic Energy Commission (AEC), the laboratories were and are operated by private contractors such as universities, large companies, and sometimes entities set up specifically for the purpose. This arrangement is unique to the U.S. In all other nations, the governmental body equivalent to the AEC runs their laboratories directly. The contract for the operation of Argonne National Laboratory was held by University of Chicago, the birthplace of Argonne. With interludes when the management was shared by the University with other universities or corporations, but mostly alone, the University of Chicago is the contractor for Argonne to this day. The university then and now administers the laboratory with a very light hand in day-to-day matters, but is very concerned that the research be of the highest possible quality, and concerned with the selection of the top management, in particular with the selection of laboratory director.

Part of the quality assessment was a system of review committees, which conducted annual reviews, typically lasting two or three days, of the work of each of the major technical divisions. These were University of Chicago review committees, but made up of eminent scientists and engineers from other universities and from other national laboratories. Each Argonne division was reviewed by its own committee of specialists, who gave the university a report of their findings. One of the subsidiary benefits of this was that the Argonne people who made the presentations came to know principal people from the other labs.

And the work of the other laboratories was highly regarded at Argonne. Their principal scientists and engineers were treated accordingly, with respect and often with considerable curiosity as to their progress on the subjects they were working on. Again, it has to be remembered that at this time there was a lot that was new; important quantities and phenomena remained to be established. Often the work was the first done in the world on a subject. The earliest review committee I presented my work to was in 1965, when Dick Lewis (later killed in a plane crash as a national champion in aerobatics) and I had completed a set of measurements that followed on from work the year before at Argonne on the Doppler effect in fast reactors. This is a reactivity effect fundamental to safe operation of such reactors, and the Argonne work (along with work at Atomics International by Stu Carpenter and colleagues) went some way to establishing, once and for all, the magnitude of it in different materials and conditions. I was enormously impressed by the committee and by the level of the discussion with men who had written the textbooks I had studied.

We tended to write up the work as quickly as we could and then move on. Perhaps one main paper submitted to a scientific journal or an "Argonne greenback" would be written if the work merited it, along with perhaps a four-hundred-word summary or two, for the brief twenty-minute presentations at the semi-annual American Nuclear Society (ANS) meetings. The ANS meeting sessions then were full, often crowded, with people standing, for the papers on new work. Doppler work was new and received attention. The questioning in all the sessions was lively, perhaps with several audience members in a joint discussion with the presenter and the session chairman. New insights and information occasionally made for agitated discussion. This was new knowledge being established and it was exciting. Then back to the Laboratory, back at work, and on to the next topic. Argonne had the responsibility for the fast breeder reactor and there were a lot of characteristics that were not known with certainty, and phenomena whose magnitude weren't yet known at all.

Our work day had no particular start and no particular end. Formally, the responsibility of the scientific staff was to show up at work some time during the day, and if you did you were accounted present for the day. My work in those years centered around experiments on the "Zero Power Reactors." These were temporary reactors built up of pieces like children's building blocks that could be assembled and disassembled quickly to give the information needed on key reactor characteristics. Where I was then, in Illinois, there were two such reactor assemblies for fast breeder work, and a couple more for light water reactors that were more conventional looking. The custom was to give the full responsibility for a reactor to the appropriate experimenter, who was then assigned time on the reactor and the freedom to arrange the schedules of all support personnel necessary to get the experiments done as expeditiously as possible. There were always other experimenters waiting in line to get their "time on the reactor."

We arranged for blocks of time that ranged from perhaps a week or so up to, once in my case, four months. During those periods I would come in at the regular 8:30 starting time, get routine things done during the normal working day, and prepare the reactor for the real work in the evening. The meaningful measurement work would start just after five and run as long as it had to. Midnight, or two or three in the morning was usual. Then home for a few hours and the day started again. The measurements were best suited to a schedule that had no interruption. When the set of measurements were complete, usually in a few weeks, we relaxed. A week or two of short days and then we were off on another topic again.

3.1.2 Life at the Laboratory in the Shaw Years and After

That was the Argonne way, as it was then. It began to change in a few years. The freedom to do unfettered research and the fraction of time that could be devoted to actual research became more constricted. There were two waves of this. The first came in the Shaw years, with his version of Rickover's rough methodology in managing technical projects. Its effect on the breeder program overall we outlined in the previous chapter. The second wave came later as an inexorably increasing burden of paperwork, bureaucratic oversight with little discriminating thought given to its effect on the very purpose of the laboratory work. Elaborate attention to routine safety matters was demanded, along with incessant training exercises for less than crucial activities, and the like. The technical staff of the lab, the creative people, felt the effects of these things, but differently in the two eras.

In both cases, to the scientific staff the actual bureaucratic impositions were more of an annoyance than anything else. The reduction on the quality and value of the research was something else. The laboratory did its best to maintain an atmosphere for good research and development work, and in the main, it succeeded. The reporting requirements, when there was nothing worth reporting or when the meaning of results was as yet unclear, were handled with the care appropriate to their degree of importance. Imposition of full "quality assurance" (QA) rituals on experimental equipment assembled for a measurement or two and then discarded was avoided when it could be, and minimized when it couldn't. This was poor usage of time, it was felt, but on the whole the scientists tended to take it in stride.

An anecdote: At one point in the Shaw era we were designing a large variable temperature zone in one of our low power reactors, to give an overall measurement of certain effects of temperature. As it was not an easy design, I had asked one of the most talented of the nuclear engineers to take it over and get it done. He rapidly determined that the stress analysis was not of quality, took a week or so and did it himself, analyzed the possible safety problems and dealt with them, and was going on to tackle more. At just that point, Shaw's representative on the Argonne site demanded that full QA requirements be imposed. My colleague came into my office and outlined the problem, stressing that it would take weeks to do what was demanded, and would slow everything after it was in place. "What do you want me to do, Chuck? I can do this stuff. It's easy. It just takes time. I get paid the same. But I sure can't get the real things done with this going on. What do you want me to do?"

But the thing that was damaging was something bigger. *It was the damage done to the ability of the lab to direct its own work that caused the principal damage in the Shaw era. It was the termination of the EBR line, so there would be no follow-up on the program of development for the EBR-II prototype and there would be no EBR-III.* It is true that at that time there was no hint of the kind of things that it turned out the EBR line was capable of. But that is almost precisely the point. Development work should have gone on. Discriminatingly directed, the things that were discovered later would have been known far earlier. To halt innovative breeder reactor development *at this still relatively immature stage of its evolution,* and turn the breeder program into what became a program basically of construction projects and commercial operating procedures, was a massive error in judgment. No follow-up of promise when R&D turned it up was thought to be necessary.

This breeder design chosen *had never been prototyped.* An entire reactor concept was adopted to solve a burn up problem that turned out to be temporary. Reactors have many important properties. Burnup is one—important, but only one. It was obvious (then) that metal fuel wouldn't last in-reactor and oxide would. The choice wasn't based on satisfactory resolution of problems found in research on other important facets of reactor behavior *as it should have been.* No research had been done on the implications of oxide fuel in sodium-cooled reactors when the choice was made. The loop configuration was used on water reactors and it was okay; it took no thought to carry it over to the sodium reactor. Then, when real research was done on the important characteristics of the resulting reactor, the characteristics simply had to be lived with. It would work, of course—three other countries have operated such designs in the years that followed. But it was nowhere near what sodium-cooled technology was capable of.

The flaws in Shaw's management varied in their effects on my colleagues and myself. It depended very much on where you were in the organization and what your technical specialty was. In the technical areas like chemical engineering and metallurgy whole programs of research were terminated as unneeded. The effect was to move talented and experienced people out of research work on breeder reactor properties, and to freeze development. Bethe's critical comment on this I quoted in the previous chapter. There was no replacement. Argonne had the right people. But they went on to other things.

In engineering areas and in technical management, particularly in operations areas, the changes were often severe. But in the breeder reactor specialty areas where Argonne was acknowledged to be unique, the effects even on the management of the divisions were minimal. In reactor physics, where some truly important breeder core characteristics were still very much in doubt, the discriminating judgment of a division director like Avery mattered, and Shaw and Shaw's people left this work pretty much alone. The same was true of reactor safety, where the characteristics of oxide in a sodium-cooled reactor were leading to fears of truly violent explosions, albeit in highly unlikely situations, but possibilities which had to be faced. Only Argonne could do convincing work on this, and Avery again by this time was in charge of such work.

An anecdote: In the cooperative arrangements between the U.S. and U.K. of those years, the AEC had a U.S. liaison officer on-site in one or other of the British nuclear laboratories; a technical representative, I think he may have been called. The assignment of the one in Britain at the time, Bill Hannum (later in charge of physics, then breeder safety, for the AEC and DOE, and still later, a colleague of mine at Argonne on the IFR) was coming to an end. I thought another two-year spell in Britain might be fun, so I went in to AEC Headquarters to talk about it. I was interviewed for a full day by the full complement of RDT leadership, except Shaw himself, who was absent that day. At the end of the day I got back to the man I started out with, an older man who predated the Shaw period, and who was in charge of physics at the time. During the day, I had been told with some urgency by one of Shaw's officers that Shaw would expect me to call him "any time of the day or night" when something went wrong in the British program. My physics contact was white-haired, professorial, very much an old school gentleman, and he asked me if I'd had any special instructions. When I told him this one, he snorted, "Forget it, Chuck, the British are our colleagues. You're no James Bond." I was spared any conflicted decision-making, when approval for the U.K. assignment came through, by the offer I accepted instead of a considerable promotion at my own lab.

Important for obvious reasons, budget cuts to free up money for other DOE projects and programs came fairly often in the seventies and early eighties. They probably affected the lab and its staff as much as anything. So there was certainly change, but the changes were not totally disruptive. Scientists and engineers were not really that much affected, nor, I think, did they think a great deal about such things. Their treatment as professionals they took as normal. This attitude speaks to the success of the lab's traditions, I think, in its treatment of its technical staff and its expectations for them. Their work went on without concern for their jobs.

The central role of Argonne in breeder reactor development, though, was diminished a lot, and that was what truly mattered. Bearing the responsibility alone for important programs brings out the best. When Argonne no longer built the experimental reactors aimed at developing an entire technology with a sense of urgency, a lot was lost. That is what the IFR program returned to Argonne, at least in part, and the best of the technical people thrived under those conditions. Every day brought new excitements. That is the atmosphere in a healthy first-rate laboratory, doing what it does best.

3.1.3 Argonne in the IFR Years

When the IFR program became established, people knew the goal of their work and it was a big goal indeed. They could see why their work was important. "Like the old days," it was said. Problems had to be solved—now. Others depended on them. Everyone working on IFR development could say why it mattered, in a single sentence. They knew what they were part of, why it was important, why they were doing it. I tried always to remind people whenever they'd been part of something important that had just been done. They too would make history.

I had seen the years go by with extremely expensive, barely perceptible progress in Milt Shaw's slow, step-by-perfect-step proof of every element of an imperfect concept. No demonstration at scale ever got built. It wasn't the concept that concerned me at this point—the oxide concept was gone—*it was the time it had taken up.* I could understand Zinn's impatience to get on with development, his urgency: *The pace of development really matters.* I knew which model we had to follow. Get on with it. Solve problems as we go along.

But we had so many areas to develop. We could do only so much. We had to grow the program as we went along; we didn't start out with much. The funding was always a question. Chang juggled that between the many areas that needed development. Basically, it had to come from our operating budgets. Normally, where there is construction there's a construction budget that is set apart from the normal funding for personnel, equipment, and so on. We didn't have this luxury, and budgets were tight. I knew where the big expense was going to come. We had to demonstrate the fuel cycle, all of it; fuel fabrication, spent fuel processing, and the waste processes. The processing was the key, though. We were going to have to bring the old EBR-II Fuel Cycle Facility hot cell building, by now not much more than an unused derelict, back into operation.

It would need construction; it had to be brought up to sufficiently modern standards that it could be safely put back into operation. The Argonne practice: People mattered, facilities didn't. Facilities for development work had to meet the more rigid standards of today, but that's all; they didn't need to be an art form. The thirty-year-old Fuel Cycle Facility at EBR-II was going have to go back into operation. Another practice: The most difficult job goes to the best man you have, the man with talent and drive, no matter his educational background. So Mike Lineberry got the job; Mike had already taken on the management of the hot cells a year or so before because I wanted to make sure that the newer, operating cell, HFEF, would be ready to take plutonium when the new fuel came out of the reactor for examination. Mike, a Cal Tech PhD physicist with a broad background, worked with Chang, Ray Hunter at DOE, and our budget people, and got that old facility back into operation on an almost invisible budget.

In the DOE, our counterpart was the estimable Ray Hunter. Admirable for his knowledge (his background went all the way back to the 1960s fast reactor plant, Fermi-1, of Detroit Edison), his ability to find ingenious solutions to budget problems, and his straightforward up-front dealings, Ray was really the other half in this. He wanted to get real things done every bit as much as we did. He found ways, and enough funding, just enough, came: And into operation the "refurbished Fuel

Cycle Facility" went. We pushed its development in every area as fast as we could, with Ray's sympathetic help.

An anecdote: Mike Lineberry had watched Chang present one of his intricate budget formulations to Ray in a meeting called on that delicate subject. Chang finished. There was a long silence. Finally, Mike said, Ray spoke, "Yoon," he said, "I know there's a pea under there somewhere. Is it up to me to find it?"

People, I think, took extra pride in working on the IFR. It was "the old days." I think a lot of the staff knew it. In Idaho, a smaller, closer community, technical people outside of Argonne knew about "the IFR Program" and many followed its progress. When it was over, we could take this cold comfort: We had run as far as we could, as fast as we could. Could one do more?

Figure 3-3. Till with his division directors and staff at Argonne–East

3.2 The Argonne Experience (Chang)

I was born in the year Enrico Fermi achieved the first controlled chain reaction in Chicago Pile-1. At some point, possibly when President Eisenhower's Atoms for Peace program reached Korea in the late 1950s. I convinced myself that a nuclear career was my destiny. The Korea Atomic Energy Research Institute was founded and a formal Nuclear Engineering department at Seoul National University was established in 1959. In those days in Korea, college admission was based on an entrance examination and the cutline for the Nuclear Engineering department was the highest among all the colleges and departments. I was a high school senior. Nuclear engineering became my dream. However, cold reality came as we were graduating—there were no career opportunities just then in the nuclear field in Korea. About one third of our class drifted away to other disciplines. I decided to follow my dream, and after a brief stint in mandatory military service I came to the U. S. for my graduate studies. Eventually, I settled here.

I was first introduced to Argonne in 1966, when I attended a week-long student conference at Argonne during my master's program at Texas A&M University. I thought then it might be nice to work at Argonne someday. When I finished my PhD program at the University of Michigan in 1971, I wrote to Argonne, but there were no openings at the time. So instead, I started my career at Nuclear Assurance Corporation (NAC) in Atlanta. NAC's main activity then was collecting fuel cycle data and providing supply and demand analysis services to utilities and fuel cycle suppliers. They had hired a couple of experienced engineers from GE and wanted to expand into a new service in nuclear operations reliability assurance. The idea was to collect the plant operating data, unusual occurrence data, maintenance data, etc. via Texas Instruments terminals from all operating plants, and provide trend analyses. Capability for reactor operators to query the system for recommended actions based on prior experience in other plants was also to be provided. Although the benefit of having available the data from all operating plants was obvious, the effort had to be abandoned when no one wanted to be the first to sign on.

An anecdote: Years later, when the TMI-2 accident happened, I went down to the library to check a loose-leaf publication of unusual occurrence reports to NRC from all operating plants. There I found that the same sequence of events that had taken place at TMI-2 had also taken place in the very similar Davis-Besse plant some time before, but with difficulty the plant had been brought under control. I often wondered later whether the TMI-2 could have been avoided if the nuclear operation reliability network had indeed been implemented. The operator could have queried such a system, or might even have received a live response from the other plant operators.

When Argonne positions opened up in 1974 I was excited by the prospect; I felt I had finally found a home worth devoting an entire career to. (As I learned later, I was one of the wholesale hires Till initiated that year to build up his division.) This was the boom period for nuclear. There were forty some reactors in operation, the new reactor orders were piling up every month and soon the total reactor orders would amount to well over two hundred. The Atomic Energy Commission also had ambitious plans. There were two major programmatic environmental impact statements under preparation: a Generic Environmental Impact Statement for Mixed Oxide, in anticipation of a wide spread reprocessing and recycling in LWRs, and a Programmatic Environmental Statement for Liquid Metal Fast Breeder Reactors

(LMFBR), which assumed a very rapid deployment of a large number of fast reactors by the year 2000.

My first assignment at Argonne was in core design studies for the national advanced fuels program. Carbide and nitride fuels were being investigated primarily for their potentially superior breeding performance to the reference oxide fuel. These studies were good grounding for fast reactor design the design tradeoffs involved. Soon I started directly working for Till in reactor strategy studies, the beginning of a long-lasting professional relationship. He has been my mentor throughout my entire career at Argonne and greatly influenced my philosophy on things nuclear and beyond.

The two-year period, 1977 to 1979, in which the International Nuclear Fuel Cycle Evaluation (INFCE) was carried out was very hectic—we worked long hours on calculations and analysis. With some forty countries involved, there was a lot of travel as well. Argonne took the leading role in the Fast Breeders Working Group. There was a parallel domestic program as well, the Nonproliferation Alternative Systems Assessment Program (NASAP), which was broader in scope, and like the INFCE effort, involved other institutions and laboratories. The interactions with other organizations on the merits of both technical and institutional issues were intense. Because we had built the capability to do calculations on all reactor types, we were inundated with technical feasibility reviews for all kinds of alternative reactor and fuel cycle concepts. This was an eye-opening experience for me. It broadened my perspective on ways to think about reactor design principles. It gave me a framework for how best to assess and compare various alternatives, which was invaluable later on.

It was in this period that Till was put in charge of Argonne's reactor program. Funding for breeder development across the nation was still substantial. Even laying aside the CRBR project funding, the fast breeder R&D program budget was several hundred million dollars a year. The breeder R&D programs at Argonne were carried out in the various divisions corresponding roughly to the AEC and later to its successor governmental organizations. There were separate divisions for reactor physics, safety, materials, component technology, and chemical engineering. EBR-II was also a separate division. The R&D work was coordinated by governmental national technology management centers on safety, fuels and materials, component technology, and so on. At Argonne this had the effect of scientists or engineers working in one division not having much knowledge of the programs of other divisions, or interactions with their staff and the technology they were working on. The coordination was done not at the laboratory but by the government.

Till wanted to integrate the breeder program elements at Argonne, and make feasible a return to the EBR line of design development if the opportunity were to present itself. He asked me to form a Systems Engineering and Integration Group. That was not a particularly descriptive name for the activities of the group. CRBR was still on-going, and an industry team was designing a follow-on larger prototype breeder reactor using the same technology so there was little point in drawing attention to any alternative efforts, no matter how small. But this was an exciting endeavor for me. The design effort was led by original EBR-II design engineer—Ralph Seidensticker, Jim Burelbach, Bill Kann, Wally Simmons, Ken Kuczen, Ernie Hutter, and more, drawn from various divisions, augmented by contributions from then current EBR-II staff—John Sackett, Gerry Golden, Roy McConnell, Ralph Singer, and others. Technical leaders from technology programs—John Marcheterre, Paul Huebotter, Dave Lennox, Mike Lineberry, and others were deeply involved as well.

Morale was very high. We were extremely enthusiastic about what we were doing. For the first time in many years, we felt like we were on a mission. In essence we were designing EBR-III, although we never called it that. We tried to incorporate as much as possible of the lessons learned from EBR-II and all the newer technology innovations that had arisen over the years.

One outcome of this effort was a deepening conviction that the pool-type design must be the choice for future fast reactors. As will be discussed in detail later in this book, it offered both cost reduction and the potential for a much greater degree of passive safety for the reactor. At this time, the loop-type design was the reference line of development; FFTF and then CRBR were loop designs. By mid-1982 we had put together a conceptual design of a 1,000 MWe pool-type LMFBR plant and we made the rounds of the reactor vendors. Westinghouse, who was the primary contractor for CRBR, wasn't interested. GE at the time was working on their innovative design, which later evolved to their modular fast reactor design, PRISM, and wasn't interested either. Rockwell International, though, was thinking about a pool configuration, and they were enthusiastic about forming a team with us to jointly develop a large pool-type plant design.

The Rockwell management team, Wayne Meyers and John McDonald, were very supportive and their engineering team, led by Ernie Baumeister, Jay Brunnings, Dick Johnson, and others was superb. It was pure enjoyment to work with them and I developed a very healthy respect for their capabilities. They were perhaps the best of all the collaborative teams I experienced in the fast reactor field. The final report on Large LMFBR Pool Plant was issued in the fall of 1983. This was truly a joint effort between Rockwell International and Argonne National Laboratory, but there were major contributions from Bechtel, GE, Combustion Engineering, and Chicago Bridge & Iron. The CBI contribution was in evaluating the constructability of a pool plant, and they provided very valuable lessons on designing large vessels and their components with constructability an important component of the design. When the IFR program was initiated the following year, the pool was a settled issue; it was to be a key feature along with metal fuel and pyroprocessing.

The ten-year IFR era was a very exciting time indeed at Argonne. Although there were political challenges from the beginning, on the technical front the new discoveries and pioneering nature of the development brought an era of high morale, dedication, and enthusiasm in everyone involved. There was very little extra funding for a project this ambitious, but we made up for the shoe-string budgets with effort, imagination, creativity, and drive.

A good example, right at the beginning, was the means for plutonium fuel fabrication. The EBR-II driver fuel contained no plutonium, and to start irradiation testing of the new IFR fuel alloy, we had to have a fabrication capability for plutonium-bearing fuel. There was no such capability by then at the Laboratory, so the first option we looked at was to rely on the plutonium fabrication facility at Los Alamos National Laboratory. Their schedule was not ours. Schedule was a very big concern to us. We needed to get on with it, and quickly. I asked Leon Walters, who was heading the fuels effort of the IFR Program, to estimate how long it would take to construct a new plutonium glove box fabrication facility at Argonne-West. He came back with a three-month schedule estimate. We proceeded with that plan, rather than pursue the LANL option. At the mid-point of this construction project, the DOE program manager visited the site to review its progress. He commented that there was no way we could make the schedule, and I had to agree because the work site was chaotic. But the project was completed on time, actually with a few days to spare. Walters told me that everyone was so motivated to complete the project on schedule that they worked especially effectively, including weekends throughout. A plutonium fabrication capability, starting with nothing, had been put in place in three short months.

EBR-II driver fuel at that time was fabricated in a small room in the old Fuel Cycle Facility building. The DOE safeguards office didn't think highly enriched uranium was secure there and provided funding specifically for construction of a fully safeguarded fuel manufacturing facility. I wanted to take advantage of the funding offered in this situation to alter the fabrication equipment design to increase the batch size from five kg to twenty kg so we could demonstrate the scalability of the process to higher throughputs. The fabrication people had been accustomed to the lower throughput and were somewhat reluctant initially, but eventually agreed to a twenty-kg batch capability with two crucibles of ten kg each, limited by criticality constraints. This turned out to be invaluable later, when this new Fuel Manufacturing Facility (FMF) proved able to supply metal fuel not only for EBR-II, but for FFTF as well. The Semi-Automated Fabrication (SAF) line at Hanford, originally built to supply MOX fuel for CRBR, wasn't complete, and it still required substantial funding. Supplying metal fuel from FMF required little funding. DOE made the obvious decision to convert the FFTF core to metal fuel and several lead assemblies were irradiated successfully. The core conversion to metal fuel was terminated by the later decision to shut down FFTF permanently.

Inherent passive safety potential was recognized from the onset of the IFR Program. However, it was the initiative of the EBR-II management team itself, led by John Sackett and Pete Planchon, who planned and executed the long series of plant characterization tests necessary to safely lead up to a full-scale demonstration of the ability of the reactor to survive loss of cooling accidents, without control or safety system operation. They brought this work to culmination in the landmark tests of April 1986 (described in detail in Chapter 7). At the time DOE was funding both the Power Reactor Innovative Small Module (PRISM) design by GE and the Sodium Advanced Fast Reactor (SAFR) design by Rockwell International in competition with each other. Because of our joint effort with the Large Pool Plant, it hadn't taken much to convince the SAFR design to adopt metal for its reference fuel, while GE had remained somewhat reluctant to do so. But after the inherent passive safety tests, the PRISM project manager, Sam Armijo, visited Argonne to announce their decision to switch to the metal fuel as their reference fuel as well.

It was possibly in pyroprocessing that the most exciting new discoveries and advances were made. Conventional wisdom has a development program progressing from laboratory-scale to bench-scale to engineering-scale to pilot-scale, and so on. We did not have the luxury of completing all the necessary development before proceeding with hot (radioactive) demonstration with EBR-II spent fuel in the refurbished Fuel Cycle Facility (FCF). Les Burris, who had been involved in the early EBR-II melt-refining, lead the pyroprocessing development effort with other early pioneers, like Bill Miller, Bob Steunenberg, Dean Pierce, and others. The fundamental process development had to go on at the same time the design of the electrorefiner and other equipment systems for hot demonstration in FCF was done. Fabrication, testing and eventual qualification followed, all while the process itself was being developed. Early results were vital to success. The important thing was to keep the work moving along in all the necessary areas. The equipment could be altered and corrected if process development made it necessary.

The refurbishment of the FCF was a challenge of a different nature. The facility had been constructed long before the requirements now formalized in DOE orders had been established. Bringing the facility into compliance with the DOE orders on a shoestring budget was the challenge. Some of us initially questioned its very feasibility. Mike Lineberry and Bob Phipps, who together held this responsibility, overcame repeated nerve-wracking hurdles. Harold McFarlane secured the necessary environmental permits, and Bob Benedict assured the technical readiness of the facility and process equipment systems. Refurbishing the FCF itself while at the same time proving and installing pyroprocessing equipment systems based on entirely new principles was an engineering feat, one of the principal technical accomplishments of the IFR Program.

During the early stage of the IFR program, Sadao Hattori of Japan's Central Research Institute of Electric Power Industry (CRIEPI) learned of the program and visited Argonne-West for a briefing and tour. Impressed with the potential of the IFR, he arranged a full day presentation of the IFR program in Tokyo in 1987. A key person in the audience was Ryo Ikegame, executive vice president of Tokyo Electric Power Company and the chair of the LMFBR steering committee of the Federation of Electric Power Companies. He represented the utility industry, and his support meant a signed contract with DOE in 1989 for a joint program on the IFR technology. The original scope was limited to the research on pyroprocessing. Two years later the contract was expanded to include Japan Atomic Power Company, representing all other utilities. The scope was expanded to include the fuel cycle demonstration in the refurbished FCF. A separate contract was also signed with Tokyo, Kansai and Chubu Electric Power Companies to investigate the feasibility of applying pyroprocessing to LWR spent fuel. Another new contract with Power Reactor and Nuclear Cycle Development Corporation (PNC), a predecessor of the Japan Atomic Energy Agency, was also agreed to just before the IFR Program was canceled in 1994. Altogether, these agreements represented an over $100 million contribution from Japan. These contracts were terminated when the IFR program was canceled.

The few years we collaborated with the Japanese utilities were among the highlights of my career. Given the situation with nuclear energy in the U.S., I truly believed that the IFR with pyroprocessing might be first commercialized in Japan. Ikegame had vision. He also commissioned a fact-finding utility panel headed by Yoshihiko Sumi, executive vice president of Kansai and president of JAPC later, to visit our facilities at both Argonne sites. Yoichi Fuji-ie, who served as the chairman of the Atomic Energy Commission of Japan, was also a strong supporter. He used to bring Japanese delegations to tour Argonne-West facilities. "Seeing is believing" is not an empty bromide in these situations. Over the years when utility representatives, whether from Japan or U.S., toured EBR-II and FCF, they usually went away convinced of the technology.

When I received DOE's Lawrence Award for my technical contributions to the IFR development, the award ceremony ironically took place one week after the DOE's announcement of the program cancellation. My acceptance speech included the following:

> *"When you do not have natural resources, then the recycling of nuclear fuel becomes a security issue. When you are faced with the high cost of reprocessing, the IFR pyroprocessing becomes your dream solution. When you are criticized from all directions about plutonium recycling, then the IFR's inherently proliferation-resistant fuel cycle, which never separates out plutonium, becomes your dream solution. When you are faced with the hundreds of thousands of years of containment requirement for the waste,*

then the IFR's ability to recycle the long-lived actinides and burn them in the reactor becomes your solution. When you are faced with public concern for safety, then the IFR's demonstrated walk-away safety becomes your solution. The technology ought to speak for itself."

I believed just that, then and now. The technology speaks for itself. I also thought then that the IFR should become the technology of choice for all aspiring fast reactor programs. But that did not happen. When I introduced the IFR technology to our European colleagues, one European fast reactor leader commented, "Our horse is in the middle of a river. We will sink or swim with it. We cannot switch horses in mid-stream." The answer to the question "If the IFR is so great, how come France, Japan and others have not adopted the technology yet?" would seem to be this. If the IFR technology is to be fully demonstrated, the U.S. must accept this as our responsibility.

3.3 Summary

The Argonne National Laboratory's efforts to maintain a working environment for its scientists and engineers that encouraged creativity and rewarded scientific success, and minimized bureaucratic non-essentials, were surprisingly successful over the years. In the early years, when the laboratory had very nearly complete responsibility for fast reactor development, the scientists and engineers were breaking new ground with every successful experiment and analysis. They were the nation's breeder reactor program, they had the responsibility, and they conducted themselves accordingly. Wide latitude in pursuing the common goal of contributing to a firm understanding of important reactor phenomena was taken for granted. It was an extraordinarily productive period, and Till saw the last few years of it when he arrived at Argonne in 1963.

The changes that took place in the mid-1960s took from Argonne much of the responsibility for breeder R & D, and removed all responsibility for direction of the breeder program as a whole. That ended the early period. For the next twenty years, the work at Argonne was directed, in greater or lesser degree, depending on the field, and with exceptions, by the AEC and its successor agencies in Washington. The exceptions were principally in the specialty fields of reactor physics and reactor safety, where the depth of Argonne's expertise and experimental facilities could neither be matched nor effectively reassembled elsewhere. The first decade of this period, the "Shaw Years," were particularly difficult for the technical management of the laboratory, but it can be honestly said that while the working scientists felt some impact, they were in large measure shielded from such difficulties. Then and later, as more bureaucratic requirements were imposed, the laboratory coped, always trying to keep the scientific staff working as many hours possible on their scientific work, not on non-essentials. In essence, the Laboratory tried, and in large measure successfully, to bend to such demands but not to break.

When the IFR program began to gain momentum in the mid-1980s, much of the excitement of the early days slowly returned. The laboratory once again directed its own program, and it was working on directions that were new and soundly based. They were Argonne directions. Important discoveries once again were being made across the board, in fuels, in spent fuel processing and recycling, in safety, in all the necessary fields, and it was exciting to the staff. Let no one say the pace of development isn't important. Within the very constrained budget of that IFR period, a fraction of the national breeder development budget a decade before, we moved along as fast as we possibly could. Rapid pace is vital to excitement about, and to success in, development. With new discoveries and new knowledge coming almost month by month, the IFR people could see for themselves that once again they were a part of something big. They were making history.

CHAPTER 4

IN THE BACKGROUND

"Energy security does not stand by itself but is lodged in the larger relations among nations and how they interact with one another... The renewed focus on energy security is driven in part by an exceedingly tight oil market and by high oil prices, which have doubled over the past three years. But it is also fueled by the threat of terrorism, instability in some exporting nations, a nationalist backlash, fears of a scramble for supplies, geopolitical rivalries, and countries' fundamental need for energy to power their economic growth. In the background—but not too far back—is renewed anxiety over whether there will be sufficient resources to meet the world's energy requirements in the decades ahead."

Daniel Yergin [1]

4.1 Introduction

In the face of gathering evidence that the world is facing increasing constraints on fossil energy supplies and of repeated calls for international action to reduce greenhouse gas emissions, today's apparent complacency about energy supply is not easy to understand. It is accepted, possibly even in the most rigidly anti-nuclear circles, that fast reactor fuel is unlimited and carbon-free. Only a few decades ago our nation planned ahead for assured energy supplies. From the 1960s onward to 1977, when the Carter administration derailed it, the fast reactor was given the highest priority among non-fossil sources of energy because of the ever-lasting nature of its fuel. In the thirty-plus years since, under the huge demands of the modern day fossil fuel supplies have depleted further. Nuclear energy and the eventual need for the fast reactor have been all but ignored in energy planning in the U.S. At best, nuclear energy is viewed as a useful but relatively minor contributor to the national energy supply, comparable in importance as well as in size to hydro. As a consequence, from the late seventies onward there has been no viable plan to assure adequate energy supplies for the nation in the future.

Fifty years ago, in the early days of nuclear power, fossil fuels, domestic, abundant and cheap, provided all the energy of every kind the nation could need. And to the present day, though now with the help of huge imports, they have continued to do so. The view that this will continue much as it does today seems to be generally held, even as facts increasingly contradict it. The facts today lead directly to a contrary and very simple conclusion. If we are to maintain our present prosperous manner of life, large amounts of nuclear-generated electricity will be essential. There is no other non-fossil source with the necessary magnitude. And this need is not far off. The outlook for fossil fuel sustainability is worse than commonly thought.

Conventional oil production, the free-flowing crude oil that is easy and cheap to recover, is approaching its peak worldwide. It will soon begin a long decline, if, in fact, it hasn't already. Debate continues about the details of the precise timing of the peak, and its shape, but the petroleum geologists and others who have insisted that such a production peak is coming soon and may already have occurred are now being joined by increasing numbers of prominent figures inside and outside the oil business.

The National Petroleum Council, the Advisory Body to the Secretary of Energy, in 2007 issued a report on America's oil and gas position to the year 2030. [2] The membership of the Committee included the CEOs of several of the major international oil companies and informed observers and analysts expert in oil matters. Significantly, the effort was chaired by Lee Raymond, the long-time CEO of ExxonMobil, and the architect of the Exxon-Mobil merger, a man noted for hard-headed practicality. The report was carefully worded. While stopping just short of an outright endorsement of the imminence of the Peak Oil picture, it states that "the capacity of the oil resource base to sustain growing production is uncertain," that "accumulating risks" threaten "continued expansion" of oil and gas from conventional sources, and that "several outlooks indicate that increasing oil production may become a significant challenge as early as 2015." Importantly, the uncertainty is stated to be based on the rate and timing at which "significant quantities of unconventional oil enter the supply mix," and the industry's ability to overcome increasing risks to supply. Thus non-conventional sources are looked to, not the free-flowing oil of the present, for the necessary increases in total liquids, and even then trouble may come by 2015.

The National Petroleum Council report in a sense led the way. Its authoritative membership gave credence to the repeated statements of geologists who for some years had warned of the approaching crisis of oil production reaching its peak. Today it's almost common wisdom. There is a rising chorus from every corner of the energy business stressing the need to face the implications of the imminence of peak oil. The International Energy Agency's World Energy Outlook [3] makes the point that the era of cheap oil has now passed and the IEA's chief economist, Fatih Birol, has recently stated repeatedly that "the age of cheap oil is over." [4]

Saudi Arabia is considered by all to be the key to the future of world oil production. It has the greatest daily production, varying somewhat with demand, but close to ten million barrels a day, and it has the largest estimated (by the Saudis) reserves. Today world production stands at about eighty-five million barrels a day. If world production of conventional oil is to increase, it is agreed that that is where it must come from. Saudi production has not increased significantly since 2003, even with the encouragement of steeply rising prices. When the world peak is reached, supply can no longer meet the ever-increasing demand. Prices can be expected to rise steeply as competition for the available supply increases.

The principal supply argument involves the effect of adding in "other liquids" and non-conventional oil production in postponing the peak or even raising the maximum production rate somewhat, as more broadly defined. The "other liquids" to be added in are the light hydrocarbons, above methane in molecular weight, useful as liquids, and present in substantial quantities in "wet" natural gas. Several percent of total oil consumption can be aggregated in the totals from these. Non-conventional oils are the heavy oils, principally in Venezuela, and tar sands, principally in Western Canada, which are hydrocarbons far up the carbon chain in molecular weights. They are plentiful, but sharply limited in the rate they can be recovered as useful additions to the totals of oil production. Finds in deep water, such as the recent discoveries off Brazil, and in other inhospitable locations like the far north, become economic to extract as prices rise and technology improves. But the total amount all of these can add to present production seems to be no more than a few percent.

The outlook for natural gas historically has always been closely related to that for oil. A very substantial fraction of gas produced worldwide is "associated gas" —that is to say, gas contained in reservoirs along with the liquid oil (incidentally, providing the driving force for lifting oil in the bore hole) and therefore a byproduct of oil production. The obvious difficulties in transport of a gas make much natural gas production fairly local, certainly continentally local, and gas far less geographically flexible than oil in world markets. Pipelines have limitations over extreme distances and between continents. Liquefied natural gas (LNG) has constraints of a different kind, but they too are severe, so only a very small fraction of gas produced is transported in this way. Gas is still simply flared off in places where transport is too difficult or too uneconomic, or where it has simply not yet been provided for (such as Saudi Arabia). In some cases, gas is re-injected as storage for the future, or to enhance the driving force for increased recovery of the oil itself.

In the U.S., substantial drilling for dry gas not associated with oil production over the last decades has made it the major source of domestic gas production. The U.S. today consumes about sixty-three billion cubic feet (bcf) of natural gas per day —an energy equivalent of about 10.5 million barrels of oil per day, or about half of the amount of oil that the U.S. consumes each day. Of that 63 bcf per day of natural gas consumption, we import about 1 bcf in the form of LNG, and we import about 8 bcf per day from Canada via pipelines. Thus we are about 98.5% self-reliant on natural gas supply from North America and about 86% self-reliant on natural gas supply from the U.S. Contrast that with oil, where we are much less self-reliant; only about 41% North American self-reliant and only about 27% self-reliant from U.S. sources. [2]

This picture of a large supply of natural gas has been strengthened by the increase in production from a number of the gas shale fields, at the same time that the recession (and global financial crisis) hit; as a result, there has been more gas available than needed, and the price has therefore dropped considerably. This, in turn, has led to a considerable reduction in the number of rigs that have been drilling new wells. Even with this, however, there are concerns about the ability of wells in the gas shale to produce to the targets that are being set up. For instance, the very high costs for the wells and technology required to create them has meant that, as a result of having relatively short lives, only 28% of them have returned a reasonable profit. The subject remains controversial. Some see shale gas allowing substantial increases in domestic gas production; others argue, based on the rapid declines found after initial well production, that the present enthusiasm for the prospects of shale gas is overblown. [2]

Gas production is thought to be sufficient to postpone the peak in total hydrocarbons for spans of time anywhere from a year or two up to a decade or two. But gas is not a direct substitute for oil and it is simply not plentiful enough to replace declining oil production for long, even where it can substitute for oil.

The outlook for coal is the cloudiest of all. Coal, particularly in America where large deposits exist, has for many years been cited as the ultimate fuel of last resort. Estimates of coal lasting from two hundred fifty to four hundred years "at present production rates" have been, and still are, common. And coal has been described as plentiful globally as well. The biggest deposits are in the U.S., China, Russia, and Australia; coal is by no means uniformly distributed globally. China is using rapidly increasing amounts of coal, to the point where their reserves will soon be strained. Indeed, despite being the world's largest coal producer, China is now a net coal importer. The U.S. has already mined much of its more easily recoverable anthracite and even the bituminous coal of Illinois. The softer lignite coals of Wyoming and Montana, of lower heat content, require strip mining on an ever increasing scale. Coal, for this reason, and for the carbon dioxide generated when it is burned, is of environmental concern. The principal point about coal, however, is that the amount of coal actually recoverable is very poorly defined. The coal resource itself is poorly defined: many of the current numbers date back to 1970s when the first global estimates were made. And the resource numbers, even if they were accurate, are deceiving in their estimates of the amount that can actually be mined economically. For a variety of fairly obvious reasons, the coal that can be recovered is only a fraction of the resource in the ground. The current guesses are that coal production will peak globally in the range of 2025 to 2050. This is based only on actual physical constraints. If constraints due to carbon dioxide atmospheric fears seriously enter the energy policy arena, coal will be limited to an even greater degree.

All in all, carbon and hydrocarbon availability will also certainly peak and diminish in the next twenty-five years or so. Populations and energy demand will, on the other hand, continue to increase, exponentially if the past is a guide. All the "alternative energy" sources, as defined by environmental groups, with some possible but limited exception in bio-fuels, will come to nothing on the scale of energy replacement required. Physical limitations guarantee this.

Nuclear energy has no such physical limitations. Its predicted role, however, remains marginal in all accepted mainstream predictions of future energy supplies. Again, the reasons are obvious, and directly attributable to the successful anti-nuclear campaigns of organized environmental groups.

The only result of this path, unaltered, is increasing shortage of energy—life-changing and draconian. A perusal of the current literature will demonstrate that this kind of future is looked upon with equanimity, even with enthusiasm, by those who push "alternative energies." It is a utopian recipe for global disaster. And it will not happen. Civilizations, nations, will do their best to maintain their energy supplies and will do whatever is necessary. The most realistic, and peaceful, avenue is plentiful electricity supplied by nuclear power. It will not substitute directly for all other forms of energy. But it will provide limitless electrical power. Electricity is very adaptable, and if you have it abundantly a lot of substitution can be done. A great deal of nuclear capacity will be required. Military might is not required. Attention is.

Comfortable views of nuclear power as a small element of the global energy picture are common enough. The recent report, "Nuclear Power Joint Fact-Finding" by the Keystone Center [5], is an excellent example of this. In essence, they extrapolate the situation today, where nuclear energy is a useful but small contributor to U.S. energy overall, and surveying the present state of the nuclear industry in the U.S., conclude that there will be difficulty in maintaining even this. There is no evidence of any alarm at this. The important, the all-important point, is evaded completely: The carbon-based energy system that we have relied upon completely for our nation's well being is now endangered. This is not to happen far out in the future; masked somewhat by the recession, it is happening now. Real additions of large magnitude to energy supplies are absolutely essential.

The alternative is conflict—military action, if history is any guide, as nations fight for energy resources; or pushed further, some scenario of the doomsayers whose theme is the inevitability of the collapse of civilization. The situation is serious, and it is made more serious by the fact that it is not generally recognized as such and little of use is being done. And if nuclear power is to fill the role that must be assigned to it, the IFR or something very similar is needed. In any event, the principal reactor type must possess breeding characteristics very similar to the IFR.

We will turn now to the apparent facts and the evidence underpinning them, and then go on to our principal purpose and the subject of this book, the technical features of the IFR itself. But first, we will examine further the evidence for its need.

4.2 Energy Today

Our energy supplies are threatened by a combination of factors, ominous for the welfare of all. A civilized nation must assure its energy supply. Instead, our nation has acted as if energy is secondary to a variety of other goals. In fact, the last four decades have seen one action after another seemingly almost purposefully aimed at choking off our own energy supply. Today, all of this, in no very well defined way, does seem to be fueling a growing general sense of unease about energy.

One very important thing can be done immediately. After the decades of experience we've now had with reactors, this shouldn't be controversial, but history says it will be. Today's need is simply to start a broad construction program of new Light Water Reactors, to begin to assure the nation an abundant supply of electrical energy, no matter what the scarcity of fossil fuels turns out to be, or what the national policy is for their use. Not one or two reactors, but a broad program, similar to the reactor construction undertaken in the early 1970s. Assuring electrical energy supply is a very important accomplishment, and one that will not be easy. For there is no nuclear power plant construction today at all, *nor has there been since construction was completed on plants ordered in the 1970s.* A "nuclear renaissance" was discussed with increasing seriousness in the latter years of the last (Bush) administration, but the new administration has given nuclear low priority; apart from a two-reactor plant proposed by Southern Nuclear, and apparently looked on with favor by the Obama administration, the possibility of new plants may even be receding. Killing the Yucca Mountain repository was at best not helpful. No positive face can be put on this national mistake.

There are perhaps as many as a few dozen new Light Water Reactors in various stages of planning by U.S. utilities. Some may still be contemplated. The two most advanced proposals, by Southern Nuclear, may be going ahead. These reactors are an evolutionary improvement over the reactors of the present generation of nuclear power plants. For at least two decades, the present plants have provided 20 percent of America's electricity. (Compare this to the fraction of 1 percent provided by the favorites of the day, solar and wind.) The organized opposition to nuclear power made investment in nuclear plants too financially risky in the mid-1970s. That opposition remains, and at least part of it is now in controlling positions in the present administration. The legal tools used then to wreck utility planning are still largely usable today. Some changes have been made, but any sustained construction remains unlikely.

Nevertheless, the possibility of new plants is being talked about. If such plants do come into being, they will be based on improvements to the present reactors in a number of areas. The designs exist and are referred to as "Advanced" Light Water Reactors or "Generation III." The present generation of reactors, designed in the early 1970s, and built through the late 1970s and 1980s, had expected lifetimes of forty years, and their licenses were granted on this basis. Their performance, by and large, has been superb, and with the care in maintenance they are given, there is no reason to take them out of service. Components needing replacement have been routinely replaced, and after further licensing review many have already had their licenses extended to sixty years. It can be expected that most of the rest will follow as their forty year licenses expire. Their capital cost now paid off, their costs are for operation and maintenance only, so they produce electricity very cheaply indeed. And they do it very reliably, around the clock, year after year.

These reactors, and any new generation of reactors, will run for decades. So the common wisdom these days often deprecates the need for fast reactors. In addition to greatly extending fuel supplies, fast reactors have a number of other desirable features, not the least of which is that among the possibilities for large amounts of energy generation, the fast reactor is unique in that we know both how to build them and how they will perform. Although there is reason today to get on with them, as the situation stands today they remain a project for the future, a dim and distant future, to be left to generations to come. It is said that with reliable new-generation Advanced Light Water Reactors, the need for the characteristics of fast reactors is many decades away. Fast reactors, after all, merely recycle fuel, and what is the need for that? There's lots of uranium today, it can be shown, to fuel the reactors we have, and as we are making no plans for many more, to fuel the reactors we will have.

The weakness in the argument is obvious of course. If nuclear power is to be kept as a minor contribution to our energy it won't take a great deal of uranium. But further it is said if nuclear power is used somewhat more, in the rest of the world, perhaps, more uranium will be found. There'll be lots to keep nuclear plants going —if there aren't too many of them. And nuclear won't be needed anyway, if "new alternative sources" of energy are developed.

But is this so, really? Does this really sound right? What are these "new alternate sources"? What is *their* promise? And what will be the strain on all resources globally? The U.S. is a powerful nation, but it does not stand alone, not today in this global economy. The LWR, and the Advanced LWR when installed in numbers, is the workhorse reactor technology not only in the U.S. but around the world. This reactor uses uranium prodigiously. Less than 1 percent of the uranium mined is used; the rest is discarded. It cannot use more. Fast reactors can extend this a hundredfold. Virtually all of the nuclear fuel is thus consumed.

Ultimately, resources must be conserved. Recycling, using appropriate reactor technology, is essential if nuclear is to contribute significantly to future global energy demands. Only fast reactors have the basic physical properties to extend resources significantly. There isn't a choice; a fast neutron spectrum is required and so is plutonium fuel. That's what nature decrees. With the small amount of uranium these reactors require, even very low-grade resources become economic. In fact, it is perfectly possible that the few parts per million of uranium found in sea water can be used economically in a fast reactor. The uranium resource then becomes infinite. That is the fundamental fact. All of this has been known for years, from the earliest beginnings, in fact. It is the fundamental rationale for fast reactors.

Scientific and engineering realities constrain the amount and the timing of nuclear growth. The time required for development to be completed of any truly new and different reactor type—not simply an evolution of present types—is at least thirty years. This is probably so even if there is strong national will to proceed. The deeply entrenched opposition built up over the years, much of it purposefully by anti-nuclear groups serving political ends, continues unabated today. However, the reaction to the spike in oil price in recent years suggests that when the need does come, it will come with irresistible force, and political opposition will make way for it. But the realities of the sequence of design, licensing, construction, operating experience, scaling up, and finally construction program expansion, will consume the years needed for new technology introduction in magnitude. Yet very little is being done today.

4.3 National Energy Considerations

Our subject is the technology of a specific type of nuclear reactor, the Integral Fast Reactor, and the incentive for its further development. The incentive is straightforward: the fast reactor, alone among non-fossil sources of energy, can compete with fossil fuels, the huge sources of present-day energy, in the magnitude of rates of energy supplied, and far surpass them in the magnitude of total energy possible and in the very long time its fuel resource will last. That is the justification for development of the fast reactor—its principal reason for being—and while other attributes enter, none are as important. It can create fuel to supply energy to supplant the present sources, and it can minimize the waste that results. And it is our contention that IFR technology provides the best form of the fast reactor.

But the question that follows immediately is, will it have to? We need to go a little deeper into the present energy situation, both in the world and in the U.S., and examine estimates and trends that allow a look into the future. The future we are talking about is not some distant time; action or inaction now will affect it importantly.

Energy fuels national economies; it sustains all civilized life. Against this is the reality that the source of most such energy is in the earth's crust; it's a limited endowment of the fossils of past life. That is the basis for 90% of the energy generated in today's world. These fuels are beginning to need help—new sources are needed. The question is how much and how soon. The situation for each fossil fuel is somewhat different.

4.4 The Relative Rarity of Carbon-Based Resources

What Daniel Yergin termed "in the background" are the hard facts of the present status of the world's energy resources. It is impossible to overestimate their importance. Fossil fuels produce the world's energy today. Unlike minerals, this resource was not endowed at the creation of the earth and distributed more or less uniformly around the world. This is a resource of finite amount that had to be created by living organisms. And it needed fortuitous combinations of geological and ecological circumstances that are rather rare in geological history. Under these conditions and no other, occasional deposits of carbons and hydrocarbons were created, trapped, and left for use today. But considering the extent of the earth's surface, although familiarity makes them seem commonplace, such deposits must be regarded as unusual and rare.

Their finite nature, in the face of ever rising populations, makes shortage inevitable. All fossil fuels are in finite deposits and all are measurable in amount. Population and the consequent demand for energy are growing exponentially, and there is little to suggest this is to change. When an exponential demand is to be met by a finite supply, at some point the supply will be insufficient. That is a simple statement of mathematical fact. The question, of course, is when. Will world energy supply reach that point decades or centuries into the future, is it just a few years away, or is it happening now? A great deal depends on the answer.

World crude oil production depends heavily on the output from a remarkably small number of fields. Oil fields have lifetimes measured in decades. The huge ones were discovered decades ago. There are some forty thousand oil fields in the world today, but only 360—these are the aging giant fields, each of which once held more than five hundred million barrels of recoverable oil—supply 60% of today's low-cost crude oil. Only 120 of them supply nearly 50%. Just fourteen fields, which are on average close to fifty years old, produce 20%, four of them supply 11%, and one, the world's only super-giant, Ghawar in Saudi Arabia, itself supplies fully 5% of world's production of crude oil.

The giant oil fields are found first. Because of this, there have been few such discoveries since 1980 even with improvements in exploration. A few have been found in deep water in the Gulf of Mexico, Brazil and West Africa, and in the Caspian Sea, but very, very few compared to the discoveries up to the 1960s. The giants the world has depended on for decades are all either proven to be in decline or are thought to be approaching it. Eighty percent of the world's oil comes from fields over twenty-five years old. Oil production today outweighs new discoveries by a large factor, quoted variously as between three and nine.

The one super-giant, the Ghawar field in Saudi Arabia, alone produces over four million barrels a day. It too is old. The sustainability of its present production is the subject of much debate, but there is good reason to believe that it may be approaching, or is actually in decline. There is nothing remotely like it to replace it. The other three fields that make up the four responsible for 10 percent of the world's production, in addition to Ghawar, are Burgan in Kuwait, which supplies about two million bbl/day, and Canterell in Mexico and Daqing in China, both of which supply about one million bbl/day. None are thought to be capable of increases, and all are either suspected to be, or have actually proven to be, in decline.

Liquids associated with natural gas add some production, and the relatively small remainder is "non-conventional oil." Non-conventional oil is high cost, from hostile locations, deep water, or from heavy oils, tars and bitumens. The resource base of the latter is large—bigger than for conventional oils. But the massive scale of recovery operations, limitations due to environmental concerns (both direct damage during extraction and broader greenhouse issues), and the smaller net energy difference between the product and the production operation, make these sources far inferior to the flows from the giants. They will be important during, and after, conventional oil production declines, but the production rate inevitably will be limited. There is some thought that the very large resources of shale oil will be developed under a high priced scenario. One thing is clear: production rates will be very limited, and the resource amounts are large. So if shale oil production is feasible there will be a very long, but ultimately limited supply of oil to augment other more ample sources of energy. This oil picture is troubling. The bare facts are enough to raise concern.

The magnitude of our oil use, two thirds of which is imported, makes energy independence for the U.S. impossible. Domestic fossil fuels have been abundant and cheap historically in the U.S. and many other countries, have provided all the energy of every kind that this nation needed, and have fueled what has been the world's most vibrant economy. In recent decades, increasing shortfalls in domestic energy production have been masked by increasing imports of oil, and also, to a degree, of natural gas from Canada. Oil imports have become immense. Much is made at the moment of the need for off-shore drilling, and of "alternative energies" deriving from the sun each day. In the routine of politics it's said that policy changes regarding one or the other, but not both, will bring renewed energy independence for this country. It can be said flatly and with complete certainty that the magnitude of our oil imports makes this impossible. Our oil imports alone are fully one sixth of the total oil production of the entire world. No discoveries off-shore can substantially offset such amounts, nor can all the sun-based alternatives that could in any way possibly be marshaled.

Present energy supply practices cannot be sustained indefinitely. There is nothing in prospect to replace energy imports of these magnitudes, nothing that can begin to match magnitudes of this kind. The view that things will continue much as always, with plentiful energy fueling the American economy, must confront facts that appear to tell a much different story.

U.S. oil imports are a substantial fraction of all oil on the world's market. If the very magnitude of the amount of fossil fuels consumed today is the most important fact, it is followed closely by the dramatic changes in the distributions of production and consumption of oil and natural gas. Oil production in the U.S. peaked four decades ago. At that time the U.S. produced a quarter of the world's oil and was close to self-sufficient. U.S. production has steadily declined since; it is now less than ten percent of world production. The Alaskan field added a temporary "bump" to the decline, but it too is now in steep decline. Imports have increased to the point where the U.S. alone imports a third of all the oil available on the world's markets after indigenous usage in the producing states is taken out. With world economies growing, fueled by oil, it is not hard to see trouble ahead—world oil production cannot increase apace. Recent rises and market volatility in oil and gasoline prices begin to suggest what lies ahead.

Two very different views of the world's future oil production have challenged each other in the last a decade or so, but may be coalescing somewhat at present. The so-called "economist's view" of oil production is that as oil prices increase, the amount of oil produced will rise to meet future demand. Typically forecasts are based on this assumption, resulting in forecasts of continued production growth, with no end in sight. At the other extreme is the "peak oil" view, held by a number of oil exploration geologists, but by no means all of them, that the world's total endowment is now well enough known that a peak in world oil production can now be foreseen; once we're there, like the peak in U.S. domestic production in 1970, decline is inevitable and irreversible. An excellent summary was recently provided by Campbell, founder of the Association for the Study of Peak Oil and Gas (ASPO) and a very experienced oil exploration geologist. [6] The IEA chief economist,

Fatih Birol, has recently been quoted at length to this effect [4] and IEA projections moved toward this view for the first time in World Energy Outlook 2008. [7]

There is consensus now on "conventional oil" production: it's now at its highpoint. It's not often stated this way, but there now seems to be broad agreement that production of "conventional oil," the free flowing "light sweet" crude that is easy and cheap to recover, has probably reached its peak level worldwide. This is a truly startling development. It is new and it has grave implications. "Peak oil" theorists used to be routinely ridiculed, but the fact is that increasing numbers of oil industry insiders, at the highest levels, are now also saying the same thing: the point of maximum production is approaching. Their comments attribute the peak to a variety of factors, but the fundamental point is an approaching inability to meet current demand growth. In this principal fact they are in agreement with the peak oil theorists. Controversy and debate continues about the details—the oil industry people speak of it as a plateau, a long-duration peak, and the "peak oil" people forecast more rapid decline based on their observations of past oil fields in decline. But the very fact of it now seems to be largely accepted. The principle of exponential demand meeting production rates that are slowing, leveling, and declining is not altered by details. Its implications remain the same. It is true that the current recession, higher prices for oil, and the resulting lesser usage of oil to fuel a no-longer-robust economy obscure the realities somewhat at the moment. But postponing the crisis point is not the same as preparing a solution for it.

Natural gas is linked to oil. It has been suggested that a "natural gas bridge" is possible when oil production falls, "bridging" the gap between oil scarcity and some new non-fossil source of energy, typically wind or sun. Peak gas, however, is linked to peak oil in a fundamental way. World gas supplies, even today, are not assured, and will decline, loosely linked to oil. Demand projections for world electricity forecast annual growth rates approaching 9 percent or so; all assume, either explicitly or implicitly, that "abundant and cheap," as well as "environmentally friendly" natural gas will take the increasing load. No practical credence can be given to suggestions that wind farms or other new, dilute, and variable "alternative energy sources" will make a meaningful contribution. Without cheap gas, the "gas bridge" to "alternative energy sources" collapses. The other end of the "bridge" exists in imagination only.

Most U.S. gas comes from gas-only fields, although worldwide it is produced principally where oil is found. Gas is found in three types of formation: associated gas, the gas occurring in associated oil fields; non-associated gas, the dry gas from conventional gas fields with identifiable boundaries; and unconventional, continuous gas fields in tight formations, coal bed gas, and shale. The first two have discrete boundaries, high permeability, and consequent high recoveries. Unconventional gas fields have more diffuse boundaries, low permeability, and consequently low (and consequently more expensive) recoveries.

Most natural gas produced worldwide is "associated gas," that is, gas contained in reservoirs along with oil. And it remains the case today that the world's large reserves of natural gas are closely associated with major oil fields, and as would be expected, are found in the major oil-producing countries. Until recently, natural gas has been a byproduct of oil production. Relatively recently, gas alone has been drilled for in the U.S. Not associated directly with oil, it has been the mainstay of domestic gas production. About two thirds of the gas produced in the U.S. over the last hundred years has been non-associated gas, and practically all production today is non-associated gas.

There is also a growing contribution from continuous fields. In particular, very optimistic statements have made about the future of shale gas. We'll have to see. There is a lot of it, but production is expensive and with its low permeability current wells have shown very short lives, with production rates dropping 50% in the first few months.

Conventional gas production in the U.S. peaked in 1972, a year or two after the oil peak, declined about a third by 1983, and with substantial increases in drilling activity has slowly increased since then. Today production has leveled out, and we produce about the same amount of gas as we did in 1972. Various unconventional sources have been tapped, and we import about 15%, principally via pipeline from Canada, with a small contribution from LNG as well. But that's the point. With all the publicity new domestic gas fields are given, as the Red Queen said to Alice, "it takes all the running *you* can do, to keep in the same place."

There is no significant LNG infrastructure in place, and none seriously contemplated, to ameliorate future domestic gas shortages. Only 3% of North American natural gas consumption is LNG. From 1993-2005, the number of drilled wells in Canada quadrupled, with a production increase of only 10%. With the break in the economy and lessening of demand, there is a current production surplus and consequent lower prices. Depletion of fields in Western Canada means their supplies are less assured than they have been. At current rates of usage, estimates vary, but the range usually quoted is for thirty to a hundred years of domestic supply. These estimates depend on substantial recoveries from unconventional sources.

Given the oil supply situation, it is proposed that we use natural gas for vehicles, and displace coal with gas for electricity generation to lessen U.S. carbon dioxide emissions. This will certainly have an effect on gas prices and it is arguable whether this represents a stable situation. All in all, not the assured long term energy supply we need.

Gas is continentally local. The principal point is that natural gas is a much more local resource than oil. Worldwide, the obvious difficulties in transportation make gas production of little use in the absence of pipelines, and storage is problematic. Pipelines have continental limitations, LNG has severe limitations in the cost and in the acceptability of the necessary infrastructure, and thus, for a lack of a better alternative, considerable amounts of associated gas still are simply flared off. However, although with less certainty, the point is that the large gas reserves, like the oil associated with them, are known. Recovery is the issue. Gas production, it is now thought, can be sufficient to postpone the peak in total hydrocarbons by a few years at most.

Coal outlook is the least well-defined. Although coal is mined on every continent except Antarctica, it is by no means distributed uniformly. The biggest deposits are in the U.S. and Russia, with China, India and Australia following in that order. The US has 27% of the world's coal, and coal is always thought of as our fuel of last resort. The principal point about coal, however, is that the amounts that will actually be recoverable worldwide are very poorly defined and technology dependent. The resource amounts themselves are poorly defined, some of the numbers date back to the 1970s when the first global estimates were made. Further, for coal particularly, the resource numbers are deceiving. The amount of coal that can be recovered is certainly only a fraction of the resource in the ground. Current guesses are that coal production will peak globally in the range between 2025 and 2050, based solely on physical constraints. If constraints due to CO_2 emissions begin to seriously enter the picture, the place of coal will be limited to an even greater degree.

Carbon-based fuel will soon become increasingly unavailable. All in all, it seems evident that carbon and hydrocarbon availability will peak and diminish in the next very few decades—within the next twenty-five years, and possibly sooner. With this in prospect, it is difficult to understand the complacency with which the stagnation of nuclear power in this country continues to be accepted. Real, practical additions at magnitude must made, and soon. Yet little is being done.

4.5 Uranium Is the Key to an Orderly Transition

Among the various issues routinely used by the organized anti-nuclear groups is a supposed shortage of uranium. Similar, they say, to fossil fuels, reserves of uranium are insufficient to allow it to make anything other than a small contribution to future energy needs. This is nonsense for the most fundamental of reasons, of course: the fossil fuel endowment arises from life forms, whereas uranium is a mineral, a basic component of the earth's crust. Present geological knowledge of uranium distribution is adequate for us to be certain that a huge energy contribution is possible with nuclear power. Still, a review of the considerations and the time scales will be useful in bringing out the main points.

First, remember, natural uranium as mined is 99.3% composed of the isotope U-238 and 0.71% of U-235. Only the U-235 is fissionable (there's a small contribution from U-238) in conventional light water reactors. Their fuel is enriched initially to a U-235 concentration of 3% to 5%. The energy produced is about equally split between fission of the U-235 and of the Pu-239 produced in operation by neutron capture in U-238. The heat from these reactions is used to drive steam turbines in a cycle whose conversion efficiency is about 33%. The fuel is replaced when the U-235 concentration has decreased to about 1%, which corresponds to a residence time in the reactor of about three years. A 1 GWe (1,000 MWe) light water nuclear plant refuels about twenty tonnes of fuel a year; it actually consumes only a tonne or so, with the rest currently going to waste. (A tonne is a thousand kg, or about twenty-two hundred pounds.)

It is worth noting that a coal-fired plant of comparable size consumes about ten thousand tonnes of coal per day, millions of tonnes per year. (The basis of the difference is very fundamental: Typical chemical processes emit about twenty electron volts of energy per reaction, whereas a nuclear fission reaction releases two hundred million electron volts, and therefore yields about ten million times more energy than do chemical processes; thus the huge difference in the amount of fuel required.) Nuclear energy is very concentrated: a very little fuel gives a very large amount of energy.

The overall picture for uranium is typical of a fairly common mineral in the earth's crust. It has an average crustal abundance of about 2.7 ppm, about the same as zinc. There is an estimated forty trillion tonnes of uranium in the earth's crust. To date, we have mined less than one ten-millionth of this. World consumption of uranium currently is some seventy thousand tonnes a year. To give a feel for contribution to power cost, a price of $100/lb ore, more or less the current spot market price, contributes about 0.55 cents per kWh, about 25 percent, to the price of nuclear generated electricity when used in LWRs.

At prices of $100/lb recently, a considerable amount of new exploration has resulted in an increase in both the known and the estimated amounts of uranium. The worldwide uranium resources data are jointly compiled by the OECD Nuclear Energy Agency and IAEA, updated every two years. [8] The "Identified Resources" reported in 2009 was 5.404 million tonnes as compared to the 4.743 tonnes reported in 2005. The "Identified Resources" category consists of "Reasonably Assured Resources" and "Inferred Resources." In addition, "Undiscovered (Prognosticated and Speculative) Resources" are estimated at 10.4 million tones at around this $100/lb recovery price.

The substantial increase in identified resources is the result of renewed exploration effort, and the increase in activity has continued. The exploration expenditures during a decade starting in the mid-1990s have averaged around $100 million per year. However, in recent years the exploration expenditures have shot up to about $1 billion per year. The identified uranium resources amount to approximately a hundred years' supply with current consumption and current technology, and the undiscovered resources—at current consumption levels—calculate to another two hundred years.

Deffeyes and MacGregor present an interesting table of uranium deposits in the earth's crust. [9] They give an estimate of the distribution of uranium in different types of rock. For example, Table 4-1 illustrates that shales and phosphates contain, at concentrations of ten to twenty ppm, eight thousand times more uranium than current minable ore bodies. These rocks are theoretically minable, and would give an energy gain of fifteen to thirty.

Table 4-1. Uranium Deposits

Type of Deposit	Estimated tonnes	Estimated ppm
Vein deposits	2×10^5	10,000+
Pegmatites, unconformity deposits	2×10^6	2,000-10,000
Fossil placers, sand stones	8×10^7	1,000-2,000
Lower grade fossil placers, sandstones	1×10^8	200-1,000
Volcanic deposits	2×10^9	100-200
Black shales	2×10^{10}	20-100
Shales, phosphates	8×10^{11}	10-20
Granites	2×10^{12}	3-10
Average crust	3×10^{13}	1-3
Evaporites, siliceous ooze, chert	6×10^{12}	.2-1
Oceanic igneous crust	8×10^{11}	.1-.2
Ocean water	2×10^{10}	.0002-.001
Fresh water	2×10^6	.0001-.001

Metals abundant in the Earth's crust in economic concentrations are likely to be found in deposits as minerals. To be useful, the metals need to be reasonably extractable from their host minerals. Uranium compares very well with base and precious metals. Its average crustal abundance of 2.7 ppm is comparable with that of many other metals such as tin, tungsten, and molybdenum, and many common rocks such as granite and shales contain uranium concentrations from five to twenty ppm. Further, uranium is predominantly bound in minerals which are not difficult to break down in processing.

Metals, which occur in many different kinds of deposits, are easier to replenish economically, since exploration discoveries are not constrained to only a few geological settings. Currently, at least fourteen different types of uranium deposits are known, occurring in rocks of a wide range of geological age and geographic distribution. There are several fundamental geological reasons why uranium deposits are not rare, but the principal reason is that uranium is relatively easy both to place into solution over geological time, and to precipitate out of solution in chemically reducing conditions. This chemical characteristic alone allows many geological settings to provide the required hosting conditions for uranium resources. Related to this diversity of settings is another supply advantage—the wide range in the geological ages of host rocks ensures that many geopolitical regions are likely to host uranium resources of some quality.

Society has barely begun to utilize uranium. The significant nuclear generation built by the late 1970s gave one cycle of exploration-discovery-production, driven in large part by late 1970s price peaks. [10] Because of constraints on nuclear construction driven by anti-nuclear campaigning, the amount of nuclear worldwide has been limited sufficiently that this initial cycle has provided more than enough uranium for the last three decades and several more to come. Clearly, it is at best premature, and at worse nonsense, to speak about long-term uranium scarcity when the entire nuclear industry is so young that only one cycle of resource replenishment has been required.

Related to the youthfulness of nuclear energy demand is the early stage that global exploration had reached before declining uranium prices stifled exploration in the mid-1980s. The significant investment in uranium exploration during the 1970-82 exploration cycle would have been fairly efficient in discovering exposed uranium deposits, due to the ease of detecting radioactivity. Still, very few prospective regions in the world have seen the kind of intensive knowledge and technology-driven exploration that the Athabasca Basin of Canada has seen since 1975.

This fact has huge positive implications for future uranium discoveries, because the Athabasca Basin's history suggests that the largest proportion of future resources will be as deposits discovered by more advanced phases of exploration. Specifically, only a quarter of the 635,000 tonnes of U_3O_8 discovered so far in the Athabasca Basin could be discovered during the first phase of surface-based exploration. A sustained second phase, based on advances in deep penetrating geophysics and geological models, was required to discover the remaining three quarters.

The extraordinarily rich deposits of Cigar Lake, at uranium concentrations one hundred times greater than the 0.2% concentrations commonly mined, added 577 thousand tonnes in one large increment. [11] Concentrations like this were totally unexpected.

The immaturity of uranium exploration is shown by discoveries of this kind, and it is by no means certain that all possible deposit types have even been identified. Any estimate of world uranium potential made only thirty years ago would have missed the entire deposit class of unconformity deposits that have driven production since then, simply because geologists did not know this class existed.

There is no persuasive evidence that at the current levels of nuclear power worldwide there is any danger of running out of uranium for centuries, even if the present light water reactor technology continues to be used. The known ore amount is large on this scale and the prospects for further economic discoveries are very favorable. On this basis alone, the persistent misrepresentation of the abundance of uranium resources, with the assertion that the world is in danger of actually running out, is demonstrable nonsense. Promotion of the view that limited supplies of natural uranium are the Achilles heel of nuclear power as society contemplates a larger contribution from nuclear power is disingenuous at best. Very small amounts of uranium fuel provide very large amounts of energy.

Only when nuclear is called upon to substitute wholesale for declining fossil resources will proper, efficient use of the uranium resource come strongly into play. As we shall see, the combination of resource usage and appropriate handling of nuclear waste will make the breeder reactor, of the right form, mandatory for answering the world's energy needs.

4.6 The Importance of China in All Such Discussions

China is implementing an energy program which, along with large numbers of coal plants, will bring online fifty to sixty new nuclear power plants by 2020. The Chinese economy has been growing at an average rate of 8% per year, with electricity demand growing twice that fast. The Ministry of Electric Power has estimated that 15-20% of China's present energy demand cannot be met, and that a hundred million Chinese have no access to electricity. To keep up with its rate of economic growth, China estimates that it will have to double its electricity-generating capacity every decade. And at 385 GWe of current online capacity, China even now has an electric grid system second only to that of the United States, and is now the world's largest total energy consumer.

Three quarters of China's electricity is currently coal-fired, but continued expansion of coal-fired capacity to meet the growing demand is not an option. Forty percent of China's railroad capacity is already dedicated to hauling more than one billion tons of coal per year.

China is now systematically pressing on with a multifaceted program that will make it a world leader in nuclear energy technology. China has imported commercial power plants from Russia, France, Canada, and the U.S. for the immediate benefit of nuclear energy, and to train its own cadre of engineers and operators. Today, China has thirteen reactors operating and more under construction, but today nuclear energy accounts for only about two percent of its total electricity output.

The Chinese government plans to choose one reactor design (and supplier) for its next group of nuclear plants to standardize its nuclear operations, rather than continue with widely varying designs from different suppliers. The goal is to have an increase of nearly eightfold in nuclear capacity, up to eighty thousand MW by 2020, from ten thousand MW today. Due to the size of China's electric system, even this aggressive effort will bring the nuclear share up to only 6% of installed electric-generating capacity. This program requires that at least five or six new reactors come on line each year, over the next ten years. By 2050, China plans to have four hundred GWe of nuclear capacity, more than the total nuclear capacity existing today worldwide. [12] This rate of growth is certainly feasible; it is similar to that in the U.S. through the late 1970s, before the organized opposition were able to stop further nuclear growth.

The China National Nuclear Corporation (CNNC) plans for quick expansion by adding more plants at existing sites, using the same reactor design as the operating units. More advanced, next-generation reactors will likely be chosen for new power plant sites. The program is of high national priority. At the same time as commercial-scale nuclear plants are being imported, domestic programs have been under way to develop indigenous conventional nuclear power plant designs. China will have an independent production capability for domestic use, and also for export. Their research and development program is intended to push forward on next-generation nuclear technologies.

A high-temperature gas-cooled reactor research and development program began in 1990s at Tsinghua University in Beijing, often described as China's MIT. A $30 million, 10-MW high-temperature gas-cooled pebble bed reactor (HTR-10) began construction in 1995, and started thermal testing in December 2000. In 2003, the reactor was incorporated into the power grid. In 2004 the HTR-10 demonstrated that it is “passively safe” under planned loss of coolant conditions. China plans to have a full-scale 195-MW version of its HTR-10 on line by the end of this decade, at an estimated cost of $300 million. China's nuclear industry plans to sell these 200-MW-sized reactors to utilities and in rural areas as modules which can be mass-produced and assembled quickly, with additional modules grouped together as electricity demand grows.

But the principal line of development is the PWR followed by the Fast Breeder Reactor for long term fuel economy. Three steps are planned: the China Experimental Fast Reactor (CEFR) at 25 MWe, UO_2 and mixed Pu-UO_2 fueled; the China Prototype Fast Reactor (CPFR) at 800 MWe, mixed Pu-UO_2 and metal-fueled, and the CDFR at 1000 to 1500 MWe, metal-fueled for the highest production of plutonium fuel. The intent is to expand the system as rapidly as possible. For the fast reactor, the self-contained, integral (IFR) fuel cycle is planned.

The rapid increase in PWR capacity will, however, stress the world's uranium markets. There, as with oil, gas and coal, China will play an increasingly dominant role.

4.7 The Energy Picture in Total

Soon the world's energy needs will no longer be dominated by the western world. In 2010 China has passed the U.S. in energy consumption. Oil production at best will have long since reached a plateau. Severe competition for imports of oil can be expected, as the two biggest users of oil are the two biggest importers and the two most powerful nations on earth.

Electricity growth will be very robust; electricity will near 50 % of total primary energy usage. Nuclear is assigned only a small part in most prognostications, non-trivial, but small and constant. But note carefully: for these predictions to be explainable, all growth had to be assigned to coal and natural gas.

The scenarios foreseen by a number of knowledgeable institutions and observers regarding peak production rates are quite similar. There is disagreement on the dates of the various peaks, but with a surprising degree of agreement considering the disparate interests of those involved. The graph below shows the main points. It is taken from the Association for the Study of Peak Oil (ASPO) 2006 Base Case Scenario. [13-14] Here they show that global production of conventional oil peaked in 2006, while all liquids (including non-conventional oil) and natural gas combined will peak in approximately 2010.

The combined peak of oil and gas will probably determine the peak—at least the first peak—in total world energy production and consumption. A similar coal analysis suggests that the global coal peak will occur around a decade after the petroleum/gas peak, so there will be a ten-year interval, starting around 2010, of relatively slow fall-off in total energy from fossil fuels, followed by a gradually accelerating decline.

The U.S., with its available coal resource base, will likely seek to produce liquid fuels from coal. This will reduce the amount of coal available to the export market. Peak oil is sometimes spoken of (e.g., by the team that produced the 2005 Hirsch report, "Peaking of World Oil Production: Impacts, Mitigation and Risk Management" [15]) as a liquid fuels crisis that will primarily impact the transport sector. Taking into account regional gas constraints and a likely near-term peak in global coal extraction, it is actually far broader—an all-fossil-fuel energy crisis with implications for electricity generation, space heating, and agriculture as well.

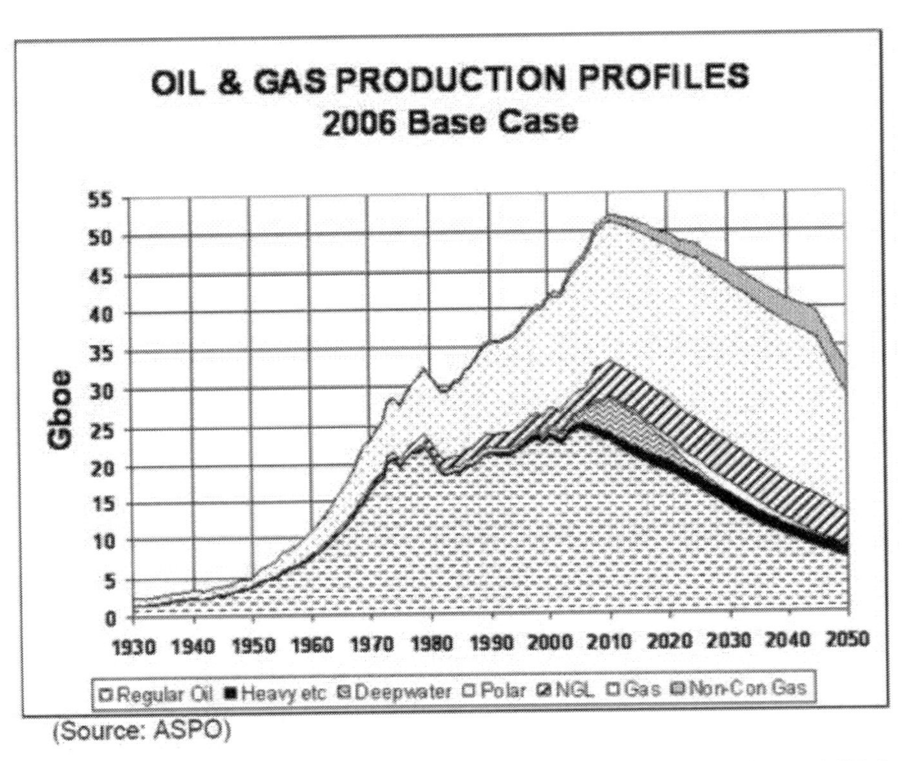

Figure 4-1. Oil and Gas Production Profiles: 2006 Base Case from ASPO [13]

Oil, natural gas, and coal together supply over 87 percent of total world energy, which stands at about four hundred quadrillion BTUs, or "quads," per year. Decline rates once the peak is past are reasonably estimated at 2.5% a year, based on field past experience. Substitution for a realistically possible 2.5 percent annual decline in all fossil fuels would require ten quads of energy production capacity from new sources each year. This is under the most optimistic assumption of no growth in energy demand. Ten quads represent roughly 10 percent of total current U.S. energy production. By way of comparison, today's total installed world wind and solar generating capacity—the result of many years of investment and work—stands at a total of less than one quad.

Tar sands and oil shale recovery are constrained by shortages of gas and water for processing. The rates of extraction will always be limited. Biofuels have low net energy gains (and so require vast areas to be devoted to biomass crops) and require substantial quantities of fresh water. Other renewables, which now produce only tiny amounts of energy—solar, wind, tidal, wave, and geothermal—have some potential for increase; however, there is no credible scenario in which these could grow enough to offset projected declines in any one of the three principal fossil fuels, much less all three together.

The ten quads each year must come from nuclear expansion. It is routinely and airily stated that expansion of nuclear power is problematic given future constraints in the availability of uranium. Properly managed, with the right reactor deployment, this statement is completely false. Cost issues are always brought up by opponents who have, through their campaigns, sought to drive the costs up. This issue is a red herring as well—the costs of nuclear plants are in keeping with other construction, and they pay off in reliable electricity over the many decade lifetime of the nuclear plant.

The findings of the 2005 DOE-funded Hirsch report [15] regarding society's vulnerability to peak oil apply also to peak coal: time will be needed in order for society to adapt proactively to a resource-constrained environment. A failure to begin now to reduce reliance on coal will mean much greater economic hardship when the peak arrives.

World fossil energy will begin to decline very soon, and there is no perfect substitute. The climate modelers and anti-nuclear activists will always point to policies with mandatory energy curtailment and societal adjustment to lower consumption levels. Policies such as these impact everything—agriculture, transport, trade, urban design, and national electrical grid systems—and everything dependent on them, including global telecommunications. Substitution of nuclear for fossil fuels is perfect for electricity. For transportation, agriculture, and other motive usages it is not—but electricity is energy and energy can be used in any number of innovative ways. No energy is no option.

Will America willingly return to the simple agrarian ways dreamed of by many in the environmental movement? This idea is influential in thinking today, while all forms of energy at least in this country are abundant, but will it withstand real scarcity? An America willingly retreating into the Middle Ages for lack of energy, while China builds itself in to an industrial powerhouse. Does this seem even remotely likely? Those who project with apparent satisfaction very limited nuclear power for America while all depleting resources show an increasing inability to sustain their historical role often have real political influence They must face these facts. Will they?

4.8 Summary

The outlook for the continued fossil fuel supplies necessary for a prosperous economy, indeed for civilization itself, is not comforting. All will peak in the coming decades: free-flowing oil already has, and the others will follow. The only real questions are the imminence of the peaks and the rate of post-peak decline.

Nuclear is the only non-fossil source of energy that has the necessary magnitude on which to base our civilization as fossil supplies dwindle. One type of nuclear reactor is not bound by uranium supply. The fast breeder reactor in effect creates its own fuel. The Integral Fast Reactor (IFR) was almost completed when its development was cancelled by a political decision. We now come to the main chapters of this book, where we describe the IFR and the underlying technical basis of its characteristics.

References

1. Daniel Yergin, "Ensuring Energy Security," *Foreign Affairs*, March-April 2006. http://www.foreignaffairs.org/20060301faessay85206/daniel-yergin/ ensuring-energy-security.html.
2. National Petroleum Council, "Facing the Hard Truths about Energy," A report of Advisory Council to the Secretary of Energy on America's oil and natural gas position to 2030, July 2007.
3. International Energy Agency, "World Energy Outlook 2010," OECD/IEA, 2011. http://www.iea.org/work/2011/Roundtable/WEO_Moscow_2feb11.pdf.
4. "High Oil Prices Here To Stay," *Huffington Post*, March 11, 2011. http://www.huffingtonpost.com/2011/03/02/high-oil-prices_n_830166.html.
5. "Nuclear Power Joint Fact-Finding," the Keystone Center, June 2007.
6. C. J. Campbell, "Open Letter to the Guardian," November 2009. http://www.peakoil.net/Campbell.
7. International Energy Agency, *World Energy Outlook 2008*, OECD/IEA 2008.
8. OECD/NEA and IAEA, *Uranium 2009: Resources, Production and Demand*, 2010.
9. K. S. Deffeyes and I. D. MacGreger, "World Uranium Resources" *Scientific American*, 242, 66-76, January 1980.
10. C. MacDonald, "Rocks to Reactors: Uranium Exploration and the Market," *Proceedings of World Nuclear Association Symposium*, 2001.
11. Canadian Nuclear Society, "Uranium Mining in Northern Saskatchewan," http://www.cna.ca/curriculum/cna_can_nuc_hist/uranium.
12. Word Nuclear Association, "Nuclear Power in China," updated September 15, 2011. http://www.world-nuclear.org/info/inf63.html.
13. The Association for the Study of Peak Oil and Gas (ASPO), Newsletter No. 74, February 2007. http://www.energiekrise.de/e/aspo_news/aspo/newsletter074.pdf.

14. W. H. Ziegler, C. J. Campbell, and J. J. Zagar, "Peak Oil and Gas," Swiss Bull. Angew. Geol. Vol. 14, No. 1-2, 2009. http://www.angewandte-geologie.ch/Dokumente/ Archiv /Vol14_1_2/1412_7Ziegleretal.pdf.
15. R. L. Hirsch, et al., "Peak of World Oil Production: Impacts, Mitigation & Risk Management," 2005. http://www.netl.doe.gov/energy-analyses/pubs/Oil_Peaking_ NETL.pdf.

CHAPTER 5

CHOOSING THE TECHNOLOGY

In this chapter we outline the bases for the most important choices of materials and configuration for the IFR. They are unique to the fast reactor. They complement each other. "All the pieces fit together" in the words of the great American physicist, Hans Bethe. They bring out the best features of each of the components of the IFR system—the reactor itself, the process for recycling the spent fuel, and the technology for disposal of the actual waste.

5.1 Aims and Considerations

Nuclear power in the U.S. was at a standstill when the IFR development began in 1984. The professional anti-nuclear activists had had their way. Safety, in the wake of the Three Mile Island-2 accident, now greatly concerned the public. Waste had been made an issue based on its "forever" nature. The growth of civilian nuclear power was portrayed as leading to the proliferation of nuclear weapons, based on India's successful weapons test in 1974. These assertions, and more, had some limited truth in them. Nuclear power was still being developed, and some of the technologies of the time were in the initial stage of that development. But the dangers had been exaggerated greatly, and purposefully, and they had successfully frightened a public that had earlier been very supportive of the growth of civilian nuclear power.

In November of 1983, Congress had cancelled the Clinch River Breeder Reactor (CRBR) project, once and for all. The CRBR was to give the nation its first large-scale demonstration of the practicality of the breeder reactor. Instead, through two administrations it had become a political football. First cancelled in 1977 by a Carter administration that included many anti-nuclear activists as part of its all-out attempt to limit the growth of nuclear power worldwide, it had been kept alive by Congress. But in 1983, only partly completed, it was no longer to be funded. The only technology that could promise both electrical power without limit and fuel supplies also without limit would no longer be a priority for the nation.

Argonne had been a part of the development. It was certainly not a major player in CRBR design and construction itself, but it had fairly important development programs supporting it. The principal development laboratory for the breeder from the beginnings of nuclear power development, Argonne had taught scientists much of new possibilities in breeder development. Argonne still had the capability to do development, with a highly trained and motivated staff and large facilities for doing the necessary experiments. And we were certain of the importance to the nation and to the world of success in development of this huge, indeed unparalleled, source of electrical energy. On this basis, then, we began the development of a more perfect from of the breeder reactor, a more perfect form of nuclear power.

The problems of CRBR itself were to the last degree political. But the limitations of the CRBR line of development of the breeder had contributed to its vulnerability too. It needed large and expensive reprocessing facilities to allow its fuel to be recycled. Such facilities were identical to those used to purify plutonium for nuclear weapons—vulnerable to misuse where they did not already exist. Where they did, with one or two exceptions, they had been used in weapons fabrication. Safety was also a concern: calculations indicated that the most serious kinds of accidents possible, unlikely though they were, could lead to explosive energy releases that were containable but with difficulty. Further, CRBR waste would have lifetimes of hundreds of thousands of years, no different than spent fuel left untouched. And the breeding characteristics of the technology, although adequate, were never more than that, and were not close to the best that fast reactor nuclear characteristics would allow.

Although the primary purpose of the IFR development program was to provide an alternative technology giving unlimited electrical power, we intended to attack all the problems found in CRBR technology and more. We would do it by proper choices in the technology itself. We would eliminate—and where we could not eliminate, at least ameliorate—the concerns that seemed legitimate to us about the present forms of nuclear power. Where legitimate, we wanted to see what could be done about them.

Just a handful of choices—the reactor configuration, the materials to be used, and the technology for processing the fuel—lock in place the characteristics a reactor system will have. In appearance the IFR will be much like any other nuclear power plant: several acres of buildings housing service facilities of various kinds, an electrical switchyard array, cooling towers, a reactor building enclosed in stout containment, and particular to the IFR, a small fuel processing facility. But it will be a much different system than reactors of today. It will be a "breeder:" it will "breed"—create via nuclear transmutation of uranium—more new plutonium fuel than it consumes. It will recycle its fuel over and over until at last it will have used its fuel a hundred times more efficiently than reactors of the present day. And it will do this with less toxic waste, with much simpler management requirements, and more safely as well. The IFR choices allow this.

The first and most important choices of the materials for any reactor are those for the fuel and coolant. Operational and safety characteristics are set largely by these choices. Enhanced or diminished by purposeful design, the inherent properties of the materials always dominate and limit the scope of the design. For the IFR, a metallic uranium- plutonium-zirconium alloy was chosen for the fuel. It would need to be developed. Earlier EBR-II work would be its basis, but it would be a new fuel type. Metallic liquid sodium would be the coolant, its disadvantages far outweighed by its advantages, evident and undiminished from earlier days.

5.2 The Fuel Choice

5.2.1 Metal Fuel

The IFR metal alloy fuel was the single most important development decision. More flows from this than from any other of the choices. It was a controversial choice, as metal fuel had been discarded worldwide in the early sixties and forgotten. Long irradiation times in the reactor are essential, particularly if reprocessing of the fuel is expensive, yet the metal fuel of the 1960s would not withstand any more than moderate irradiation. Ceramic fuel, on the other hand, would. Oxide, a ceramic fuel developed for commercial water-cooled reactors, had been adopted for breeder reactors in every breeder program in the world. It is fully developed and it remains today the *de facto* reference fuel type for fast reactors elsewhere in the world. It is known. Its advantages and disadvantages in a sodium-cooled fast reactor are well established. Why then was metallic fuel the choice for the IFR?

In reactor operation, reactor safety, fuel recycling, and waste product—indeed, in every important element of a complete fast reactor system—it seemed to us that metallic fuel allowed tangible improvement. Such improvements would lead to cost reduction and to improved economics. Apprehension that the fast reactor and its associated fuel cycle would not be economic had always clouded fast reactor development. Sharp improvements in the economics might be possible if a metal fuel could be made to behave under the temperature and radiation conditions in a fast reactor. Not just any metal fuel, but one that contained the amounts of plutonium needed for reactor operation on recycled fuel. Discoveries at Argonne suggested it might be possible.

Metal fuel allows the highest breeding of any possible fuel. High breeding means fuel supplies can be expanded easily, maintained at a constant level, or decreased at will. Metal fuel and liquid sodium, the coolant, also a metal, do not react at all.

Breaches or holes in the fuel cladding, important in oxide, don't matter greatly with metal fuel; operation can in fact continue with impunity. The mechanisms for fuel cladding failure were now understood too, and very long irradiations had become possible. Heat transfers easily too. Very little heat is stored in the fuel. (Stored heat exacerbates accidents.) Metal couldn't be easier to fabricate: it's simple to cast and it's cheap. The care that must be taken and the many steps needed in oxide fuel fabrication are replaced by a very few simple steps, all amenable to robotic equipment. And spent metal fuel can be processed with much cheaper techniques. Finally, the product fuel remains highly radioactive, a poor choice for weapons in any case, and dangerous to handle except remotely.

Important questions remained—whether uranium alloys that included plutonium could be developed that had a high enough melting point and didn't harm the fuel cladding, while at the same time retaining the long irradiations now possible for the uranium EBR-II fuel. Early metal fuel had swelled when irradiated—the reason it had been discarded. But the swelling problem had been solved for all-uranium fuel. EBR-II had been operating with fairly long burnup uranium metal fuel for over a decade. Long-lived metal fuel resulted from metal slugs sized smaller in diameter than the cladding that allowed the metal to swell within the cladding. If properly sized, the metal swelled out to the cladding in the first few months of irradiation, and when it did, it exerted very little stress on it. After that, the fuel would continue to operate without any obvious burnup limit nor any further swelling.

Before the metal swelled sufficiently to give a good thermal bond with the cladding the necessary thermal bond was provided by introducing liquid sodium inside the cladding. The compatibility of liquid sodium with uranium metal allows this. As the fuel swells, sodium is displaced into the empty space at the top of the fuel pin, provided to collect fission product gasses. The bond sodium is important. It provides the high conductivity necessary to limit the temperature rise at the fuel surface and therefore the temperature of the fuel itself. The swelling itself, it was found, is caused by the growing pressure of gaseous fission products accumulating in pores which grow in size in the fuel as operation continues. But as swelling goes on, the pores interconnect and release the gasses to the space above. At less than 2 percent burnup the point of maximum swelling is reached, and the interconnections become large enough that sodium enters the pores. This, in turn, has the effect of restoring heat conductivity, which then acts to minimize the fuel temperature rise in the fuel.

The soundness of the basic uranium design had been established by thousands of uranium fuel pins of this design that had been irradiated without failure in EBR-II. But now, metallic uranium-plutonium would need to be designed to accommodate swelling. Would the plutonium content cause swelling behavior different from uranium alloy? And, more worrying, plutonium forms a low-melting-point eutectic (mixture) with iron, below the temperature required for operation. A new alloying element would be necessary to raise the eutectic melting point. Zirconium was known to be helpful in that. Zirconium also suppressed the diffusion of the cladding elements, iron and nickel, into the fuel. Iron and nickel form a lower melting point fuel alloy; worse, they form those alloys in the fuel next to the cladding. Zirconium solves these problems. Ten percent zirconium was chosen as optimal, because higher amounts gave fuel melting points too high for the techniques we intended to use to fabricate the fuel. Ten percent gave fuel with adequate compatibility with the cladding, and a high enough melting point to satisfy operating requirements, and could be fabricated with simple injection-casting techniques.

Thus the fuel would be a U-Pu-10Zr alloy. But would it work? Ten percent burnup, about three years in the reactor, was our criterion for success. We would have one set of tests initially, and everything depended on its success. In the event, the fuel passed 10% with no difficulty. It got close to 20% before it was finally removed from the reactor. There were no failures (such as burst cladding). The very first IFR fuel assemblies ever built exceeded the burnup then possible for oxide fuel in the large programs on oxide development of the previous two decades. Metal fuel which included plutonium had passed the test. All the benefits from its use were indeed possible. The program could then turn to a thorough sequence of experiments and analysis to establish, in detail, its possibilities and limitations.

5.2.2 Plutonium

The IFR fuel cycle is the uranium-plutonium cycle. In this, non-fissile uranium-238 is converted slowly and inexorably to fissile plutonium-239 over the life of the fuel. If there is a net gain in usable fuel material, the reactor is a breeder; if not, the reactor is a called a converter (of uranium to plutonium), as are all present commercial reactors. But all reactors convert their uranium fuel to plutonium to some degree. Water reactors convert enough that about half the power the fuel eventually produces comes from the plutonium they have produced and burned in place. A significant amount of the plutonium so created also stays in the spent fuel.

A large and lasting nuclear-powered economy depends on the use of plutonium as the main fuel. The truth about this valuable material is that it is a vitally important asset. Its highly controversial reputation has been built up purposefully from the activists, with little countervailing public awareness of its "whats and whys." Its very existence is said to be unacceptable. In this way, breeder reactor development was stopped in the U.S. and today continues only fitfully around the world. The fact that present reactors fueled with uranium convert uranium to plutonium very efficiently indeed, creating new plutonium in yearly amounts comparable to the best breeders possible, is lost in the rhetoric. But facts are facts. The principal plutonium-related difference between breeders and converters is that breeders recycle their plutonium fuel, using it up, cycle after cycle, so the amount need not grow. Present reactors leave most of the plutonium they create behind as waste. For efficiency in uranium usage, there is little incentive to recycle it; perhaps a twenty percent increase in uranium utilization is achievable, at a considerable cost to the fuel cycle. (Other reasons, such as waste disposal, may make reprocessing of thermal reactor fuel attractive, but not the cost benefits of plutonium recycling.)

However, it is plutonium that brings the potential for unlimited amounts of electrical power. Plutonium no longer exists in nature except in trace amounts. Its half-life is too short: 24,900 years. The earth's original endowment decayed away in the far distant past. It has to be created from uranium in the way we just described. Plutonium is a metal. It's heavy, like uranium or lead. It is chemically toxic, as are all heavy metals if sufficient quantities are ingested, but no more so than the arsenic, say, common in use for many years. It is naturally radioactive, but no more so than radium, an element widely distributed over the earth's surface. Its principal isotope, Pu-239, emits low-energy radiation easily blocked by a few thousands of an inch of steel, for example, and it is routinely handled in the laboratory jacketed in this way. It is chemically active, so in fine particles it reacts quickly with the oxygen in the air to form plutonium oxide, a very stable ceramic. If this is ingested, either through the lungs or the digestive system, as a rule the ceramic passes on through and the body rids itself of it. A popular slogan by the anti- nuclear organizers is that "a little speck will kill you." Nonsense—a little speck of the ceramic plutonium oxide will not react further, and will generally pass through the body with little harm.

Plutonium has been routinely handled, in small quantities and large, in laboratories, chemical refineries, and manufacturing facilities around the world for decades. There have been no deaths recorded from its handling in all this time. A study of the wartime Hanford plutonium workers gave the unexpected result that these people on average lived longer than their non-plutonium-exposed cohort group. This was explainable as the likely result of better and more frequent checkups because they were involved in the study, but at the very least there was certainly no shortening of lifespan.

The last point is plutonium's use for weapons. The very fact that Pu-239 is fissile makes this a possibility, as it does also for the two fissionable isotopes of uranium, U-233 and U-235. But plutonium for a time was exceptional because it could be chemically separated from the uranium that it was bred from, and it did not require the large, expensive diffusion plants necessary for the separation of the fissile U-235 isotope of uranium. But this ease of acquisition argument changed with the development of centrifuges. Now the fissile element U-235 can be separated from bulk uranium with machines. And instead of a stock of irradiated fuel, a chemical process, and facilities for handling, machining, and assembling a delicate implosion device, as one must have for plutonium, for uranium one has a nearly non-radioactive natural uranium feed, centrifuges that can be duplicated to give the number needed, and a nearly non-radioactive product, easily machined and handled, which allows a more simply constructed weapon. Plutonium can no longer be singled out as more susceptible to proliferation of nuclear weapons than uranium. The fact is that uranium now is probably the preferred route to a simple weapon in many of the most worrying national circumstances. Iran's current actions are a case in point.

Weapon-making is complicated by the presence of radioactivity. Plutonium processed by an IFR-type process remains very radioactive; it must be handled remotely, and delicate fabrication procedures are correspondingly difficult. Uranium is so much easier. This is not to imply that the large and sophisticated weapons laboratories like Los Alamos or Livermore could not use such isotopically impure reactor-grade plutonium; it is sufficient to say they would not choose to do so with much more malleable material available. And the beginner would certainly avoid the remote techniques mandatory for IFR plutonium.

This is the situation: Plutonium, as used in the IFR, cannot be simply demonized and forgotten. It is the means to unlimited electricity. The magnitude of the needs and estimates of the sources that might be able to fill those needs lead to one simple point: Fast reactors only, taking advantage of the breeding properties of plutonium in a fast spectrum, much improved over any uranium isotope, can change in a fundamental way the outlook for energy on the necessary massive scale. Their resource extension properties multiply the amount of usable fuel by a factor of a hundred or so, fully two orders of magnitude. Fine calculations are unnecessary. Demand can be met for many centuries, by a technology that is known today, and whose properties are largely established. This technology is not speculative, as are fusion, new breakthroughs in solar, or other suggested alternatives. It can be counted on.

5.3 The Coolant Choice

Liquid sodium was the choice of coolant from the beginnings of fast reactor development, because the neutron energies must remain high for good breeding and sodium doesn't slow the neutrons significantly. (Water does, and so nullifies breeding.) But sodium has other highly desirable properties too—it transfers heat easily and removes heat from the fuel quickly; it has a high heat capacity which allows it to absorb significant heat without excessive temperature rise; its boiling point is far too high for it to boil at operating temperatures, and importantly, even to boil at temperatures well above operating; and finally, although a solid at room temperature, it has a low enough melting point to stay liquid at temperatures not too far above that. In addition, there is no chemical reaction at all between the sodium and the structural materials making up the core (such as steel and zirconium). It is chemically stable, stable at high temperatures, stable under irradiation, cheap, and commonly available.

Further, as a metal, sodium does not react at all with metal fuel either, so there is no fuel/coolant interaction as there is for oxide fuel exposed to sodium. In oxide fuel, if the cladding develops a breach such reactions can form reaction products which are larger in volume than the original oxide. They can continue to open the breach, expel reacted product, and could possibly block the coolant channel and lead to further problems. Metal fuel eliminates this concern.

For ease of reactor operation, sodium coolant has one supreme advantage. Liquid at room pressures, it allows the reactor to operate at atmospheric pressure. This has many advantages. Water as a coolant needs very high pressures to keep it liquid at operating temperatures. A thousand- to two-thousand-psi pressure must be maintained, depending on the reactor design. Thick-walled reactor vessels are needed to contain the reactor core with coolant at these pressures. The diameter of the vessel must be kept as small as possible, as the wall thickness necessary increases directly with diameter. With the room-pressure operation of sodium coolant, the reactor vessel, or reactor tank as it is called, can be any diameter at all; there is no pressure to contain. And leaks of sodium, if they happen, have no pressure behind them, they drip out into the atmosphere, where generally they are noticed as a wisp of smoke. The important thing is that there is no explosive flashing to steam as there is when water at high pressure and temperature finds a leakage path.

Its principal disadvantage is that it is highly chemically reactive with oxygen, in water or in air. It must not be exposed to either so it must be maintained in an inert gas environment. Argon, a relatively common noble gas, which itself is completely non-reactive and heavy enough to blanket surfaces and keep them blanketed, is the obvious choice to do this. Its opacity is little more than a nuisance; techniques have been developed over the years to deal with it.

5.4 The Reactor Configuration Choice

The reactor configuration—that is, the arrangement of the vessel containing the core and the necessary piping—is important too. The piping that carries sodium coolant exposed to radiation in the core, the primary sodium as it is called, mustn't ever leak to the atmosphere. Primary sodium is radioactive; short-lived Na-24 is formed, which has a fifteen-hour half-life and decays with two hard gammas, at 1.38 and 2.75 MeV. Gamma radiation at energies this high is penetrating and hazardous to humans. Its half-life is short enough that it dies away reasonably quickly. But as sodium burns in air with a heavy white smoke, radioactivity would spread from a leak. A pool configuration eliminates this possibility. It keeps all primary sodium and its associated piping inside a double-walled tank. Radioactive sodium is never exposed to the atmosphere, if the primary system does leak it merely leaks sodium back into the pool. The heat in the primary sodium is transferred in a heat exchanger inside the tank to a secondary cooling circuit. Only non-radioactive sodium from the secondary cooling circuit is brought out of the vessel. This piping may develop a leak, but there can be no spread of radioactivity from it. Radioactivity from sodium leaks is a non-existent problem in the pool reactor configuration.

The pool configuration is a conscious choice, just as the fuel and coolant materials choices are. The reactor tank is sized large enough to accommodate all the primary system components. The core itself, the primary piping, and the primary heat exchanger (where the heat is transferred from the radioactive primary sodium) are submerged in the pool of primary sodium. The tank boundary has no penetrations; it is a smooth walled tank, and it in turn sits in another larger diameter tank. This guard vessel provides double assurance that there will be no leaks to the room. Unpressurized, a leak of sodium from the primary vessel would go into the space between the two vessels. That space is "inerted" with argon gas, and instrumentation is provided to monitor the space for any leaks into it. (There were none in the thirty-year lifetime of EBR-II.)

It should be noted that of the two possible reactor configurations, pool or loop, each is suited to one particular coolant type. The water-cooled reactor, because of its high pressures, needs a small-diameter reactor vessel and the loop design is almost mandatory. The sodium-cooled reactor, because of its low pressure coolant can have any sized vessel. The primary coolant is radioactive, so it's best to have primary components, piping, and connections inside the primary tank. The pool is a natural choice, and it was the choice of Argonne's designers of EBR-II in the late 1950s. The loop design, of course, is possible, and in fact it became the choice for the U.S. breeder development in the late sixties and seventies, and several of the breeder reactors built around the world were given this configuration, but for a number of reasons it is not the natural choice for sodium cooling.

As will be seen in the chapter on safety, sizing the pool to provide enough bulk sodium to absorb the heat of accident conditions adds some remarkable extra safety properties to the system. It allows safe regulation of the reactor power even under conditions where an accident has disabled the control and safety systems. In such an accident the massive pool of sodium provides ballast—heat can be absorbed until the natural reactivity feedbacks of a metallic-fueled core come in strongly enough to reduce the reactor power to harmless levels.

These "natural reactivity feedbacks" reduce reactivity as the core expands from the increased temperatures of an accident. Neutron leakage is much more important to reactivity in a fast reactor than a thermal reactor. In a fast reactor, neutron cross-sections are small and neutrons typically travel tens of centimeters before being absorbed, compared to distances of fractions of a centimeter in thermal reactors. The core dimensions are small too, so a large fraction of the neutrons are born close to the boundaries and many leak from the reactor. Small increases in the core diameter due to temperature increase in turn increase neutron leakage in the axial direction and give reactivity reductions sufficient to reduce reactor power.

5.5 The Spent Fuel Processing Choice

The fast reactor fuel cycle is different from that of the thermal reactor in its needs. This is so even if plutonium from the thermal reactor is recycled after reprocessing its spent fuel. The fast reactor, because of its fast neutron spectrum, has less (unproductive) neutron capture in plutonium 239. (The ratio of unproductive capture to fission decreases sharply with increasing neutron energy.) This is the reason that breeding is becomes possible, of course, as the small amount of Pu-239 capture leaves the excess neutrons necessary for U-238 neutron capture to make more than enough Pu-239 to replace the Pu-239 burned in operation. This also has the effect of breeding relatively pure Pu-239, as it is capture in Pu-239 that leads to Pu-240, so the low capture amounts means that the fraction of Pu-240 in the bred plutonium will not build up as fast as it does in a thermal reactor where Pu-239 neutron capture is greater.

An ideal processing scheme for a fast reactor is one good enough to assure that all the actinides recycle back into the fuel for the reactor, not just plutonium and uranium. Because of the high burnups of fast reactor fuel—20% as we have noted will not be exceptional—significant quantities of highly radioactive actinides will build up. The concentrations of the other actinides are far less than those of the plutonium isotopes, but nevertheless they are a significant contributor to the reactivity of the fuel. All are fissionable in a fast spectrum—so much so that they in fact make good fast reactor fuel. In contrast to thermal spectrum reactors, where they are a reactivity poison, the higher actinide isotopes do not need to be stripped from the fuel in recycling and go into the waste. And they, not the fission products, are the principal very long-lived radioactive products in spent fuel. These are the so-called "man-made" elements. Generated by neutron capture in the fuel not resulting in fission, they maintain sufficiently intense radioactivity for the waste to be a significant hazard for unimaginably long times—hundreds of thousands of years. But in the IFR, as they are fuel, they burn, limiting their quantity in recycling, and reducing by at least two orders of magnitude the amount that otherwise would be in the waste. As they generate heat, as well as lasting a very long time, this has important implications for long-term waste storage and disposal.

But while a processing scheme that's good enough to do the job is necessary, it's equally true that it's desirable not to have one that's too good. A pure plutonium product removes any technological barriers to weapons use. The fast reactor can produce quantities of fairly pure Pu-239 in the uranium normally surrounding the core—or the "blanket" as it is called—and a processing technology that also separates a very pure product is precisely what is not wanted. It is not a coincidence that the PUREX process used in some nations to reprocess thermal reactor fuel produces a pure product—it was developed for precisely this purpose, and has been used in all the successful weapons programs of the world to the present time.

As we get into the chemistry of the pyroprocess in a later chapter, we will see that there are basic scientific reasons that make any kind of pure product very difficult, and probably impossible, to obtain. (It is almost true to say that success in gathering even the very impure product represents something of a technical triumph!) The pyroprocess, if successfully developed at scale, is ideal for the fast reactor.

The process is not ideal for large throughputs, but it's certainly capable of scale-up for larger throughputs than those currently contemplated. And the amount of fuel loaded into the fast reactor is less than in a thermal reactor of the same power by a factor of five to ten. So the two considerations match: The fast reactor fuel amount is much less than that of a thermal reactor of the same power and the process can comfortably handle these smaller amounts. The current scale is suited to the fast reactor at the hundreds of megawatts to few gigawatts scale of operation.

The high fissile content in fast reactor fuel makes criticality a continuing concern in the presence of an effective moderator like water. Aqueous reprocessing plants have suffered unplanned criticalities in the past. The need for dilution and reduction of fissile concentrations, critical for aqueous processing, is very much less for the IFR process. A well-designed facility severely limits potential moderating materials, and no dilution at all is needed. Hydrogen is the material that moderates neutrons most effectively and it is therefore the material most effective in causing criticality in small masses of fissile material. Hydrogen-bearing liquids are the working fluids, the basis of the PUREX process. They do not exist at all in the IFR pyroprocess. Thus the fast reactor's concentrated fissile content is matched by a process that maintains these concentrations. No dilution is necessary to address criticality concerns. Again, the two facts match.

The PUREX process is developed. Its development is sunk cost. Plants have been expensive. The Japanese PUREX plant for reprocessing thermal reactor fuel cost in the neighborhood of twenty billion dollars. More new plants will be large and expensive, it is certain, and they are not ideally suited to the fast reactor. Pyroprocessing on the other hand, is relatively cheap, as we shall see in the chapter on economics, and the process, as we have seen, is suited to the characteristics of the fast reactor. The virtues of PUREX for the purposes for which it was developed stand on their own. Its sterling qualities for military plutonium have been amply demonstrated over the years. It is inevitable that comparisons be drawn between pyroprocessing and the established technology. That technology is aqueous reprocessing, personified by PUREX. Yet they are quite different, and although this is debatable, it remains our belief that the established technology is poorly suited to the fast reactor, and by no means ideal for civil use in general.

This is an example of the penalty incurred when the unique characteristics of the fast reactor are not recognized in the basic choices of materials and technologies. Choices perfectly well suited to the thermal reactor, and to the naval reactor variation of the thermal reactor, penalized the fast system when they were simply carried over to it in the late 1960s. In the core, as burnup proceeds, the higher plutonium isotopes and other actinides build up, but they do not do so rapidly or in great amounts in the blankets. The onus for dealing with proliferation and diversion issues that may arise must be put on the processing technology.

That is a principal reason why the process portion of the IFR fuel cycle is so important. And it is a reason why PUREX reprocessing is undesirable for civil fast reactor use. Pyroprocessing is advisable for this reason alone. The IFR process utilizes a form of electrorefining to separate the actinides from the fission products. It is directed specifically to the recovery of the uranium as one product; the transuranics, including plutonium, as another; and the fission product wastes as a third product.

The electrorefiner is a vessel about a meter in diameter and in height. It has electrodes appropriate to the process, a molten salt electrolyte, and it operates at 500°C. The transuranics are transported to one kind of cathode, in which liquid cadmium is the key, and uranium to another, a steel rod. The anode is the spent fuel itself chopped into short segments. The segments dissolve gradually into the electrolyte as electrically-charged ions suitable for electrical transport to the relevant cathode. Various elements transport at different voltages, and by selecting the voltage, just the elements wanted can be gathered at the appropriate cathode. The uranium can be gathered in fairly pure form, but voltages for the higher actinides are close enough together that they are collectable only as a group, and the product also includes some uranium. Uranium and the actinide elements heavier than plutonium are always present in the electrolyte and it isn't possible to separate pure plutonium in these circumstances.

At the voltages used, a large fraction of the radioactive fission products remain in the electrolyte to be removed later and stabilized in a long-lived waste suitable for disposal. The metal fission products, some forty elements in all, structural materials and noble metals, do not stabilize in the electrolyte. They, along with steel hulls of the chopped fuel, are removed mechanically and put in final form for waste disposal.

These then are the bases for the principal choices for the IFR, the materials and the configuration. In the chapters to follow we will go into the how the IFR technology performs, the experimental programs, and the results, the analyses, and the implications. We will present the knowledge that came from these programs, what is known, what is fairly certain, and where necessary, what work remains to be done.

5.6 Summary

The material and configuration choices appropriate to the fast reactor are different from those for thermal reactors—the dominant commercial reactors of the present day. The fuel, coolant, and reactor configuration chosen largely set the characteristics that a reactor can have. Our choices were metallic fuel, liquid sodium coolant, and a pool configuration. These would not be the choices for a thermal reactor, but the combination of metal fuel and metallic coolant brings out the best in the fast reactor, and along with the pool, they yield a whole set of desirable characteristics. These choices define a reactor type that can truly be called revolutionary in its possibilities for an energy system for the future. The development required will be described, item by item, in the chapters that now follow.

CHAPTER 6

IFR FUEL CHARACTERISTICS, DESIGN, AND EXPERIENCE

Although based upon the decades of experience with related fuel designs in EBR-II, IFR fuel is a new fuel. It is unique to the IFR; it has not yet been used elsewhere in the world. It was developed and extensively tested in EBR-II in the IFR decade from 1984 to 1994, and to some degree further in FFTF in the latter part of this period. Its traits are responsible for many of the positive characteristics of the IFR system. It provides ease in reactor operation and is largely responsible for the unusual reactor safety characteristics, the simplified fuel cycle, and the more satisfactory waste management qualities of the IFR. This fuel choice is where it all starts.

The qualities needed for a fast reactor fuel to perform well in reactor operation will be examined first. We then briefly examine the plusses and minuses of candidate fuels other than oxide or metal, concluding they do not have sufficient advantage over oxide to justify further development. We then turn to metallic fuel, the choice for the IFR, and describe its history and the reasons it was replaced by oxide in the demonstration fast reactors of the 1970s, when, at the same time, development work at Argonne was diminishing its weak points and evolving into a superior fast reactor fuel possessing many advantages. We describe the keys to this success, describe its performance, and cover the depth and breadth of the experimental results establishing its range of characteristics as IFR fuel.

We look then at the detailed structure of the fuel as it changes under irradiation, showing why the fuel behaves as it does. The satisfactory performance of fuel under irradiation at steady high power is first and foremost, of course, but fuel should have other characteristics as well. In the real world, reactor power changes, responding to circumstance, and we look at the robustness of the fuel in responding to such changes. We look at the ability of the fuel to go on operating safely even if cladding is breached or punctured, without contaminating the reactor or otherwise raising concerns. And finally we note two other important features of the fuel that are covered in detail in later chapters: it eases the requirements on the non-reactor portion of the fuel cycle, and it provides the basis for improved protection against serious accidents as well.

6.1 What Makes a Good Fast Reactor Fuel?

The ability to withstand the combination of high temperature and intense radiation is the characteristic of most importance in a fast reactor fuel. It means long life, and that, in turn, eases the load on the rest of the fuel cycle. The ability to generate high power without excessively high internal fuel temperatures is important too. Fast reactors generate their power with less tonnage of fuel than their thermal reactor counterpart. Paradoxically, the higher the enrichment, the less fuel you need. Instead of the three or four percent in the thermal reactor, enrichments in a fast reactor are in the range of twenty to twenty-five percent. The basic cross sections of all materials are small for fast neutrons, so the fraction of the fuel material that is fissile must make up for its smaller cross sections. With a lot of fissile material in each pin the power produced per pin must then be correspondingly high to minimize the total amount of fissile material required for a given power production. So the fuel must withstand high power densities as well as withstanding long irradiation times. The two are related, but they are not identical.

The most basic feature is time in the reactor. The longer the fuel can remain in the reactor, the smaller the amount of fresh fuel needed. And for a system that recycles its fuel, like the IFR, this means the flows to spent fuel processing are smaller, which means smaller and cheaper processing and new fuel fabrication. Breeding new plutonium in the core means that the fissile content of the fuel stays fairly constant as the fuel burnup proceeds, so the flow of fissile material can be minimized only by long fuel residence times in the reactor.

In theory, processing could be avoided completely by designing for very long fuel residence times indeed, up to the entire lifetime of the reactor, and periodically such systems are proposed. Generally the power densities allowed must be low. The phenomenon that normally limits fuel lifetime in a fast reactor is irradiation damage to the fuel cladding, and this directly limits the power density. Even with high breeding capability so adequate reactivity is maintained, the fuel lifetime is limited by the cladding. The intensity of the neutron flux and the length of the exposure to the flux—the "fluence" as it is called—on the fuel cladding is what is important in limiting fuel life, not the fuel material itself. The result is that the power produced by such a reactor will be correspondingly low, or alternatively, a very much larger core, and thus a larger reactor vessel and components, will be needed. There isn't any way around this tradeoff. The construction cost of nuclear systems combined with low power production is deadly to plant economics.

At the other extreme—and more realistic economically—was the scheme implemented by the designers of Experimental Breeder Reactor-II in the early 1960s. As the metal fuel of those days couldn't withstand long residence times, their idea was to keep fuel in the reactor as high a percentage of the time as possible by rapidly processing it at low burnup and immediately returning it to the reactor.

To do this, an extraordinarily simple spent fuel process was used, done remotely in a high radiation environment. It refreshed and re-fabricated the fuel for return to the reactor in a matter of weeks. In this way the total inventory of fuel in the system could be kept to a minimum.

The process had to be simple and it was, amounting to melting the fuel in a crucible after removing the cladding. The fission product gases released and were collected, other fission products were removed when they reacted with the crucible, and the bulk of the material then went to fabrication to be recycled back into the reactor. But as the melted fuel reacted to some degree with the crucible, several percent of the product was left behind. The process was variously called "melt-refining" and "pyroprocessing." It worked well—some five recycles were completed in a few years—but from the beginning it was realized that this was not a complete process and substantial improvements would have to be made for commercial viability.

The EBR-II fresh fuel contained almost no plutonium; a little was bred into the fuel in-reactor, but only a very little. The commercial fast reactor cycle would have to deal with plutonium, and lots of it; plutonium is the fissile material used in the fast reactor fuel cycle. No simple adaptation of melt-refining would increase the plutonium content of the processed fuel to the needs of fresh fuel, so the cycle as it stood was not suitable for a commercial power-generating system. But the idea of a simple fuel cycle for a fast reactor was born there, and later became one of the principal elements of the IFR. Long residence with correspondingly high burnup combined with simple processing is the right combination, and that was what was to be developed for the IFR.

Next, if the reactor is to breed efficiently, in addition to the coolant the fuel itself should slow the neutrons as little as possible. The fuel material should therefore contain little in the way of light elements. Oxygen is light enough to slow neutrons significantly, and the oxygen in oxide fuel brings the breeding characteristic down significantly. This became a concern in the Clinch River Breeder Reactor demonstration plant, partially completed in the 1980s, when it had difficulty meeting its stated breeding goal. Obviously, no light material at all, as in pure uranium or plutonium, is best for breeding.

A variety of issues impacting on operation are affected by the choice of fuel. When a fuel pin develops a leak what happens next determines whether the entire primary system will become radioactive from contamination by fission products if operation continues. A safety problem is possible too if the fuel material extrudes out and blocks a coolant channel sufficiently to deprive fuel of proper cooling. Shutdowns to remove faulty pins waste valuable power production time. Further, a good fuel choice will have a lot of leeway in the power it can generate without failing by melting or for any other reason. In particular, overshoots in power arise in a number of incipient accident situations, and the fuel should be capable of accommodating them without difficulty.

Best of all is a fuel, coolant, and configuration that will shut the reactor down with no ill effects in the face of incipient accidents with no need for a human or engineered safety system response. This is possible if the reactor power inherently adjusts to match the amount of cooling, whatever the amount is. This can be achieved. And going beyond even the major accident possibilities to the ultimate, where the core may be damaged, what will be the result? In the worst possible accidents, power increases could cause substantial volumes of the reactor core to approach fuel pin melting. The last thing you want is to have the fuel behave in a way that leads to a further large spike in reactor power—as an explosion in fact, violent enough to breach the reactor containment and release radioactivity to the environment. What is needed is a response that reduces reactivity sufficiently to terminate the power increase without damage, or in the extreme where fuel is damaged, to have very little explosive energy released. Damage in this case is contained in the reactor vessel, and the containment building remains as backup.

These are the kind of considerations involved in the fuel choice. Such considerations underlie the experimental programs that have been undertaken around the world on the various possible fuels for fast reactors.

6.2 What Are the Candidate Fuel Types?

Metal fuel was the choice for all the early fast reactors: EBR-I, EBR-II, Fermi-1 in the U.S., and the Dounreay Fast Reactor, DFR, in the U.K. Its advantages were obvious. Its nuclear characteristics are superb; they give a neutron energy spectrum leading to the best possible breeding performance. This was a particularly important consideration in these early days when uranium was thought to be scarce, and would be today in a rapid expansion of fast reactor generated nuclear power. It is entirely compatible with sodium coolant, a liquid metal, so there is no reaction whatsoever between the two. And it is simple and cheap to fabricate; a single casting operation suffices.

However, it was soon found that it swelled badly under irradiation. After a short time it burst the cladding—the principal reason that Argonne adopted a quick turnaround fuel cycle for EBR-II. No amount of work hardening of the metal or softening of it by annealing could stop the swelling. Metal fuel swelled. It allowed only very limited burnup. The best achievable was about 3 percent, a fraction of that needed for an economic fuel cycle. So in the mid-sixties it was abandoned as the choice for future fast reactors.

Further cementing the decision was an event in the Fermi-1 reactor in 1966. Fuel had melted during a startup when flow to one of the subassemblies had been blocked by a structural component that had come loose. The subassembly partially melted, affecting its immediate neighbors as well. The operators continued to increase power without realizing that a subassembly wasn't being properly cooled. The lesson drawn at the time was that the higher melting point of oxide would be safer; it was thought it would have more margin-to-failure, and perhaps, lesser implications if fuel did fail. And the ease with which the switch to oxide was accepted was abetted by the fact that oxide was now the established choice in the rapidly expanding commercialization of the LWR.

However, oxide has its disadvantages too. It has a limited ability to conduct heat. It is an insulator, really, and it has limited specific heat as well, (a measure of the ability to absorb heat with limited temperature rise), which means that fuel temperature is quite sensitive to power. It will withstand very high temperatures, however, and the temperatures at the centerline of the fuel pins reach 2,000°C or more. The wide range in temperature tolerance is a positive trait in normal operation, but the high temperature can increase the energy releases in possible accidents. And it is not entirely compatible with sodium. But it does withstand radiation very well and it does not swell. Its breeding properties are only fair, as will be explained below, but when it was chosen at this time it was felt that they were good enough for a start, and better breeding fuels could be developed later. On this basis, in the late 1960s and early 1970s a number of first-of-a-kind oxide-fueled sodium-cooled fast reactor demonstration plants were constructed in several countries around the world.

These reactors, and a few that came later, were fueled with mixed uranium-plutonium oxide fuel. Mixed oxide fuel experience has been favorable. [1-2] A comprehensive oxide fuel data base now exists that can support future commercial reactor licensing efforts. High burnup potential, beyond 20% burnup, has been demonstrated. The consensus worldwide is that oxide fuel is fully developed, and it is certainly the *de facto* reference fuel for fast reactors. Its characteristics, good and bad, are very well established.

Other ceramic fuel types have also been tried. Carbide and nitride in particular received some attention in the mid-1970s. The U.S. fast reactor development program began a significant program on these fuels, motivated by improvement in breeding. These fuels, while still ceramic, do breed better and allow new core loadings to be built up faster than in the reference oxide fuel. At this time, concerns about breeding had arisen in the U.S. fast reactor demonstration reactor project, CRBR, then underway. And in the early seventies, a rapidly expanding fast reactor economy was still confidently planned that would follow light water reactors as the next generation of nuclear power. So an "advanced fuels program" to test the performance of the carbide and nitride fuels in fast reactors was begun.

Of the two candidates, carbide was favored. Effective transfer of heat generated in fuel requires either a very close fit of the fuel to the cladding—difficult to obtain—or some substance put in the imperfect interface between the ceramic fuel and the steel cladding. The fuel must be thermally bonded to the cladding. The carbide fuel pins tested used helium or sodium as thermal bonds. In early stages of the irradiation program, helium bonding was selected as the reference. Most of the irradiation database is for helium-bonded carbide fuel. It turned out that carbide fuel's swelling under irradiation is greater than that of oxide, and greater than had been thought. It is also brittle, and considerable cracking was observed under irradiation. [2-3] This combination led to concern that cracked and brittle carbide fuel might puncture the cladding after some time in reactor. This "fuel-cladding mechanical interaction" as it was called, of some concern in oxide, was thought to be worse in carbide.

In the end, carbide fuel performance in-reactor was no better than oxide, and the breeding potential was not sufficiently improved to warrant continuing with it. The final straw was a greater anticipated difficulty in reprocessing carbide. It should be mentioned that a full carbide core has been demonstrated in the India's Fast Breeder Test Reactor, with the fuel reaching a peak burnup of 10%. For their 500 MWe Prototype Fast Breeder Reactor, however, they too adopted oxide because of difficulties in reprocessing carbide. [4]

The nitride fuel irradiation database was less than for carbide fuel. Nitride fuel has less swelling and a lower fission gas release than carbide fuel, but it exhibits even more extensive cracking and fragmentation during the startup and shutdown transients. [2,5] It is considerably easier to reprocess than the carbide fuel. But its fabrication is more difficult and more expensive. Also, the principal isotope of nitrogen, N-14, has a significant cross section for neutron capture. It acts as neutron poison and produces biologically hazardous C-14. The problem is avoided if nitrogen is fully enriched to N-15, but only at significant cost. The neutronics characteristics of nitride, such as its breeding capability, are similar to practical Zr-alloyed metal fuel, but its fabrication is much more difficult, and cracking and fragmentation remain issues not faced by metal fuel. Development of both carbide and nitride was terminated simply because there seemed to be little incentive to go further.

6.3 The Basis of Metal Fuel Development

The technical outlook changed dramatically for metal fuel in the decade following the decision to adopt oxide as the reference fast reactor fuel. The facts of metal fuel performance had changed, but this drew little attention at the time and in the wider world was assigned even less importance. EBR-II had continued to use metal fuel. Recycling of the fuel by the melt-refining process had been stopped by the AEC, but the reactor operation had continued. EBR-II was assigned the role of an irradiation facility, testing oxide fuel for a Fast Flux Test Facility (FFTF) that was to be built, and later was built, at Hanford, Washington. A wide variety of tests were planned to prove the oxide fuel proposed for it. EBR-II would irradiate such fuel, but its actual operation was with metal fuel. Because of the very limited burnup at the time, new fuel would be needed every few months, and in the absence of recycling this represented a real and unwelcome cost.

As a cost reduction measure, EBR-II metallurgists began development to improve the fuel burnup capability. The important discoveries came in the late 1960s, and the irradiation experience necessary to prove their importance accumulated through the 1970s. [6-8] It turned out that the basis for high burnup was simple. Cease trying to solve the swelling problem by metallurgical means—as mentioned, cold work and annealing had been tried extensively—rather, concentrate on the design of the fuel pin itself. In particular, because metal fuel is so compatible with sodium, sodium can be used inside the fuel pin as a thermal bond between the metal fuel and the metal cladding. In this way, heat is easily transferred from the fuel to the clad and hence to the coolant and no tight fit of the fuel to the cladding is needed. And not only does the sodium thermal bond make a tight fit to the cladding unnecessary, because it is a liquid it allows any amount of clearance that the designer may need to simply *allow the fuel to swell inside the cladding.* As fuel swells sodium is displaced into a plenum formed by continuation of the cladding above the actual fuel column. Instead of trying to constrain the swelling with strong cladding, the new design allocated ample room for fuel to swell.

To be practical, though, there had to be a limit to the swelling. Eventually the swollen fuel must reach the cladding, and what then would happen? Following a suggestion by Barnes [9] in the U.K., it was thought that the mechanism for swelling might well be the fission gases trapped in the fuel. It was hoped that the force for swelling might lessen as swelling went on and when it reached the cladding the fuel might no longer have the rigidity to seriously stress it. This turned out to be exactly the case. Fuel swelling is driven primarily by the growth of fission gas bubbles as burnup proceeds and the fission gases gather at nucleation sites and grow as bubbles. When the fuel swells about 30% in volume, the bubbles begin to interconnect sufficiently to provide passage for fission gas to release to the surface of the fuel. At this point the irresistible force for swelling diminishes, so much so that further swelling is easily constrained by the cladding.

The resulting fuel is a porous material with two very useful characteristics, both needed for long burnup. It is a weak fuel exerting only a weak force on a strong cladding, and it can accommodate the eventual accumulation of a great deal of solid fission products without swelling further. The plenum above the fuel column is sized so the fission gas pressures on the cladding are kept to reasonable levels. Very long burnups become possible.

The improved EBR-II fuel design did not differ greatly from the earlier design. The fuel composition was the same. Schematically, the design itself looks the same. The important additional clearance and the lengthened plenum were simple changes, but they made the difference. A schematic of the EBR-II metal fuel is shown in Figure 6-1. The fuel slug, the gap filled with sodium, and the plenum can be seen. The fuel slug can be full length or it can be in segmented in pieces one on top of the other; the performance is the same. Since they are easiest to fabricate, most fuel has been in segmented pieces.

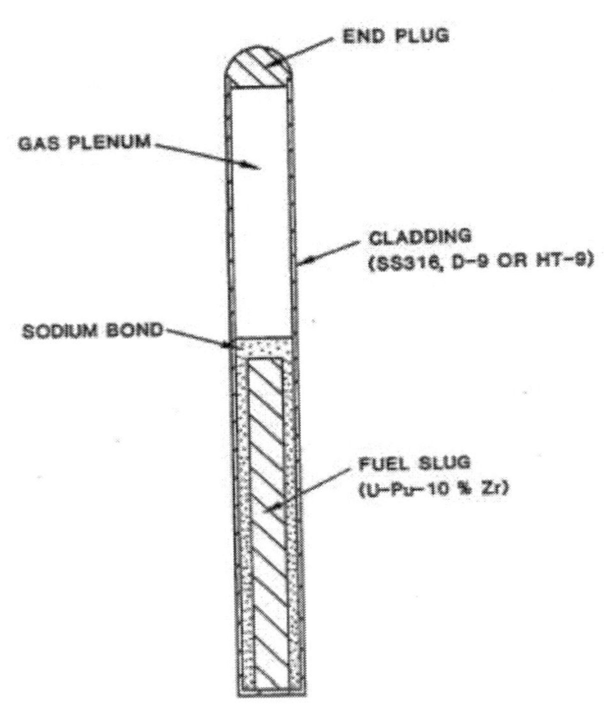

Figure 6-1. Schematic of metal fuel

Ease of fabrication generally may not seem to be that big a factor, but it is hugely important where plutonium, particularly recycled plutonium, is to be involved in the fabrication. Recycled plutonium builds up considerable amounts of the higher plutonium isotopes. Highly radioactive, hands-on fabrication is unwise if not impossible, yet hands-on fabrication is almost mandatory when fabrication is complicated. The high power densities in the fuel of a fast reactor demand excellent heat transfer, and in the absence of another heat transfer medium like sodium, very tight fits of fuel to cladding are necessary. In the FFTF, the individual oxide fuel pellets were ground to size to give the fit necessary for adequate heat transfer, and the higher plutonium isotopic content of the fabricated fuel was limited by the hands-on access needed to accomplish this. Even in LWR recycle, where the content of higher isotopes is less than for fast reactor recycle, radioactivity stays so high that it is difficult to use hands-on fabrication, but to some degree it is possible. For the fast reactor recycled fuel it isn't. Remote fabrication is necessary, and if many steps are necessary it is expensive and difficult. Oxide fuel fabrication requires many steps. Remote handling of the many process steps will be necessary for recycled fast reactor oxide fuel, no matter how it is processed. It will be difficult, it will be expensive, and it will not be easy to operate and maintain

The fewer the steps and simpler the steps are, the better. Very large savings come from that. The IFR metal fuel adapts naturally to remote fabrication because the fabrication of the fuel slug is a simple casting operation and insertion into the cladding is done simply as well. At EBR-II the fuel was cast by injecting molten fuel into molds, glass tubes in fact. The resulting fuel slugs are easily fitted into the cladding tubes, as they have ample clearance because a substantial gap is needed in any case to accommodate the swelling. A measured amount of sodium fills the gap initially. The end cap is welded in place and the pin is done. Ideal for remote equipment, and inexpensive to assemble, this is a real plus for metallic fuel for fast reactor application.

It should be mentioned, too, that in scaling up from EBR-II methodologies there must be an improvement from the use of glass tubes as molds. In EBR-II fuel fabrication the glass was simply broken off, leaving the fuel slug. The fuel contaminates the broken glass, and it represents a waste stream neither desirable nor necessary. There was some experimentation at EBR-II along these lines in late years of the IFR program. Probably the most interesting was in using an electromagnetic field to shape the slug, but the 1994 termination stopped all work before anything real could be developed and evaluated.

6.4 Irradiation Experience: A Very Long Burnup Fuel

The first EBR-II fuel (Mark-I fuel) was designed with an 85% "smear density." In cross section then, the fuel occupied 85% of the space inside the cladding. After the discoveries about fuel swelling, the next fuel design (Mark II fuel) had more clearance with about 75% smear density. The burnup then gradually increased. Eventually the bulk of Mark-II design fuel achieved 7–8% burnup. In the initial Mark-II fuel, the cladding was indented at the top of the fuel column to restrain the fuel from moving axially. The sharp indentation became a failure point at about 8% burnup. A design change to a spherical indentation solved the problem and much higher burnups became possible. Some of these fuel pins reached 18.5% burnup. Over 40,000 pins had been irradiated successfully by the early 1980s.

These were non-plutonium-bearing pins. The metallic compositions of both the Mark-I and Mark-II fuels were 95% uranium and 5% "fissium" alloy. Fissium was the name given to an artificial fission product alloy mimicking the composition resulting from the recycling by melt-refining. The noble metal fission products recycled back with the recovered uranium had been found to improve the fuel properties. [10] The equilibrium composition of fissium was approximately as follows. The fissium addition continued in all Mark-I and Mark-II fuels after the recycle operation ceased in 1969.

Molybdenum	2.46%
Ruthenium	1.96%
Rhodium	0.28%
Palladium	0.19%
Zirconium	0.10%
Niobium	0.01%
Total	5.00%

By 1984 when IFR development began, it was known that properly designed metallic uranium-based fuel could withstand very long burnups. The basic principles for such design had been established. Sodium thermal bonding allowed room for fuel to swell inside the cladding, swelling decreased after a time, and—very importantly—the allowable initial fuel volume left the fuel density high enough to satisfy both reactivity and breeding requirements. It was a practical fuel. Although this was not expected to be needed, smear fuel densities could go as low as half the uranium metal density and still give a practical fuel. Oxide fuel when fully dense does not, in fact, have a density much greater than this. The first step in developing a fuel with characteristics suitable for the IFR was to prove beyond doubt the long burnup capability of metal fuel. By the early 1980s this had been done.

But this was for the basic uranium fuel of EBR-II. A fuel with the plutonium contents needed for the IFR reactor had still to be developed. The IFR initiative was beginning. The IFR, as all other fast reactors, would be based on the uranium-plutonium fuel cycle. (Thorium/uranium-233, the other possible choice, is far less suited to the neutron energies of a fast reactor.) The IFR fuel alloy, in addition to the U-238, must contain a substantial fraction of plutonium—perhaps up to one third as much plutonium as uranium. It was known that plutonium lowers the temperatures at which metallic fuel can operate in steel cladding—a eutectic (a mixture) of plutonium and iron with a lowered melting point forms at the fuel/cladding interface. But it was also known that zirconium, replacing fissium, might give a fuel alloy with the necessary protection against the eutectic formation. Earlier irradiation tests of various alloys had indicated that zirconium exhibited exceptional compatibility with cladding, and very importantly too, it significantly increased both the melting point of the fuel alloy and the fuel-cladding eutectic temperature. However, a very substantial addition of zirconium would be required, 10 percent by weight or, as zirconium is only 40 percent of the weight of uranium, fully 25 percent of the atoms in the fuel would be zirconium. The effect of this had to be tested. Would the uranium swelling behave as it had before with so much additive in it?

As the Mark-II fuel assemblies reached their irradiation limits, the EBR-II core was gradually converted to a new uranium fuel (Mark-III), U-10w%Zr, and for the cladding D-9, a swelling-resistant stainless steel, or SS-316 stainless steel, the more usual steel cladding. Later, Mark-IV fuel with HT-9 cladding, an even lower-swelling steel cladding, was introduced. (Minimization of structural steel swelling under irradiation with resulting dimensional changes is always a consideration—in fast reactors the high neutron energies take their toll.)

The 10% zirconium addition was the reference alloying agent for both uranium- and plutonium-bearing fuels. EBR-II began to irradiate large quantities of the uranium/zirconium fuel, while a new capability was being put in place specifically for the fabrication of plutonium/uranium alloy fuel. Called the Experimental Fuels Laboratory (EFL), it was ready in just a few months and began to fabricate the actual IFR fuel choice, a plutonium-bearing ternary fuel, U-Pu-10w%Zr. A total of 16,811 U-Zr and 660 U-Pu-Zr fuel pins were irradiated in EBR-II over the next ten years. Forty pins went all the way to 16 to 18% burnup. There were no failures at all, an impressive performance. This untried fuel on its first try had exceeded the burnups achieved by the large oxide fuel programs at that time. (Later, oxide would achieve similar burnups.)

Burnups of the EBR-II "driver fuel" (the fuel of reactor operation) using U-10Zr had had limits imposed on them in licensing, and in the tables giving the irradiation experience, the pins with 8% burnup or less are driver fuel with those limitations placed upon them. The long burnups were from test assemblies. Not only were there no failures, but the experience accumulated with both types of fuel established the burnup capability once and for all before operations were terminated in 1994.

6.5 Understanding the Long Burnup Fuel Behavior

It turned out that the behavior of the zirconium alloy metallic fuel under irradiation is very similar to that of the U-fissium. The form and structure of irradiated U-10Zr fuel, demonstrating the gas pores, is shown in Figure 6-2. The dark areas are pores, interconnected as they grew with time in-reactor and released the gases to the plenum. Maintaining the fuel smeared density below 75% is crucial for interconnected porosity. In Figure 6-3, the maximum amount the cladding is expanded (the peak strain) and the fission gas release are plotted as a function of the fuel smear density for U-19Pu-10Zr fuel at 12.5% burnup. If the smear density is maintained below 75%, the interconnected porosity allows about 85% of the fission gases to be released to the plenum and results in a minimum stress on the cladding and minimal deformation. For the 85% smeared density fuel, the fission gas release is limited to about 75%, which means more gas in pores and continued swelling, resulting in about six times greater cladding strain.

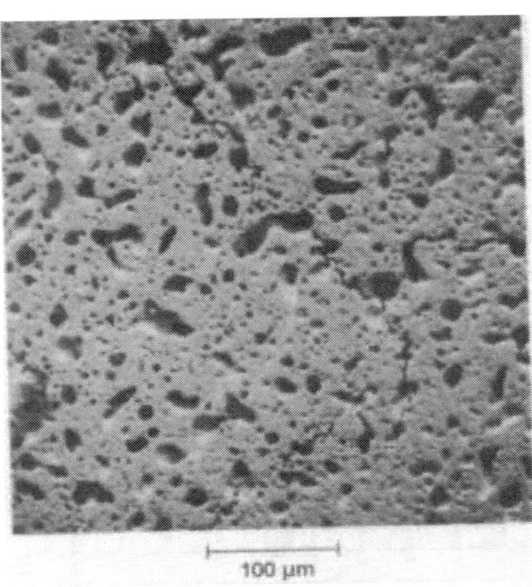

Figure 6-2. Fission gas pore structure of irradiated U-10Zr fuel

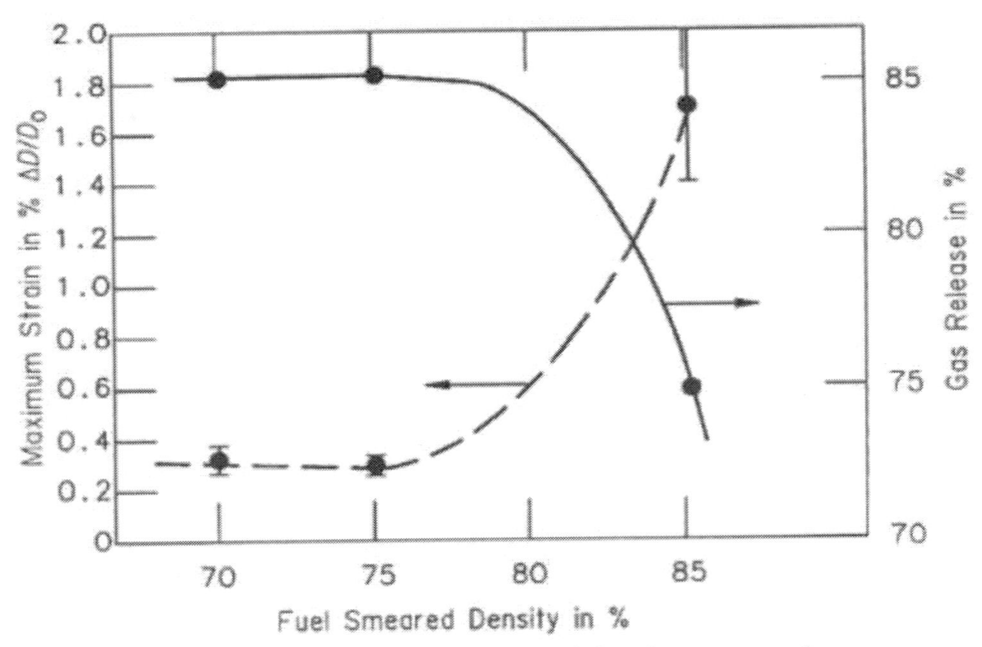

Figure 6-3. Effect of fuel smeared density on gas release and cladding diametral strain

It turned out that the high-plutonium-content ternary alloy fuel (U, Pu, Zr) has the useful characteristic of the three constituents redistributing in early stage of irradiation into radial fuel zones as illustrated in Figure 6-4. The migration is driven primarily by temperature gradient and is predominantly a radial redistribution. Overlaid in Figure 6-4 is the radial distribution of constituent concentrations. Zirconium tends to migrate to the center and toward the periphery, and uranium migrates oppositely. Plutonium pretty much stays put. The movements tend to help performance in that the higher-melting-point zirconium moving to the center raises the melting point further in the peak temperature region, and in moving to the periphery it helps the fuel/cladding compatibility.

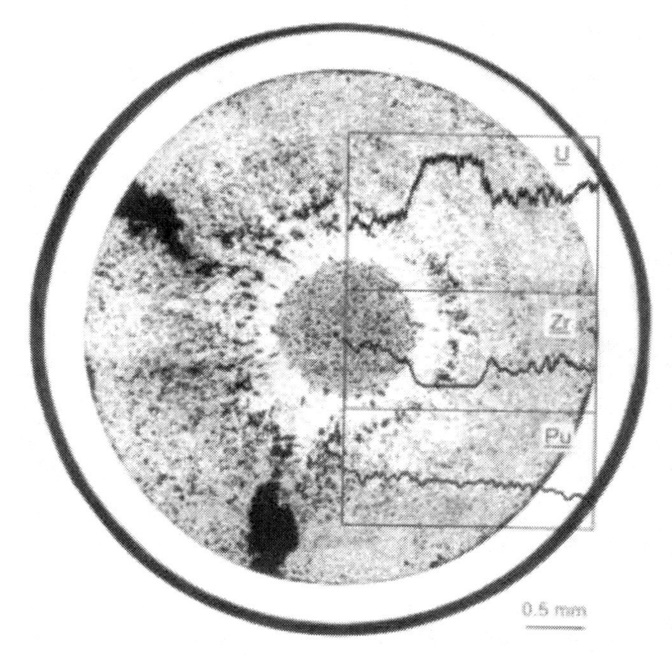

Figure 6-4. Constituent redistribution and radial zone formation of U-19Pu-10Zr fuel

Radial zone formation in early stage of irradiation markedly enhances the radial swelling. The high rate of radial swelling stresses the peripheral fuel enough to cause the crack formation shown in Figure 6-4. The large cracks eventually fill with fuel as irradiation continues. This anisotropic swelling (unequal axially and radially) results in much smaller axial growth of the ternary fuel than the U-fissium or U-Zr fuels. The axial growth of the uranium based fuel is in the range of 8–10% by the end of fuel life, but only 3–4% for the ternary fuel.

Fuel restructuring or no, the ternary metallic fuel has demonstrated the same excellent steady-state irradiation performance characteristics found in the extensive irradiation experience with the uranium-based fuel. The U-Zr and U-Pu-Zr irradiation tests are summarized in Table 6-1. [11]

Table 6-1. Metal Fuel Irradiation Tests in EBR-II

Test ID	Description	Cladding Type	Pin OD mm	Pu content % HM	Burnup at.%
X419	Lead test	D9	5.8	0/8.9/21.1	12.0
X420	Lead test	D9	5.8	0/8.9/21.1	17.1
X421	Lead test	D9	5.8	0/8.9/21.1	18.4
X423	Lead test	SS316	7.4	up to 28.9	5.0
X425	IFR lead test	HT9	5.8	0/8.9/21.1	19.3*
X427	Run beyond eutectic	SS316	4.4	0	11.5
X429	Fabrication variable I	HT9/SS316	5.8	0/8.9/21.1	14.4
X430	Advanced HT9 test	HT9	7.4	up to 28.9	11.5*
XY-24	RBCB high Pu	SS316	4.4	0/21.1	7.6
XY-27	RBCB medium Pu	SS316	4.4	0/8.9	6.6
X397	Advanced metal blanket	D9	12.9	0	2.0
X431	HT9 blanket test	HT9	9.4	0	3.8*
X432	HT9 blanket test	HT9	9.4	0	4.4*
X435	Mk-III qualification	D9	5.8	0	19.9*
X436	Mk-III qualification	D9	5.8	0	9.3
X437	Mk-III qualification	D9	5.8	0	10.3
X438	Mk-III qualification	D9	5.8	0	9.9
X441	Variable Zr: 6/10/12%	HT9/D9	5.8	21.1	12.7
X447	Mk-III high temperature	HT9/D9	5.8	0	10.1
X448	Mk-IV qualification	HT9	5.8	0	14.8*
X449	Mk-IV qualification	HT9	5.8	0	11.3
X450	Mk-IV qualification	HT9	5.8	0	10.2
X451	Mk-IV qualification	HT9	5.8	0	13.8*
X452	Fuel impurities	D9	5.8	0	6.1
X453	Fuel impurities	D9	5.8	0	8.5
X454	Fuel impurities	D9	5.8	0	9.0
X455	Fuel impurities	D9	5.8	0	9.0
X481	Pu feedstock	SS316	5.8	21.1	9.0
X482	RBCB high Pu	D9	5.8	21.1	13.5
X483	Mk-IIIA qualification	SS316	5.8	0	15.0*
X484	Mk-IIIA qualification	SS316	5.8	0	11.7
X485	Mk-IIIA qualification	SS316	5.8	0	10.5
X486	Mk-IIIA qualification	SS316	5.8	0	13.9*
X489	High Pu compatibility	HT9/HT9M	5.8	21.1/31.1	5.4*
X492	Zr sheath	SS316	5.8	0/21.1	10.5*
X496	Long life	HT9	6.9	0	8.3*
X501	Actinide burner	HT9	5.8	0/22.2	5.9*
X510	Metal fuel source pin	HT9	5.8		1.9*
X521	Synthetic LWR fuel	HT9	5.8		1.9*

*Still under irradiation when EBR-II was permanently shut down

The plutonium contents of the test assembly pins covered the range of interest.

In the column "Pu content % heavy metal" the contents are listed: (0, 8.9, 21.1) is to be read as three compositions, one with no plutonium, one with 8.9%, and one with 21.1 % plutonium.

The test matrix was designed to encompass the important variables in IFR fuel pin design. Various combinations of cladding materials (SS316, D-9, HT-9, and HT-9M), pin diameters (4.4, 5.8, 6.9, 7.4, 9.4, and 12.9 mm), plutonium concentrations (3.3, 8.9, 21.1, 24.4, 28.9, and 31.3% of heavy metal), zirconium contents (6, 10, and 12%), fuel smear densities (70, 75, and 85%), fabrication variables (fuel impurity levels), and operating conditions (peak linear power, cladding temperatures, etc.) were tested.

The test assemblies were removed at various burnup levels so that examinations could be carried out; the assemblies were then reconstituted and returned to the reactor for further irradiation. The burnup levels presented in the table are the peak values achieved. The burnup values with asterisks are for test assemblies that were planned for further irradiation when the EBR-II was shut down permanently on September 30, 1994.

As a principal unfortunate example, the X425 lead test with U-Pu-Zr ternary fuel had achieved 19.3% burnup and the X435 Mk-III driver qualification test had achieved 19.9% burnup by that time. There was no indication these tests needed to be terminated. Much higher burnup levels might well have been achieved if irradiation had been allowed to continue. This kind of information takes years to acquire. It is invaluable. It, and everything else, was thrown away in the rush to kill the IFR program and there was no inquiry by the administration as to its value.

6.6 Testing the Effects of Remaining Variants in Fuel Design: Diameter and Length

The possibility had been raised that the excellent performance experience of the metal fuel in EBR-II was due to its small pin size (4.4 mm diameter and 34.3 cm length). Perhaps metallic fuel might not perform as well in the full-length pins (90 cm or so in length) expected in commercial fast reactors. These concerns were satisfactorily resolved in tests in the Fast Flux Test Facility (FFTF) at Hanford where the fuel column length was the 90 cm typical of fuel envisaged for commercial fast reactors. Seven full assemblies of metallic fuel were irradiated in FFTF. The results are summarized in Table 6-2. One assembly, IFR-1, contained U-Pu-Zr fuel pins, which achieved a peak burnup of 10.2%. The other six assemblies were part of the qualification tests of U-Zr fuel with HT-9 cladding. This was to be the fuel type for a proposed FFTF conversion to metal fuel. All of these assemblies achieved peak burnup in excess of 10% and the lead test achieved a peak burnup of

16%. The FFTF core conversion to metallic fuel was terminated by the Clinton administration decision to shut down FFTF as well.

Table 6-2. Metal Fuel Irradiation Tests in FFTF

Assembly ID	Description	Cladding Type	Pin OD mm.	Pu content % HM	Burnup at.%
IFR-1	Lead test	D9	6.9	0/8.9/21.1	10.2
MFF-1	Driver qualification	HT9	6.9	0	10.0
MFF-2	Driver qualification	HT9	6.9	0	15.9
MFF-3	Driver qualification	HT9	6.9	0	16.0
MFF-4	Driver qualification	HT9	6.9	0	14.8
MFF-5	Driver qualification	HT9	6.9	0	11.4
MFF-6	Driver qualification	HT9	6.9	0	11.0

The potential performance issues examined were the effect of height and weight of the long fuel column on fission gas release, fuel swelling characteristics, and potential fuel-cladding mechanical interaction at the lower part of the fuel column. The examinations of the FFTF tests showed that the fission gas release to plenum was similar to that of the EBR-II fuel. No difference in constituent migration was found; axial growth was as predicted; and there was no evidence of any enhanced fuel-cladding mechanical interaction. All were satisfactory; the fuel was ready for whole core conversion to metal.

6.7 Testing the Effects of Transient Variations in Reactor Power

The next characteristic we will look is the way metal fuel stands up to big swings in power with the accompanying changes in temperature, dimension, and other properties. A strong point of metal fuel is its excellent power transient capabilities. No restriction on transient operation or load-following is needed. The robustness of metal fuel is illustrated by the sample history of a typical EBR-II fuel assembly whose entire history was tracked to provide base line data for the series of inherent passive safety tests conducted in 1986. As listed below, note the many startups, the overpowering of the fuel, and the tests causing the fuel temperatures to rise and fall rapidly, all of which strains the fuel of any reactor. But EBR-II was a test reactor; its purpose was to find out the effects of such treatment of test fuel assemblies, and in so doing its own fuel was subjected to fairly rugged treatment. Transient operation like this severely tests the robustness of any fuel, but metallic fuel copes with it easily. This is to be compared to a reactor in routine operation generating electrical power, which probably has no transients after startup, until it shuts down once a year, or even less often than that, and whose fuel is then replaced after three years.

40 start-ups and shutdowns
5 15% overpower transients
3 60% overpower transients
5 loss-of-flow (LOF) and loss-of-heat-sink tests, up to 100% power without scram

6.8 Operation with Failed Fuel; Testing the Effects

Metal fuel in sodium coolant has still another remarkable trait. It can continue to run with the cladding breached and the fuel exposed to the coolant without contaminating the coolant. Oxide fuel releases corrosion products to the coolant if the cladding breaches. But with no reaction between two metals, liquid metallic sodium and metallic fuel, in contrast to oxide metal fuel doesn't corrode. This is remarkable feature has very positive implications, no contamination of the primary system in operation with cladding breach and no possibility of corrosion products plugging coolant channels.

The tests in EBR-II of this characteristic were called "Run Beyond Cladding Breach (RBCB)." Figure 6-5 shows the result for oxide fuel with an initial breach made for the test. The opening widened due to fuel/cladding stress caused by the formation of a low density sodium-fuel product, $Na_3(PuU)O_4$, which extruded in small amounts into the coolant.

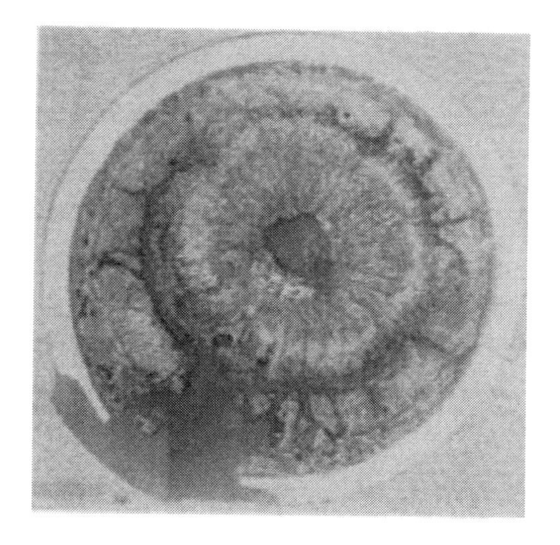

Figure 6-5. Oxide fuel (9% burnup)

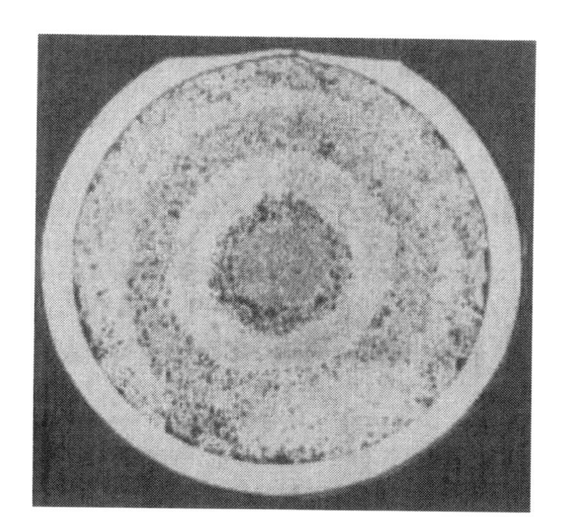

Figure 6-6. Metal fuel (12% burnup)

A similar RBCB test with metal fuel (12% burnup) is illustrated in Figure 6-6. Again an initial breach was made. There is no reaction product and the fuel loss is practically zero. The post-irradiation examination shown in Figure 6-6 is after operation in RBCB mode for 169 days. There is no indication of breach site enlargement. In another test, metal fuel operated for 223 days after the cladding breach, including many start-up and shut-down transients. The breach site remained small. It was concluded that although metal fuel is very reliable, if a fuel failure does occur the failed fuel pins can be left in the core until their normal end of life without raising any particular operational or safety concerns. This is true of no other fuel, in fast reactors or in LWRs.

The eutectic formation (melting) temperature between the fuel and the cladding is a critical parameter for metal fuel. The onset of fuel/cladding eutectic formation starts at 700-725°C range, depending on the fuel alloy and cladding types. At onset, however, not much occurs. In fact, even at a hundred degrees above the eutectic temperature the eutectic penetration into the cladding is minimal in one hour. Only at much higher temperatures, approaching the melting point of the fuel material itself on the absence of cladding, does the eutectic penetration into cladding become rapid. Eutectic formation therefore, is not a primary safety concern even during transient overpower conditions.

It does have a practical effect. Reactor designers tend to place limits on the coolant outlet temperature to assure the eutectic temperature is not reached at the fuel/cladding interface. A coolant outlet temperature limit of 500-510°C is very conservative, and up to 550°C provides ample margins to the onset of eutectic formation. The conservative design approach will probably be relaxed with more experience.

6.9 Minor Actinide Containing Fuel

Equilibrium recycled IFR fuel will contain transuranic elements near to plutonium in atomic weight, so the U-Pu-Zr fuel test matrix included fuels containing these minor actinides. The typical composition of the actinides in LWR spent fuel is about 90% plutonium and 5% each for neptunium and americium. IFR ternary fuel, U-20Pu-10Zr, will contain about 1% Np and 1% Am. For an upper bound case, a 1.3% Np and 2% Am addition was selected for the irradiation tests.

Americium has high volatility and fuel fabrication at elevated temperatures with Americium will be a challenge for any fuel material. For the IFR metal fuel, the standard injection casting technique was altered slightly to have the feedstock of 80Pu-20Am inserted into the molten fuel late in the melt cycle to minimize the time at temperature. When the Pu-Am stock was added, "sparks" were observed. The sparks were probably induced by boiling of contaminants in the Pu-Am feedstock, which contained about 3% calcium and 2000 ppm magnesium. Both of these elements boil below the casting temperature. The resulting agitation of the molten pool probably contributed to the higher-than-expected evaporation of americium. Three slugs were cast, and about 40% of americium was evaporated in the casting.

During casting, a partial vacuum (10-25 torr) was drawn above the melt for approximately twenty minutes. From data on evaporation rates of tungsten under increased pressure, it was estimated that the evaporation rate of americium could be reduced two hundred times by casting under a pressure of about 800 torr. Furthermore, based on experience with U-Zr injection casting, the vacuum level in the molds could be as high as 150 torr and the molds would still fill. (The filling technique starts at partial vacuum in the space above the melt and then pressurizes it, drawing the melt up into the glass tubes.) A judicious choice of casting parameters combined with feed stocks free of volatile impurities should make fabrication of minor actinide containing fuels quite viable.

The minor-actinide-containing test pin (U-20Pu-1.3Np-1.2Am) was irradiated in EBR-II to 6% burnup, but this test too was prematurely terminated when EBR-II was shut down in 1994. The post-irradiation examinations showed very satisfactory irradiation performance. The micrograph shown in Figure 6-7 does not exhibit any features not seen in other fuels. The radial distribution of the fuel constituents is also shown in Figure 6-7. The uranium, plutonium, and zirconium distribution is very similar to that in the standard U-Pu-Zr fuel. Neptunium does not migrate; it stays put. Americium generally follows zirconium, precipitates in pores, and tends to migrate more in porous fuel. Although this was a limited test, the information it gave was positive, indicating that recycle in IFR all-actinide fuel should raise no new difficulties from spent fuel with only the major isotopes in it.

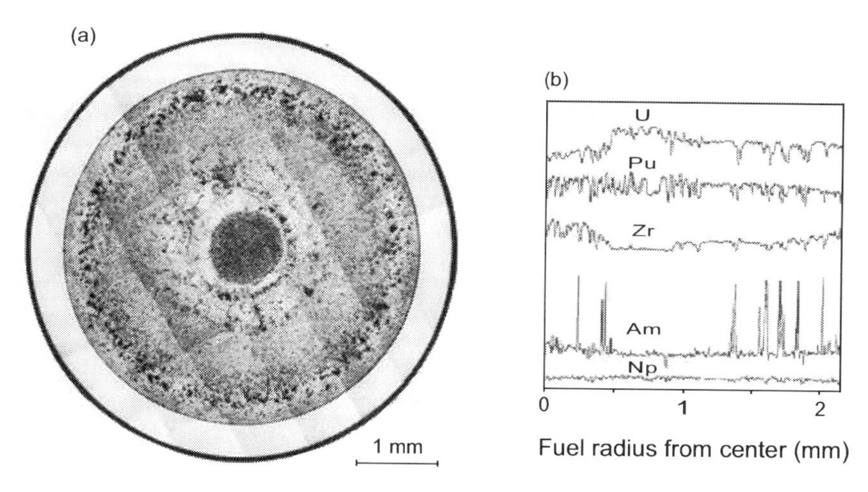

Figure 6-7. Post irradiation examinations for minor-actinide-containing fuel at 6% burnup (U-20Pu-1.3Np-1.2Am)

6.10 Other Characteristics

Metal fuel greatly eases the requirements placed on the non-reactor portion of the fuel cycle. Fabrication is simple, easy and cheap, processing is done in only a few steps, and the equipment for recycle is compact and inexpensive. Aided by small fissile material flows made possible by its long burnup, the fuel choice thus provides the bases for an economic fuel cycle. The injection-casting fabrication technique illustrated in Figure 6-8 is also amenable to remotization.

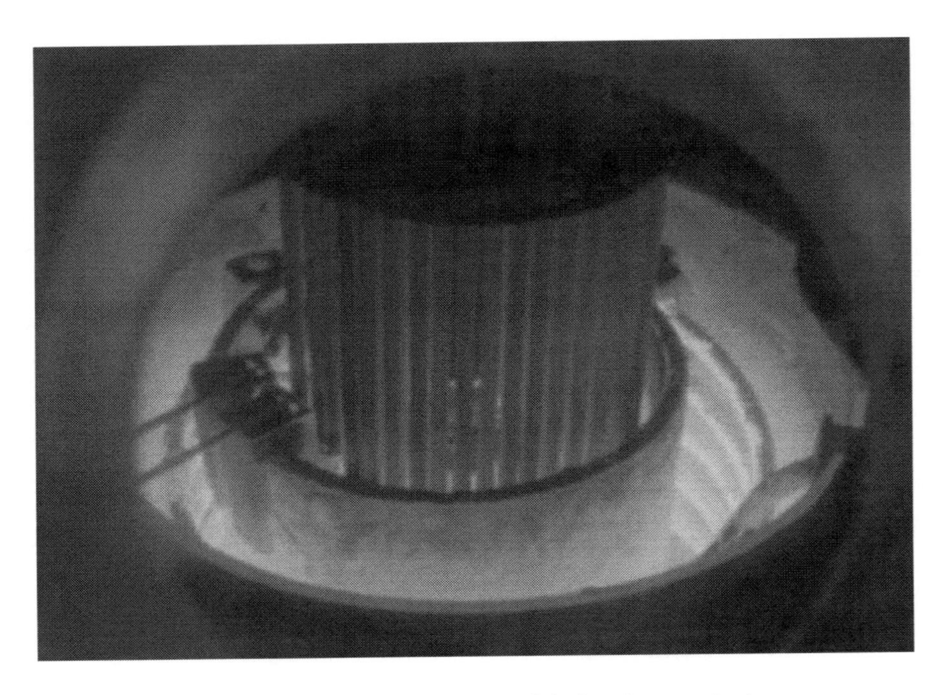

Figure 6-8. Injection-casting fabrication technique

Finally, the characteristics of the fuel increase protection against accidents with major consequences, in stopping such accidents before they can start, and in limiting the extent of the consequences if they do. Metal fuel has a relatively low melting temperature, and has been criticized on this basis. But metal fuel has about ten times better thermal conductivity than oxide fuel, so its temperature rises much more slowly with increases in power, planned or unplanned. This means more margin-to-failure, not less, in overpower accidents. The conductivity also means much lower internal fuel temperature, making possible other unique and important inherent safety characteristics. Finally, the low fuel temperature acts to limit the energy release in any serious accident where fuel melts, shutting the reactor down with limited consequences. But all this is the subject of the next chapter. There we will discuss the safety attributes of the IFR in considerable detail.

6.11 Summary

Although it is based upon EBR-II experience and the certainty of its behavior relies upon the decades of experience with related fuel designs in EBR-II, IFR fuel is in fact a new fuel. Plutonium-based, it is unique to the IFR; it has not yet been used elsewhere in the world. It was developed and tested extensively in the IFR decade from 1984 to 1994 in EBR-II, and to some degree further in FFTF late in that decade and for a year or two after. It has a number of quite remarkable traits which are responsible for many of the positive characteristics of the IFR system. In ease of reactor operation, in unusual reactor safety characteristics, in the simplified fuel cycle, and in more satisfactory waste management, all these characteristics of crucial importance are in large part due to this fuel choice.

The qualities of a good fast reactor fuel are long life in reactor, ruggedness in withstanding repeated power transients, high power generation in each pin, leeway in overshoots in power, benign behavior if cladding fails, simplicity of fabrication, and suitability to simple and inexpensive processing. And there are another set of characteristics that come in to reactor safety considerations which we will cover at length in the next chapter.

The IFR fuel can be thought of as a weak fuel material contained in a strong cladding. Fabricated to have a considerable gap initially between fuel and cladding, the metal fuel swells out to the cladding fairly soon in its life in-reactor. The swelling is caused by fission product gases collecting in bubbles in the fuel, expanding as irradiation proceeds and more gases are generated, which exerts irresistible force for swelling up to the point where the bubbles interconnect. When they do, the gases largely release to the gap and collect in the plenum provided in the fuel pin, above the fuel itself. The swelling ceases sufficiently that there is little pressure on the cladding from the fuel to the end of fuel life. This simple change in design, leaving room for the fuel to expand a measured amount, is the reason that metal fuel, with all its advantages, can now be used with confidence, when in the 1960s it was abandoned because it swelled.

Fuel lifetime in reactor under these circumstances is governed by the cladding behavior, not the fuel itself. The cladding must still withstand the challenging conditions of high temperature and intense radiation field of the high powers of a fast reactor. Its lifetime is typically three or four years. The power generated in individual fuel pins is kept high, the fuel has a lot of fissile material in it, and the more power each pin can have, the fewer the pins and the less the fissile amount needed to provide the operating power of the reactor. Metal fuel adapts easily to high power densities, as it transfers heat rapidly to the coolant and the temperature does not build up in the pins. This has two very favorable effects. The temperature at the center of the fuel pin stays close to the coolant temperature, perhaps a hundred degrees or so above it only, even at very high power densities. The fuel doesn't come close to melting even at the highest temperature point, which is its centerline close to the core outlet where the coolant temperature is highest. It will withstand very considerable power overshoots without damage, and can do it repeatedly, if called upon. It is a rugged fuel.

Further, its cladding can fail utterly and it still doesn't release radioactive corrosion products to the coolant to contaminate the entire primary system. The metal fuel and the liquid metal coolant do not react with each other. There are no corrosion products, as there is no corrosion.

The requirements placed on the non-reactor portion of the fuel cycle are eased by metal fuel. Fabrication is simple, easy, and cheap, and processing is done in a few steps, so the equipment for recycle is compact and inexpensive. Aided by small fissile material flows made possible by its long burnup, these features provide the basis for an economic fuel cycle.

Finally, the fuel choice provides increased protection against accidents with major consequences by stopping such accidents before they can start, and by limiting the extent of the consequences if they do occur. Metal fuel has a relatively low melting temperature, and has been criticized on this basis. But its thermal conductivity is about ten times higher than that of oxide fuel, so its temperature rises much more slowly with increases in power, planned or unplanned. This provides more margin-to-failure, not less, in overpower accidents. High conductivity also means much lower internal fuel temperature, in turn making possible other unique and important inherent safety characteristics. Finally, the low fuel temperature acts to limit the energy release in any serious accident where fuel melts and the reactor shuts down with limited consequences. In the next chapter we will discuss the safety attributes of the IFR in considerable detail.

References

1. W. D. Leggett III and R. D. Leggett, "A Decade of Progress in Fast Reactor Fuel," *Proc. LMR: A Decade of LMR Progress and Promise*, Washington, D.C., November 11-15, 1990.
2. D. C. Crawford, D. L. Porter and S. L. Hayes, "Fuels for Sodium-cooled Fast Reactors: U.S. Perspective," *J. Nuclear Materials*, 371, 202-231, 2007.
3. R. J. Herbst and R. W. Stratton, "LMR Advanced Fuels: (U,Pu)-Carbide Fabrication, Performance and Reliability," *Proc. Int. Conf. on Reliable Fuels for Liquid Metal Reactors*, Tucson, Arizona, September 7-11, 1986.
4. S. Govindarajan, et al., "Performance of FBTR Mixed Carbide Fuel," IAEA-TECDOC-1039, 1997.
5. A. A. Bauer, P. Cybulskis, and J. L. Green, "Mixed-Nitride Performance in EBR-II," *Proc. Symposium on Advanced LMFBR Fuels*, Tucson, Arizona, October 10-13, 1977.
6. L. C. Walters, B. R. Seidel and J. H. Kittel, "Performance of Metallic Fuels and Blankets in Liquid-metal Fast Breeder Reactors," *Nucl. Technol.* 65, 179 (1984).
7. B. R. Seidel, L. C. Walters and Y. I. Chang, "Advances in Metallic Nuclear Fuel," *J. Metals*, 39, 10, 1987.

8. G. L. Hofman and L. C. Walters, "Metallic Fast Reactor Fuels," in *Material Science and Technology: A Comprehensive Treatment*, eds., R. W. Cahn, P. Haasen, and E. J. Kramer, 10A, 1994.
9. R. S. Barnes, *J. Nuclear Matters*, 11, 135-148, 1964.
10. Charles E. Stevenson, *The EBR-II Fuel Cycle Story*, American Nuclear Society 1987.
11. Y. I. Chang, "Technical Rationale for Metal Fuel in Fast Reactors," *Nuclear Engineering and Technology*, 39, pp.161-170, 2007.

CHAPTER 7

IFR SAFETY

In this chapter we look at the all-important issues of reactor safety. What worries people most? Most people would say they are concerned most about how safe they are.

Is it safe? Mistakes are human, accidents happen. Will the reactor cope with them without hurting anybody or damaging itself? Reactors cost a lot of money and we depend on them to produce our electricity. How reliable is the IFR? How does it measure up against other kinds of reactors? And what is the evidence behind statements about its safety?

Concerns about the use of IFR technology in proliferation of nuclear weapons will be covered in a later chapter, but in short, does IFR technology add to the likelihood of proliferation of nuclear weapons? IFR processed fuel material is not usable directly in weapons; processing would be necessary by some other unrelated process, a situation not different in kind from starting with unprocessed fuel. Nor does the chemistry of the IFR process allow a pure plutonium product from the uranium/plutonium/actinide mix of irradiated fuel. So the IFR adds little or nothing to proliferation risk. We cover this area in detail in Chapter 12.

This chapter deals with the safety of the reactor itself. Starting with a description of the protection against accident consequences, we will look at the regulatory requirements all reactors today must meet, which encode good practice in providing this protection. This is basic for all reactors, and it is basic for the IFR as well. But we can go much further. We introduce the concept of passive safety, what it is, and what is needed to have it. We describe how it works in the IFR, and the features that give the IFR its unique safety properties. We then look at how this could fit into the regulatory environment, simplifying licensing. Finally, we describe the detailed experimental evidence that establishes that these properties are real in the IFR. With evidence provided by elaborate and expensive experiments carried out in large nuclear facilities, in the main built expressly for this purpose, the safety characteristics of the IFR are not subject to question—they are proven.

The important point is that IFR technology can provide an assurance of safety beyond anything possible in the water cooled reactors of the present day. As described briefly in Chapter 5 the very materials chosen for the IFR give the characteristics necessary for this. They go well beyond what is possible in other forms of the fast reactor, such as the accepted standard—the fast reactor using oxide fueling. These characteristics are intrinsic and unique to the IFR technology. In this chapter we will focus on what these characteristics are, what kind of behavior can be expected if accidents do happen, and in general why things work the way they do. We will outline the reasoning that led to the focus on such characteristics. And we will present in detail the evidence from the experimental and analytical programs that provide assurance of the reality of these assertions. There are benefits across the entire range of possible accidents, from the smallest upset to the most severe accidents that can be postulated. These benefits have not been widely understood, or appreciated, but they are important; in many severe accident situations they become very important indeed.

Because of its obvious importance, reactor safety has been exhaustively studied over the entire history of reactor development, dating all the way back to the 1940s. A vocabulary has evolved, used by scientists and engineers specializing in safety R&D and in design, whose specialized meanings allow precision in describing safety work. We will try to avoid such specialized wording, which can make for heavy sledding for the non-specialist, to try to make the subject as clear as we can. The commonly understood language may make the specialist wince occasionally, but we hope the reader will benefit.

First, let us review the modern requirements for all reactors, and the logic underlying them. Very specific goals are set for the safety of all modern reactors.

7.1 Safety Goals for All Reactors: Defense in Depth

The goals are defined to ensure adequate protection of public health and safety and of the environment. In design and in ordinary operations, ample safety margins are specified. For any accident that can be anticipated, safety systems are provided that will shut down the reactor without harm. And finally, over and above these, features are added that will minimize the consequences of accidents whose probability is low but whose consequences would be serious. This approach is called "defense in depth." Attention to quality construction, rigorous maintenance, formalized operating and maintenance procedures, exhaustive training of personnel, reliable control and safety systems, and multiple physical barriers to radiation release all play a part.

Defense in depth in fast reactors formally has three successive levels.

The first level is the design requirements for reliable prevention of accidents initiating in the first place. Large margins between normal operating conditions and failure are mandated.

The second level are the safety systems that shutdown the reactor, remove the shutdown heat, and provide emergency power. Each system is safety-grade, a high standard of quality with increased attention to quality and correspondingly greater expense, and each has a back-up system in the event of failures. The backup systems, normally idle, are continuously monitored, inspected, and periodically tested to ensure readiness. There is an independently powered and instrumented reactor shutdown system, a safety-grade emergency heat removal system to remove decay heat by natural circulation after shutdown, a second independent off-site power connection, and a safety-grade on-site emergency power supply.

The third level provides additional protection to the public for accidents that are not expected to happen ever, and for accidents that were not foreseen when the plant was designed. A second vessel surrounding the vessel containing the reactor core, called the "guard vessel," will catch primary coolant from any breach of the primary coolant system. It is sized to ensure that the reactor core remains covered with sodium coolant so it will be cooled by the emergency heat removal system even if the primary reactor vessel fails. If primary coolant leaks into the reactor building air atmosphere, or if failures of the cladding and the primary system barriers lead to release of gaseous fission products, the reactor building itself is designed to contain any radioactivity. (It might be noted here that the hydrogen explosions that damaged the Japanese reactor buildings after the quake could not happen to the IFR as the sodium coolant and structures contain no hydrogen, no H_20 as in an LWR.)

7.2 Safety in the IFR: Introducing New Characteristics

But IFR safety goes much further than this. The IFR goal is a safe result even if all the normal defenses in depth fail. A safety net is provided in case the engineered systems, against all expectations, fail. The choice of materials and configuration for the reactor provides this. It is unique to the IFR and ensures its ultimate safety.

The goal is simple: Radioactive fuel materials must be contained inside the fuel cladding and activation of materials outside the core must be negligible under all operating conditions. Further, radioactive materials in the fuel must be contained safely inside the reactor tank under all accident conditions, even the most unlikely.

The barriers going outward from the fuel are steel, the cladding of fuel itself, the primary tank, the guard tank, and the containment. Only liquid sodium coolant in the secondary coolant circuit, which is not radioactive, leaves the reactor vessel. Shielding inside the tank protects this sodium from activation. The primary coolant stays entirely within the reactor vessel and therefore is radioactive from the exposure to the high neutron flux in the core. The secondary system is so free from radiation that personnel have routine access to the area above the reactor, the heat exchangers, and the steam generators while the reactor is in operation. In loop-configuration reactor designs where the radioactive primary sodium is brought out of the reactor in piping, this is not possible.

In the IFR, then, in normal power operation, all radioactive materials are kept inside the primary tank. And under accident conditions, the fuel, cladding and coolant combination itself provides no chemical reaction between the fuel and coolant that could block coolant flow and cause trouble, nor is any other troubling chemical reaction possible such as the hydrogen generation possible where water is the coolant.

But it is the self-correcting nature of the reactor power in the IFR that gives it its most powerful passive safety features. Power self-corrects by decreasing as the reactivity decreases with increasing power. Power increases are caused by an increase in reactivity in the first place, and it is the resulting temperature increase that lowers reactivity again so power is reduced automatically. These are passive effects, reactivity feedbacks that involve no movement of the control rods and no operator action. It is this, combined with passive removal of shutdown decay heat, which prevents damage over a range of otherwise serious accidents. The cladding remains intact. There is no radioactive release even to the coolant. But even under much more unlikely circumstances, events beyond design basis where fuel may fail, the tank itself safely contains the result.

7.3 Significance to Regulatory Requirements

Important in themselves, such passive safety characteristics could very be important to regulatory requirements. They were considered in licensing in the past. Such characteristics were not possible in reactors proposed in the past for licensing. But more recently, with identification both of the possibility and the importance of such characteristics in advanced reactor designs, they have begun to be mentioned in licensing policies. The Nuclear Regulatory Commission's policy statement of 1994 on regulation of advanced nuclear power plants [1] brings in inherent safety in the following way:

> "Among the attributes that could assist in establishing the acceptability or licensability of a proposed advanced reactor design, and that therefore should be considered in advanced designs, are:
>
> - Highly reliable and less complex shutdown and decay heat removal systems. The use of inherent or passive means to accomplish this objective is encouraged (negative temperature coefficient, natural circulation, etc.).
> - Simplified safety systems that, where possible, reduce required operator actions, equipment subjected to severe environmental conditions, and components needed for maintaining safe shutdown conditions. Such simplified systems

should facilitate operator comprehension, reliable system function, and more straightforward engineering analysis.
- Designs that minimize the potential for severe accidents and their consequences by providing sufficient inherent safety, reliability, redundancy, diversity, and independence in safety systems. ... "

7.4 Evolution of Fast Reactor Safety: Treatment of Severe Accidents

The effect overall of these passive safety characteristics is what is meant by the term "inherent safety." Inherent in the properties of the materials making up the reactor and in its configuration, these characteristics keep the reactor safe.

The possible importance of passive safety characteristics can be seen in the history of safety analyses and evaluations. Safety systems have always been designed to safely handle a broad set of accidents defined for a wide variety of system failures—the so-called "design basis accidents." Design basis accidents generally assume a failure in the safety grade systems, but only a single failure, and very importantly, no failure of the control and safety systems. Such accidents are accommodated in the design and must be shown to present risks to the public within regulatory standards.

But beyond design basis accidents, accidents that involve two or more failures, simultaneous failures, of safety grade systems can be postulated. Such accidents have been studied exhaustively and are calculated to be of such low probability that they have been termed "hypothetical." However, if they did occur they would be serious; they threaten to disrupt the core, typically by melting, and if they do, because of the high fissile content of fast reactor fuel they could be serious indeed. They are termed "Hypothetical Core Disruptive Accidents (HCDAs)." They are considered to have a frequency of occurrence of less than 10^{-6} per reactor year; that is, for any reactor, for any one year, a one in a million probability. But they cannot be ruled out. Nor can they be ignored.

The possible severity of the consequences of this class of accidents has led to significant regulatory scrutiny of them in licensing reviews of past sodium-cooled fast reactors. The purpose was to identify just what the safety margins are, both thermally (temperature) and structurally, for accidents that go beyond the design basis. The consequences of a postulated accident are calculated by modeling in detail the phenomena occurring in the course of such an accident, made possible by the advent of modern computing capability. Great detail in modeling is possible. However, the very complexity of the modeling demands that confirming experiments be carried out to assure its validity. The modeling used in these calculations in the present day has been tested exhaustively against corresponding experiments, as mentioned; this has been detailed, complex, and costly, but high confidence can now be assigned to the results.

To go back a little in history, the first widely accepted analysis of extreme accidents in fast reactors was the "Bethe-Tait" accident published in 1956. [2] In this, the authors estimated the order of magnitude of the largest explosion possible when it was assumed that the sodium coolant had disappeared entirely, so the temperature rose rapidly, and the core then melted and collapsed under gravity. Far too simple to be reliable, it nevertheless began the process of bringing modeling of real phenomena into accident calculations. Modeling became more and more detailed in the following decades. In the Clinch River Breeder Reactor licensing review of the late 70s and early 80s, the modeling and calculation of HCDAs, although beyond design basis, received a great deal of attention.

A typical HCDA for an oxide-fueled fast reactor like the CRBR postulates the loss of coolant flow by station blackout (complete power failure) and simultaneous failure of both primary and secondary reactor scram systems (an assumption, no known cause identified). The resulting rapid temperature rise and boiling of the sodium coolant introduces substantial positive reactivity due to the resulting voiding (boiling) of sodium, and a prompt critical (very rapid) power excursion results in a core meltdown. Comprehensive computer codes, such as SAS4A [3], were developed to analyze such accidents. These codes trace the response of the reactor core and its coolant, its fuel elements, and the structural members to the accident, and the consequences—the energy release, and the implications of the molten core debris on the reactor vessel and surrounding concrete structures—are calculated. The purpose of course is to establish whether these consequences will be safely contained.

The effect of the calculated energy release (explosion) on CRBR containment was severe enough to be a real concern. But in the end this phenomenological modeling, supported by experimental verification, did establish the adequacy of the containment of the CRBR.

Although these HCDA scenarios have very low probability they consume much effort and expense. In licensing review they inevitably resulted in further implications for the design. Responding to licensing concern, designers tend to go the next step and add still more safety systems. For example, a self-actuated safety shutdown system with a magnetic latch that will release when the coolant temperature rises above a certain temperature (its Curie point) was considered in addition to the primary and secondary shutdown systems. Mitigation features, such as a "core catcher" beneath the core to retain molten core debris, were considered. However, in the end there is a limit to more and more safety systems. Eventually they are counterproductive in that they simply provide more ways the system could fail.

7.5 Safety Through Passive Means: Inherent Safety

In contrast to the approach described above, partly as a result of Argonne's principal role in the CRBR safety calculations and related experiments, the IFR took quite a different path. In dealing with HCDAs instead of adding engineered safety systems, we sought to emphasize possible passive safety characteristics that could actually accommodate the kinds of failures that lead to beyond design basis accidents. In other words, stop them before they start. Such characteristics, we felt, were possible with proper choice of materials for the core; the principal free choice is the form of the fuel. Sodium as a coolant has too many advantages to want to move away from it. In the end, a principal contribution of the IFR work was the demonstration that inherent safety of this kind in fact is possible—a very significant finding indeed.

Excessively high temperatures lead, in order, to coolant boiling, fuel melting, and cladding failures. But inherent neutronic, hydraulic, and thermal performance characteristics can limit such temperature rises to safe levels which avoid even the first step, coolant boiling, in beyond-design-basis accidents. And this can be done without activation of engineered systems or operator actions—"inherent safety" in fact.

Sodium-cooled fast reactors using metal fuel have these intrinsic fuel and coolant properties. They can provide the desirable attributes spelled out in the Nuclear Regulatory Commission's policy statement. The physical properties of sodium are important: The high boiling temperature (881°C) means that the coolant system requires no pressurization, as do those in light water reactors, heavy water reactors, and gas-cooled reactors. Liquid sodium has excellent thermal conductivity (about ninety times that of water) so heat flows into it easily, away from the fuel, and it has a reasonable specific heat (about one third that of water) to absorb the heat and carry out of the core. These are the properties that are exploited to yield extraordinary inherent passive safety characteristics. Since pressurization is not needed, a simple reactor vessel with a second guard vessel is sufficient, and the absence of pressurization goes a long way toward ruling out coolant loss. Further, the vessel diameter is not limited by the need to contain high pressures (wall thicknesses increase with vessel diameter) making feasible the pool design, a must for some passive features.

The characteristics of metallic fuel, like sodium coolant, that are important to passive safety include its thermal conductivity and melting point. The thermal conductivity is high, as it needs to be for several of the passive responses. The most desirable melting point is more complex. It must be high enough to provide sufficient margin to melting for safe operation, but low enough to provide passive characteristics that terminate multi-failure accidents without energy release sufficient to allow radioactive release to the environment, and preferably no energy release at all.

7.6 Handling Severe Accidents: Accidents Where the Reactor Shutdown Systems Fail

Identifying the causes of the principal severe accidents is not difficult—the reasons vary but all lead to the same result; inability to cool the reactor properly. Lack of cooling is the basis for all serious accidents, if sufficiently serious core melting results. In the language of safety specialists, these are called "anticipated transients without scram (ATWS)" events. That is to say, these are power increases that would be anticipated if certain identifiable events were to take place. Normally such events would be terminated safely by the safety systems. But the ATWS event assumes the safety systems disabled, so no control or safety system action is available to terminate them. These would be serious accidents in most reactor types.

However, such accidents would be handled completely safely through the inherent characteristics of the IFR, with no damage of any kind. The important physical factors are ample margin between operating and boiling temperatures, low centerline temperature of the fuel due to the high conductivity of metal, and sufficiently high fissile conversion in the core to limit the necessary reactivity through the burnup cycle. Each plays a part in making a safe response to ATWS events possible. We will examine these ATWS events one by one to show how the passive safety characteristics enter to stop them before serious trouble can start.

7.6.1 Unprotected Loss of Flow

In this type of accident, a loss of offsite power cuts off power to all pumps; they coast down, the primary pump with a designed-in eight-second flow halving time, the secondary with six seconds. (Easily designed in, these pump coast-down times are needed to handle the power spike in the first few seconds.) If the control or safety rods act, the reactor reduces power in response to the passive feedback mechanisms which bring in reactivity changes some tens of cents negative, a substantial reduction in reactivity. There is an overshoot in the core outlet coolant temperature as power doesn't match available coolant flow initially, but this dies away quickly in the face of the negative reactivity coming in. Combating the reactivity decrease overall is positive reactivity from the Doppler effect (a reactivity effect inherent in the neutron cross-section changes from changing temperatures, small in metal fuel) arising from the decreasing fuel temperature, which tends to hold the power level where it is. The faster the power can decrease, the smaller the temperature increase in the entire system and the easier it is to assure that there is no damage.

The magnitude of the Doppler effect can be very important in such accidents. The fact that it is small in metal, much smaller than in oxide, is the key to the safe response. Although both the magnitude of the Doppler effect per degree of temperature change and the magnitude of the temperature change itself in the fuel are greater in oxide, the principal difference between metal and oxide is caused by the much higher temperatures in oxide fuel. The high oxide fuel temperatures in operation mean that very substantial positive Doppler reactivity comes back in as the fuel temperature has to decrease many hundreds of degrees due to the feedback effects. The resulting positive Doppler reactivity slows the decrease in power. In metallic fuel the centerline temperature is only two hundred degrees C or so above coolant temperature; in oxide it is in the range of two thousand degrees. The small Doppler effect in metallic fuel allows the power to drop sharply, as very little countervailing reactivity comes in. In no channel does the coolant temperature rise to levels close to boiling. At minimum there is a margin of at least 150°C from boiling. After the power surge is over, the reactor stabilizes at a low steady power, a few percent of normal. There is no damage to the plant. The events are not sensitive to changes in parameters to any significant degree, except for the coast-down times of the coolant circulating pumps. If coast downs are extended modestly—as they were in this example—the increase in cooling capability that results at a critical time gives a significant increase in the safety margins.

7.6.2 Unprotected Control Rod Run-Out

In this type of accident, the control rod is run out of the reactor completely through operator error, or by a failure in the control system. No safety rods come in to shut down the reactor. Reactivity is added at the rate given by the run-out time. The increase in power heats the coolant bringing in feedback reactivity effects similar to those in the previous accident type. The end state is similar—power and sodium temperature stabilized at somewhat higher level, with no damage of any kind.

7.6.3 Unprotected Loss of Heat Sink

The heat produced for electricity generation is absorbed by the steam generating system in all power reactors, a substantial fraction producing electricity and the remainder "rejected" to cooling towers or to large sources of cooling water. This is just the thermodynamics of electricity production. The important point is that the steam system is the "heat sink." It absorbs the heat from the system that the reactor generates. If this is cut off, trouble starts instantly. With no heat sink, there is nothing to remove the heat being generated. Temperatures increase immediately if control and safety rods do not act to halt the fission process and shut the reactor down.

In the IFR, this loss of the cooling capability of the steam generators, with failure of the control systems to insert the rods, leads to a similar result to the previous accidents. In this case, however, there is no overshoot. The same feedback effects enter to reduce the power to a few percent of normal. This power level is maintained by the heat from fission product "decay," the heat generated from the highly radioactive but short-lived fission products. These "decay heat" levels are handled by the decay heat removal system designed specifically for passive heat removal in the absence of power. The peak temperatures in this accident can be kept far below coolant boiling and also below temperatures that could fail fuel. The requirements here are somewhat different from the previous accidents. As the overall reactor system heats up, the principal additional need is for the reactor vessel to be sized large enough for the mass of sodium in the pool to absorb the heat until the passive decay heat removal system can handle it safely. The tank is sized to accept heat over and above the capacity of the decay heat removal system for about a week. The decay heat system will handle the heat generation at that point, and the pool will then cool. The feedback characteristics will act to bring the reactor to critical again at this low power, but at a level precisely matching the heat removal capability, and it will maintain at this low level.

7.6.4 Unprotected Overcooling

Any event in the steam system that results in overcooling of sodium in the intermediate loop causes the temperature of the primary sodium entering the core to decrease. If there is no action of the control rods, the initial feedback effects will add reactivity. The accident then is similar to the unprotected control rod run-out case described above. The power increases, the temperature rises to counter the initiating effect, and an equilibrium state is reached, the reactor at low power with appropriate heat removal, and again without damage of any kind.

7.7 Experimental Confirmations: The EBR-II Demonstrations

The effectiveness of such passive safety was demonstrated dramatically in two landmark tests conducted on the EBR-II in April of 1986. Both the truly major accident events—the unprotected loss-of-flow and the unprotected loss-of-heat-sink—were initiated with the reactor at full power. These spectacular tests proved the effects of passive safety design in sodium-cooled, metal-fueled fast reactors, decisively. [4-5]

The unprotected loss-of-flow event can be initiated by station blackout. Nuclear power plants have redundant power supply sources and even if the alternate line is disabled also, the emergency power supply system on-site will be activated. If this fails too, the plant protection system shuts the reactor down. The plant protection system has redundancy too—if the primary shutdown system fails, the secondary shutdown system will be activated. All else failing, the operator can manually shut the reactor down. The unprotected loss-of-flow test in EBR-II simulated the ultimate scenario where all these safety systems and operator actions have failed and the reactor is "on its own."

The sequence of events is outlined in the graphs in Figure 7-1. With the reactor at full power the power to the primary pump was cut off. This immediately reduced the coolant flow as shown in the bottom left plot. With the reactor producing at full power, this caused the coolant outlet temperature to rapidly increase (about 200°C in thirty seconds) as shown in the top left plot. The rising coolant temperature causes thermal expansion of the core components, in particular the fuel assembly hardware, increasing the reactor size a miniscule amount. Slightly less dense than it was before, neutrons now find it easier to escape, and neutron leakage from the core increases. This reduces reactivity.

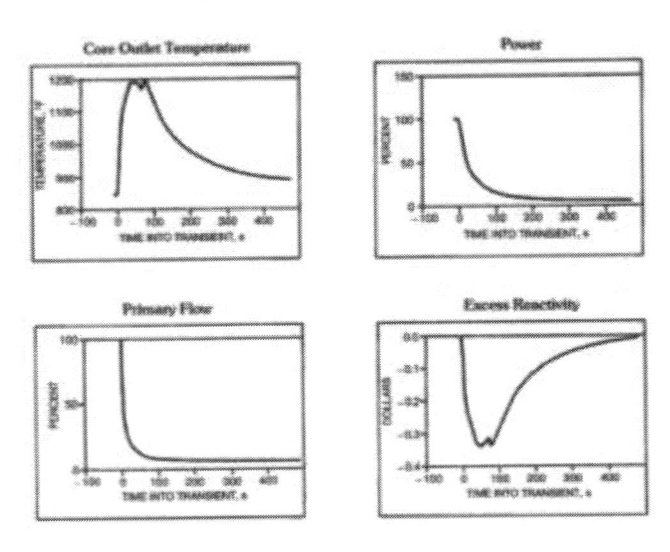

Figure 7-1. Unprotected loss-of-flow test results

During the initial tens of seconds, the mechanical pump inertia provided the flow coast down necessary to keep coolant temperatures well below local sodium boiling, enabling gradual transition to natural convection flow through the core. The negative reactivity feedback is shown in the bottom right plot; the consequent reduction in reactor power is shown in the top right plot.

As negative reactivity comes in the coolant temperature stops rising, and after some minutes an asymptotic temperature is reached at equilibrium with the natural heat loss from the system. The predicted coolant outlet temperature response during the loss-of-flow without scram test is compared with the actual data from the test in Figure 7-2. The excellent agreement shown demonstrates the ability to accurately calculate these events.

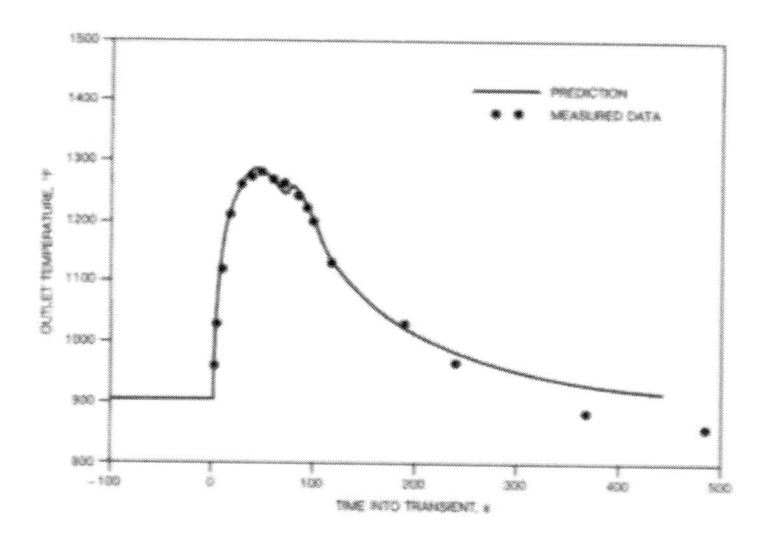

Figure7- 2. Reactor outlet temperature responses to unprotected loss-of-flow test

Following the unprotected loss-of-flow test in the morning, the reactor was immediately restarted and the unprotected loss-of-heat-sink test was conducted in the afternoon of the same day. The unprotected loss-of-heat-sink was initiated by the shutdown of the intermediate pump, isolating the primary system from any heat sink. The primary pump continued to function, transferring the heat from the core to the bulk sodium of the pool.

The sequence of the events is shown in Figure 7-3. The top left plot of Figure 7-3 shows the intermediate loop flow reduced to zero, isolating the reactor from the normal heat sink provided by the balance of the plant. The core heat is dumped to the sodium pool, its inventory large enough to absorb the heat by its pool design. The core inlet is from the pool, so the inlet temperature increases as shown in the top right plot. This turns out to be a rather slow transient—it took about 10 minutes to raise the primary sodium temperature by about 40°C. The gradual increase in the reactor inlet temperature causes the same effects as in the previous test—thermal expansion and enhanced neutron leakages—and the power is reduced as shown in the bottom left plot. The reactor outlet temperature reduces as the power drops, as shown in the bottom right plot. The reactor inlet and outlet temperatures are plotted in Figure 7-4, comparing the prediction and the actual measured data during the test, again showing excellent agreement and confirming again the ability to accurately calculate such effects.

7.8 Factors Determining Inherent Safety Characteristics

EBR-II illustrates characteristics necessary for these remarkably benign passive responses to these severe accident scenarios, but the larger IFRs possess these characteristics as well. Principal among them are:

- Sodium coolant with large margins to boiling temperature
- Pool configuration with large thermal inertia
- Metal fuel with low stored Doppler reactivity

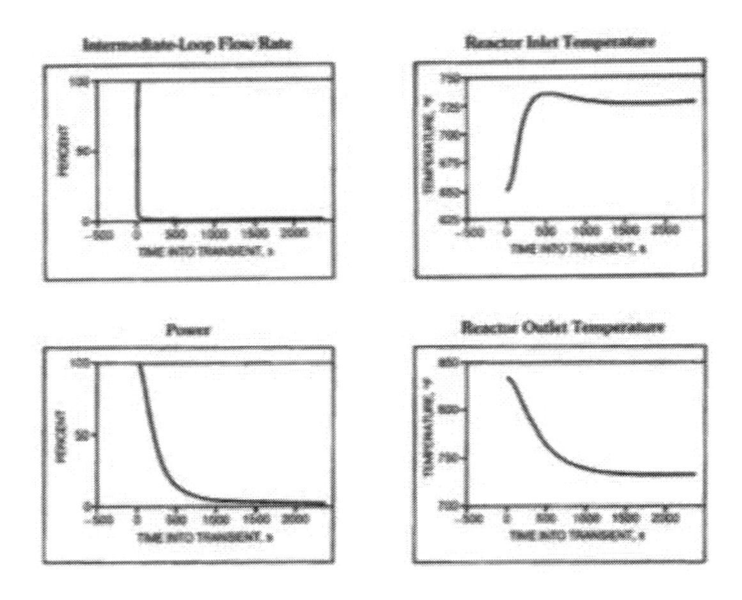

Figure 7-3. Unprotected loss-of-heat-sink test results

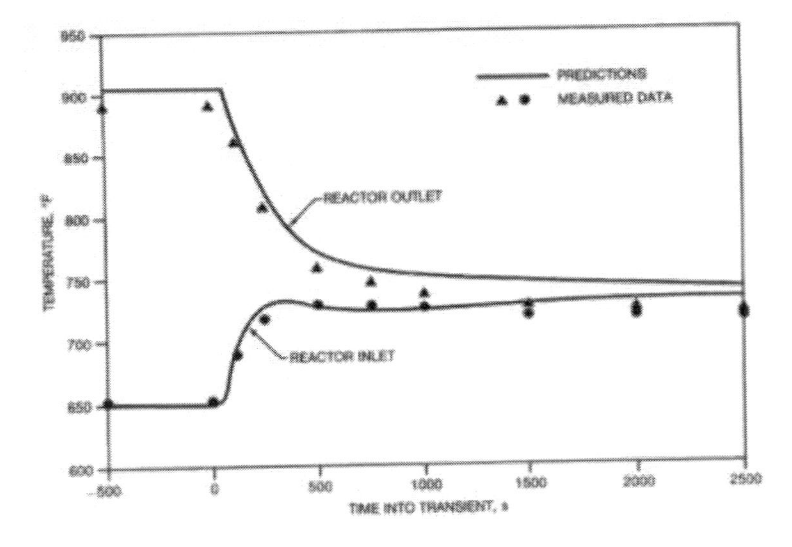

Figure 7-4. Reactor inlet and outlet temperature responses to unprotected loss-of-heat-sink test

The first point is obvious. The initial coolant temperature rise must be ridden out. The second point is necessary to provide heat dump capacity and provide time for thermal expansion of heavy structures to take place. The third point is not so obvious, but was explained above. The importance of the characteristic is perhaps best illustrated by comparison of metal and oxide fuels. The most important factor differentiating the responses in metallic and oxide fuels is the difference in stored Doppler reactivity between the two fuel types. The greater stored Doppler reactivity in oxide means that power does not decrease rapidly during the loss-of-flow without scram event. Further, when the power has been reduced to decay heat levels to counter the stored Doppler reactivity, the coolant temperature equilibrates at a much higher value. The comparison between oxide and metal cores is illustrated schematically in Figure 7-5.

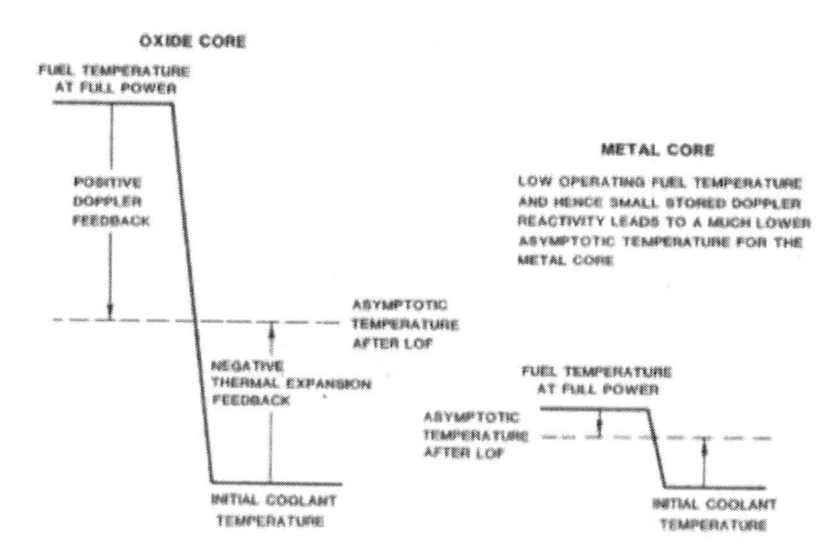

Figure 7-5. Asymptotic temperature reached during unprotected loss-of-flow event is determined by reactivity balance: comparison of oxide and metal cores

The difference between the metal and oxide cores of a typical large reactor (1350 MWe) for loss-of-flow without scram event is illustrated in Figure 7-6. [6] While the metal-fueled reactor maintains large margins to coolant boiling throughout the transient evolution, the oxide-fueled reactor with equivalent core design parameters initiates coolant boiling about five minutes into the transient, a serious problem. All other core design parameters were kept the same, so the difference in the response characteristics is due entirely to the difference in the stored Doppler reactivity.

An inherent safety test was conducted on FFTF analogous to that done on EBR-II. [7] With the FFTF core fueled with oxide, the loss-of-flow without scram test was conducted at 50% power. More importantly, it was conducted with the aid of negative reactivity feedback augmentation provided by a device designed specifically for this purpose. A gas expansion module (GEM) is a gas-filled assembly with the bottom end open; GEMs are loaded in the reflector region just outside the core. With full coolant flow, the pump head compresses the gas and the GEM is essentially filled with sodium. When the pump head is lost, the gas expands, pushing sodium out, and the empty space increases the neutron leakage and provides augmented negative reactivity during the unprotected loss-of-flow event.

Devices such as this may be desirable but cannot be described as a fully inherent safety characteristic. The integrity of the GEMs has to be monitored to assure that they too do not fail when called for. They do complicate operational procedures, since they introduce positive reactivity when the pump is turned on or its speed is increased.

A natural question about the passive safety of EBR-II is whether it is possible only in smaller cores where increased leakage is the predominant factor in reactivity changes. However, if we compare the actual unprotected loss-of-flow without scram test in EBR-II shown in Figure 7-2 and the calculated response for a large reactor shown in Figure 7-6, we see that in fact they are very similar. The temperature rise in the initial minute or so is similar in magnitude in both. The small EBR-II has much stronger negative reactivity feedbacks, so the temperature comes all the way back down to normal outlet temperature. The large reactor tends to equilibrate at a higher asymptotic temperature, but well below temperatures that might cause concern.

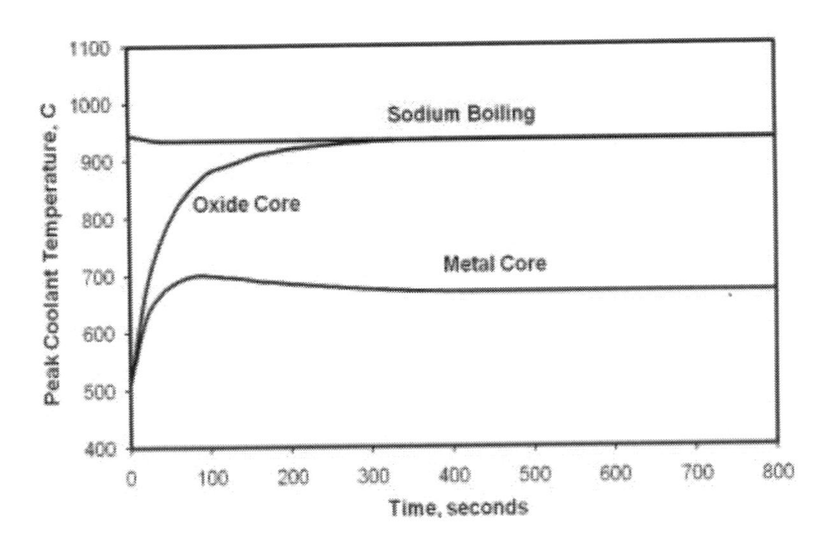

Figure 7-6. Comparison of oxide and metal fueled large reactors in an unprotected loss-of-flow event at full power

The reason for the similarity in behavior is straightforward—the feedbacks themselves are quite similar. The total positive reactivity introduced from full power to shut down power is basically the stored Doppler reactivity, which is more or less independent of the reactor size. The negative reactivity coefficient is larger for the smaller reactor, so a smaller coolant temperature rise is required to balance the Doppler than for larger reactors. However, the initial temperature rise is the important measure because it dictates coolant boiling, and the margin to coolant boiling is largely independent of the reactor size. This is illustrated in Table 7-1, where the results for unprotected loss-of-coolant and unprotected loss-of-heat-sink events are compared among three different reactor sizes with consistent equivalent design parameters.

Table 7-1. Margin to coolant boiling as a function of reactor size, °C

	471 MWth	900 MWth	3500 MWth
Loss of flow w/o scram	170	160	130
Loss-of-heat-sink w/o scram	340	310	360

7.9 Passive Mitigation of Severe Accidents of Extremely Low Probability

If the core structure becomes damaged sufficiently that the passive self-regulation characteristics no longer control the power sufficiently to prevent damage to the fuel, the IFR has other passive characteristics that limit the damage. This feature is important—fuel melting and compaction in fast reactors might reach prompt critical configurations with accompanying large energy releases. This has been the subject of exhaustive studies in oxide-fueled fast reactors to date.

In the IFR, the lower melting point of the fuel combined with its tendency to disperse when it does melt—effects that are the result of well-defined phenomena—provides a passive mechanism for dispersing the fuel so it cannot reassemble in a prompt critical configuration. This inherently limits the energy generation possible in such an accident. The tank will remain intact, and the dispersed fuel from the melted core will be contained and cooled within it. There is no release of radioactivity, therefore, in even these extraordinarily low-probability events.

Regardless of the cause (structural failures or sudden large reactivity additions from any cause) the ultimate response is fuel dispersal. Fuel dispersal in the IFR can be counted on. It introduces a large and negative reactivity injection if other negative feedbacks are not available. The goal, of course, is zero radiation release, no matter what the accident. IFR technology makes this a realizable goal.

- The integrity of the tank and core-support structure is important to both feedback and containment without radiation release. Protection from seismic events is ensured by seismic isolation systems as part of the reactor design, and protection from defects is ensured by monitoring the inherent feedback mechanisms to ensure they are in place. It is important to note that only the nuclear portion of the plant needs isolation. Loss of the balance of plant is simply the loss-of-heat-sink accident and is therefore harmless from a safety point of view.

- The potential sources of positive reactivity need to be small individually, and they need to act independently of each other. First, it should be noted that, contrary to intuition, the best configuration for an IFR is with the internal conversion high. It needs to be enough that the individual rods can have sufficiently limited worth to make inadvertent control rod withdrawal safe by passive means. No damage of any kind results from this potential accident.

- In extreme accident situations it is the early dispersal behavior of metallic fuel that halts the accident progression before positive coolant voiding reactivity can enter to seriously aggravate the situation. Interconnected porosity up to 30 percent of the fuel volume, containing fission gas at high pressure, combined with the low melting point of the fuel, disperses fuel up and out of the reactor, where it freezes on the structure above. It terminates such accidents before prompt criticality can be achieved with the large damage and releases it could cause. In ATWS events there is no release of radioactivity, the accidents are passively terminated with no damage, and in even more extreme accidents with damage to the core all products are contained safely within the tank.

- Prompt criticality, with resultant large energy release, is the cause of major accident consequences. It too is avoided by passive means. In the extremely low-probability accidents caused by unprotected rod bank run-out, where many rods and thus large reactivity is involved, or by abrupt stoppage of flow, where there is no coast-down, fuel dispersal stops the progression before prompt criticality can be achieved. Temperature and power conditions differ sufficiently in the different fuel channels to cause fuel failure at slightly different times. The low temperature dispersal provides a massive negative reactivity injection, overwhelming all other reactivity effects. This terminates such accidents with the failure of only a few assemblies. Damage to the whole core is averted, the energy released is low, there is no prompt criticality, nor are there shock waves to damage the tank. In all cases, energetic events are avoided by a large margin, and there is no radioactive release.

- With all such accidents contained within the tank, the resulting debris in the tank reaches a stable sub-critical configuration adequately cooled with no re-criticality. The debris is in the form of large irregular strands, with very large void spaces so that it cools by natural convection. Results of all analyses and experiments lead to the conclusion that the inherent properties of the fuel—a low enough melting point, its resulting mobility, its high thermal conductivity, and the form of the debris—lead to a safely stable, cooled, and subcritical debris bed, in an intact tank, with no radioactive release whatsoever.

The broad conclusion is that the physical properties of metallic fuel exploited in IFR designs provide passive response that limits damage to the point that there is no radioactive release from the reactor tank in any accident.

7.10 Experimental Confirmations of Limited Damage in the Most Severe Accidents

It is all very well to state the characteristics, but what is the proof of them? To provide the experimental verification of the calculated effects, elaborate tests were conducted in the Transient Reactor Test Facility (TREAT) on the Argonne Idaho site. This reactor can conduct severe transient overpower tests, essentially on an individual fuel channel of an IFR, that match its behavior under the various accident conditions.

7.10.1 The TREAT Tests

In unprotected overpower events of the kind we have been describing, metal fuel has characteristics that result, perhaps surprisingly, in a higher margin in power to failure and consequent fuel dispersal.

TREAT tests were conducted on both irradiated oxide and metal fuel pins. The larger margin to cladding failure for the metal fuel than for oxide is shown in Figure 7-7. Oxide fuel pins fail typically at 2.5–3 times nominal peak power, metal fuel pins at 4–4.5 times nominal peak power.

The TREAT tests also demonstrated the substantial axial extrusion of metal fuel column in the cladding even before failure occurs. When overpower reaches four times nominal peak power, about half of the fuel inventory is molten and the molten fuel distribution is cone-shaped: the bottom third of the fuel column is solid, melting starts at the center, and the molten fuel area gradually expands going up so that 80–90 percent is molten at the top. The fission gas dissolved in the molten fuel is the driving mechanism for axial expansion, extrusion really, of the fuel column. Very large negative reactivity is introduced by such axial fuel movement. The amounts of movement are far greater than that caused by simple thermal expansion, and the reactivity effects are correspondingly large.

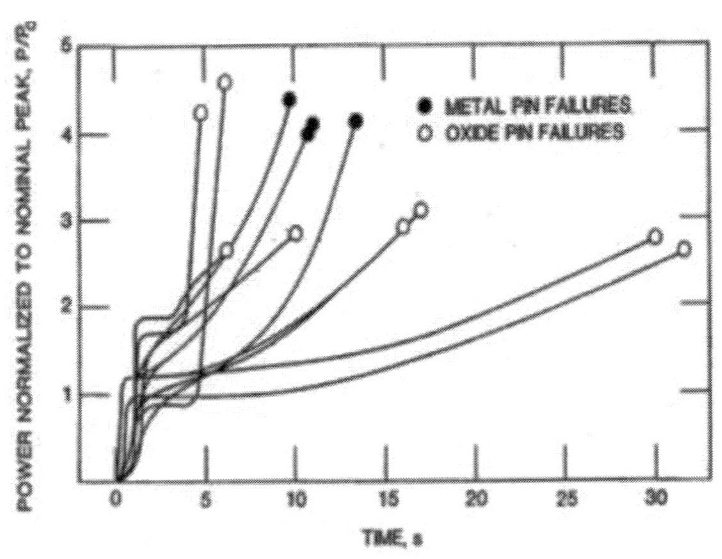

Figure 7-7. Transient overpower failure tests in TREAT show the greater margins for metal fuel

A summary of measured extrusion and calculation by a simple model is presented in Figure 7-8. At burnups below 1 percent, the amount of fission gases in the fuel matrix increases linearly with burnup, and so does the axial extrusion, which can reach 20% or more elongation. At higher burnups, fission gases escape to the plenum and the pressures throughout the pin equilibrate. The expansion of molten fuel is then driven by the temperature difference between fuel and plenum, and extrusion is limited to the few percent range. This pre-failure axial extrusion is permanent in that cool-down does not collapse the elongated fuel column. Friction of the fuel column with the cladding and the lessening of the plenum gas pressure as the gas cools along with the fuel both act to maintain the increased length of the fuel column. These experiments mimic accident conditions. Fuel movement inward toward the reactor center is not wanted at any stage of an accident, and none is found.

The mechanism for actual fuel failure under overpower conditions is overpressure of the plenum that ruptures cladding, as well as thinning of cladding by eutectic formation, or some combination of both. Because both mechanisms cause failure right at the top of fuel column or in the plenum just above the top of fuel column, upon cladding failure the molten fuel along with any fuel-cladding eutectic formed is ejected into the flowing sodium. It is swept up out of the core by the coolant flow and freezes on the cladding and assembly hardware. The TREAT tests demonstrated this further dispersal when fuel is molten. Both the pre-failure extrusion and the post-failure dispersal found in these tests provide intrinsic large negative reactivity feedbacks. They terminate overpower transients no matter what their cause.

7.10.2 The EBR-I Partial Meltdown

The EBR-I partial core meltdown back in the 1950s provides interesting and relevant technical evidence as well. It demonstrated that even under hypothetical core meltdown scenarios, metal-fueled cores have the natural dispersive characteristics described above and will avoid re-criticality—a concern always, naturally.

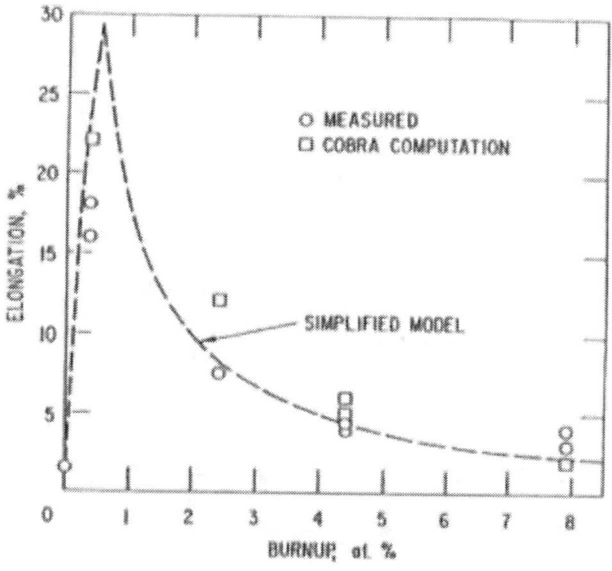

Figure 7-8. Pre-failure axial fuel extrusion as a function of burnup for metal fuel

After its startup in 1951, EBR-I had been used for a number of years to investigate the physics of fast reactors. The reactor was scheduled to be placed on standby early in 1956. As a last experiment, measurements of its transient temperature coefficients were to be made. (It is of interest to know what the components of the temperature coefficients of reactivity are.) The reactor was started at very low power and put on a short period (rapid power increase) to permit the fuel temperature to rise to 500 or 600°C. Because the object was to obtain the temperature coefficient of the fuel only, the coolant flow was shut off. The technician at the control panel was expected to use the fast-acting safety rods upon receipt of a spoken instruction from the scientist in charge. However, upon receiving the instruction, the technician pushed the button for slower-acting control rods. The scientist, when he realized the situation, reached over and pressed the rapid safety rod button and, simultaneously, the automatic high power-level trips responded to activate the safety rods. The delay in time, at most two seconds, was sufficient to permit the reactor power to overshoot to a point where partial core melting took place. [8]

The EBR-I core was small, football-sized. During the post-accident disassembly of the molten core, it was observed that considerable amounts of molten fuel had penetrated into the upper blanket coolant passages, with a maximum upward travel of approximately five inches. The lower blanket region also filled with solidified material from the fuel section. In the core region, two distinct sponge-like zones of porosity had formed. Based on observations and measurements made on the core during its later disassembly, an artist's reconstruction of a vertical section through the damaged core was made, as shown in Figure 7-9. Also shown in Figure 7-9 are what are considered to be the most reliable values of densities of the various areas.[9] Based on out-of-pile experiments, it was conjectured that the porous structure in the center of the damaged core may have resulted from vaporization of NaK (the liquid sodium-potassium coolant) entrained in the molten fuel alloy and subsequent expansion of the vapor. Furthermore, radiochemical analyses of Ce-144 distribution and mass spectrographic analyses of U-235 depletion confirmed that little mixing had occurred in the molten volume and the fuel that dispersed into blanket regions originated from the outer part of the core. Fuel dispersal was therefore was outward, reducing reactivity, which is the key point.

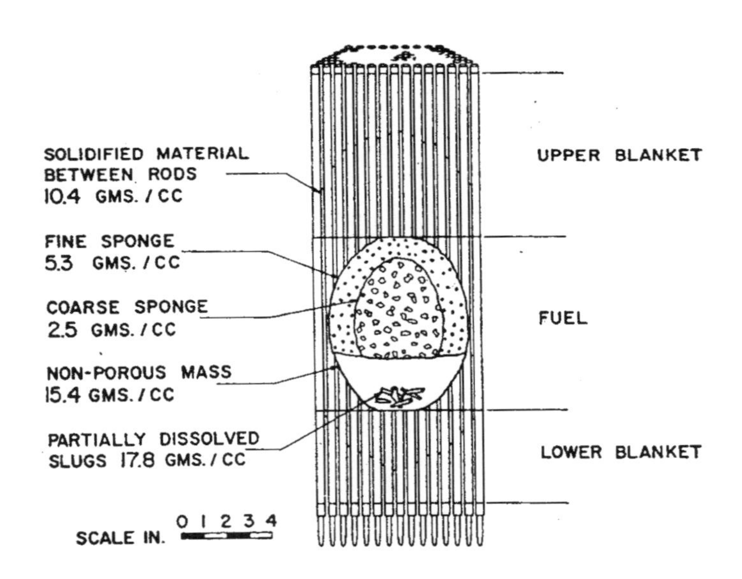

Figure 7-9. Schematic view of vertical section through damaged EBR-I core assembly

The question of whether the EBR-I power excursion was terminated by the core dispersal or by the manual scram was analyzed in detail. [10] The analysis concluded That the manual scram was the likely cause of termination, but the dispersed core would also have terminated the excursion in time to prevent prompt criticality.

7.11 Licensing Implications

What are the implications of all this in licensing? The prescriptive general design criteria codified in 10CFR Part 50 [11] were applied to licensing of the current LWRs, and with modification, the same approach was used in licensing of the Fast Flux Test Facility and the Clinch River Breeder Reactor Project. However, when the Advanced Liquid Metal Reactor program incorporated inherent passive safety in the early 1990s it wasn't clear how to take credit for such inherent safety features. The traditional approach for redundancy and diversity in engineered safety systems and severe accident mitigation systems were recommended by NRC. [12]

There have been improvements in the regulatory process in recent years in the form of an increased emphasis on risk informed decision-making. Probabilistic risk assessment is now used throughout design, safety assessment, licensing and operation. All operating nuclear power plants are now required to have a probabilistic risk analysis (PRA) of both internal and external events, and probabilistic insights are now used in some aspects of operation and regulation. The NRC is now developing a risk-informed, performance-based alternative to 10CFR Part 50 to be used in licensing of future advanced nuclear power plants [13]. The approach is based on the NRC's safety goals policy [14] but still on fundamental safety principles such as defense-in-depth and safety margins. It combines both probabilistic and deterministic (calculation of accident sequences, generally by computer codes) elements. It is technology-independent, with technology-specific requirements for particular designs.

Under the new regulatory framework, a probabilistic risk assessment would be an integral part of the design process and safety assessment, and be given a fundamental role in the licensing process. Deterministic criteria and multiple lines of defense against radioactive release would continue to be required. Under this approach, a probabilistic analysis would be used to establish the event sequences to be considered in the licensing process and to classify equipment as to its safety significance. The selected events, called Licensing Basis Events (LBEs), would be analyzed deterministically to demonstrate the conservatism of the probabilistic analysis. The allowable consequences of an event would be matched quantitatively to its frequency. This evolving risk-informed regulatory framework will recognize the benefits of inherent passive safety in a concrete, quantitative way.

Probabilistic analysis will be central to demonstrating the effectiveness of passive safety features. As mentioned previously, in licensing of sodium-cooled fast reactors to date, a great deal of attention was focused on beyond-design-basis events that lead to severe consequences. Probabilistic analysis affords the opportunity to show quantitatively that such events may have a frequency below the lower limit for consideration as LBEs (the current proposal is a frequency less than

10^{-7} per reactor-year); nevertheless, defense-in-depth considerations may still require mitigation features such as low-leakage containment.

In the context of inherent passive safety, the desirability of having negative sodium void reactivity has sometimes been raised as a goal for future fast reactors. In Russia, the regulatory criteria include such a clause. BN-600 has negative sodium void reactivity due to highly enriched uranium fueling and this property is a natural result, but the same goal has been adopted for the design of BN-800. This does require some spoiling of core geometry (non-optimum core dimensions to enhance neutron leakage). In the CRBR licensing, the positive sodium void reactivity played a key role in the analysis of the loss-of-flow without scram leading to a core disruptive accident. However, if the reactor is designed to accommodate such an event without coolant boiling, the positive sodium void reactivity in itself has no significance. What is important is that the overall temperature reactivity coefficient and the power reactivity coefficient, including all reactivity feedbacks—fuel, coolant, structural, etc.—remain negative over the full range of powers.

7.12 Sodium Reaction with Air and Water

7.12.1 Sodium-Water Reaction

The most important concern arising from sodium itself is its high chemical reactivity. It reacts violently with water and it burns in air. Compared to the other alkali metals, sodium is more reactive than lithium and less so than potassium. High chemical reactivity means that in nature it is found only as a compound.

Sodium reacts exothermically with water: small pea-sized pieces will bounce around the surface of the water until they are consumed by it; large pieces will explode. The reaction with water produces caustic sodium hydroxide and hydrogen gas, which can be ignited by the heat produced by the reaction. Because of this, contact of sodium with water or steam in the steam generator system design must be avoided.

One very conservative approach very successfully demonstrated in EBR-II assures that the barrier between sodium and water is very reliable. The steam generator tubing is made straight and double-walled. The EBR-II steam generators of this design operated without a single tube leak for their entire thirty-year life. Although early fast reactors experienced some isolated steam generator problems primarily associated with welding techniques used for dissimilar metals, fast reactor steam generators in general have been reliable. Double-walled tubing has not been the norm. Sodium itself (unlike its reaction products) is completely compatible with structural materials, so no corrosion products accumulate in the crevices of the shell side of the tubes, a phenomenon that has plagued LWR steam generators.

The sodium flowing into the steam generator isn't the primary sodium that cools the reactor. An "intermediate" heat exchanger transfers the heat from the radioactive primary sodium to a "secondary" sodium loop. Thus the sodium that goes to the steam generator, where water is present, is non radioactive. In the pool design, the intermediate heat exchanger is located in the reactor vessel itself, submerged in the pool. The purpose is to ensure that the radioactive primary system is isolated from the steam cycle so any steam generator tube leak into sodium cannot impact the primary system directly.

As an added safety measure, a steam generator relief and dump system is provided to relieve pressures in the secondary sodium system in the event of a major steam or water leak into the sodium. The system relieves overpressure through a rupture disc located on the steam generator, sized to prevent damage to the sodium system. The sodium/water reaction products are collected in a separator tank downstream of the rupture disc, where the majority of the liquids and solids are separated and the remaining gas-entrained solids and liquids are passed to the centrifugal separator tank. In this tank, the remaining liquid and solid components (all non-radioactive) are removed and the gaseous products are released to the atmosphere through the hydrogen relief stack and igniter.

7.12.2 SuperPhenix Sodium Leak in Fuel Storage Vessel

During commissioning stage of the French 1240 MWe SuperPhenix fast reactor, a sodium leak developed in its fuel storage vessel. Although this event did not lead to sodium/water reaction, it had a significant effect on operation, and the reactor was shut down for eighteen months. SuperPhenix was designed to use a large fuel storage vessel just outside of the reactor containment structure to store fresh as well as spent fuel assemblies. Fresh fuel was to be transferred from the storage vessel to the reactor vessel through the A-frame fuel handling mechanism, and the spent fuel would be transferred out through the same A-frame system to the fuel storage vessel for cooling. The fuel storage vessel was large, 9 m in diameter and 10 m deep, containing 700 tons of sodium. The vessel was cooled by sodium-filled coils mounted on brackets welded to the inside surface of the vessel wall. A safety guard vessel enclosed the main storage vessel; the annular space between the two vessels (about 15 cm wide) was filled with nitrogen; and leak detectors were mounted on the bottom of the guard vessel.

Subsequent examination following the leak detection revealed a horizontal crack some 60 cm long along a weld that attached a cooling coil support bracket to the vessel wall. Several more cracks were discovered at similar places on the vessel. After an exhaustive investigation it was concluded that the cause was the choice of the vessel material. The ferritic steel 15D3 was chosen for its high temperature properties instead of the carbon steel used in Phenix. There wasn't sufficient performance data on it in a sodium environment. Three additional factors contributed to a deleterious environment: (1) microcracks in the vessel that may have been introduced during the welding process, (2) hydrogen attack from water used for vessel testing, and (3) subsequent rust-sodium interactions. All of these contributed to attack the vessel integrity.

The fuel storage vessel was not replaced. Instead, the main vessel was removed and the guard vessel was filled with argon and used as a transfer facility between the reactor vessel and a fuel storage pool located outside the reactor building. The sodium leak event was not the reason for the final shutdown ten years later by the government; the real reasons were dictated by political considerations.

7.12.3 Sodium Fires

Liquid sodium reacts readily with air, and the oxidation reaction can be rapid and lead to a sodium fire. Burning sodium produces a dense white sodium oxide smoke. The heat, though, is much less than that of conventional hydrocarbon fires. The flame height is also an order of magnitude lower. Both allow a close approach for firefighting. The ignition temperature varies widely depending on the form of sodium, its moisture content in air, and other factors. Solid chunks cannot be ignited quickly even with a torch. A stirred liquid pool can be ignited at a temperature as low as 120°C.

For sodium fires, conventional firefighting agents are normally useless. In general, fluids cannot be used, because either they are flammable or they react violently with sodium. Only inorganic powders are used for extinguishing sodium fires. Dry silica sand, MET-L-X (fine, treated NaCl), and dry soda ash are all used. The dense cloud of aerosols does interfere with firefighting. Small sodium fires are readily extinguished but large sodium fires are difficult to extinguish.

Reactors are designed to effectively limit sodium leaks and to control sodium fires. The sodium in the primary system is blanketed with inert gas and maintained in double containment. The reactor vessel has a guard vessel, and the pipes have guard pipes around them. Leak detection monitors are installed in the inert gas in the gaps between the vessels and between the pipes. A variety of sodium leak detection systems are used. The principal technique relies on the detection of sodium aerosols in the annulus gap between the vessels or pipes. The aerosols are produced by the chemical reaction of liquid sodium with oxygen or water vapor existing as minute impurities in the inert gas atmosphere.

In the secondary sodium system, between the inert gas blanketed primary system and the steam generator, a variety of leak-detection techniques can be deployed. The methods commonly used for sodium fire detection include visual and remote television detection, electronic smoke detectors, flame photometers, atomic-absorption detectors, light-emission detectors, light-absorption detectors, and more.

[15] Provisions are also made to collect leaking sodium in steel drip trays to avoid reaction of sodium with structural concrete.

The main point to be made here is that in order to prevent the radioactive primary from sodium reacting with water or air, a non-radioactive secondary sodium system isolates the radioactive primary sodium from the steam system. As a result, potential sodium/water reactions or sodium fires can occur only in the non-radioactive secondary system. The primary system's integrity is not involved and the prevention and mitigation systems are not safety-grade systems. Their role is to protect the plant investment only.

There have been no sodium fires in primary sodium protected by inert atmosphere. We suppose there may have been unreported sodium fires which involved minute amounts of sodium in small pipes for sampling lines or purification systems, with inconsequential impacts, but we are not aware of any.

Over the thirty-year life of EBR-II, there have been several small sodium fires in the non-radioactive secondary system. They have been handled as routine matters. Typically, white smoke coming off through the pipe insulation is detected, the smoke is blanketed with MET-L-X, the sodium in the affected area is frozen and the piping cut out, and a new section is welded in. In general, sodium draining has not been required. In the most severe case, the freeze plug in the piping to permit maintenance of the bellows seal valve was insufficiently cooled and it melted, allowing sodium to flow from the open valve bonnet. The sodium ignited. The operator drained the secondary sodium to the storage tank and fire retardant was effective in putting it out.

By far the most severe sodium fire was in the (non-radioactive) secondary system of the Japanese demonstration fast reactor, Monju. During the 40% power tests in December 1995, an alarm sounded due to a high outlet temperature at the intermediate heat exchanger and a fire alarm (smoke detector) sounded at the same time, followed by a sodium leak alarm. The reactor power-down operations were initiated. When the white fumes in the piping room increased, the reactor was manually tripped after eighty minutes of initial alarms. The sodium drain started in ninety-five minutes after the reactor trip and was completed in eighty minutes. The cause of the leak was a thermocouple well that extended into the flowing sodium which bent due to flow-induced vibrations and left a one-cm-diameter opening. During the four-hour period before effective action was taken, a total of 640 kg of sodium had leaked and burned. The piping was elevated, and below the leak approximately one cubic meter of sodium oxide formed in a semicircular mound, about three meters in diameter and thirty cm high, on the six-mm-thick steel floor liner. The ventilation duct directly under the thermocouple well developed a hole extending over half its perimeter with lumps of deposit around the opening. Sodium aerosol was lightly diffused over the floor and walls of the piping room.

Monju now has redesigned thermocouple wells not subject to flow-induced vibration. But such thermocouple wells in the intermediate piping are not necessary in the first place. As a demonstration plant, Monju had more instrumentation than commercial reactors. Other mitigation features were incorporated before the restart, including compartmentalization of the secondary building, larger piping for drain tanks reducing the drain time to half, and a nitrogen injection system which as an inert gas can extinguish any size sodium fire in fifteen minutes.

It is important to note that there were no explosive energy releases or adverse effects on personnel or the surrounding environment. In properly designed sodium-cooled fast reactors, there are sufficient preventive and mitigation features to deal with sodium leakages, sodium/water reactions, and sodium fires so that the reactor safety should never be in jeopardy. The nonradioactive secondary sodium poses no more risk than is typical in common industrial safety.

7.13 Summary

The IFR stands out in its ability to cope with incipient major accidents. These most serious accidents have been increasingly unlikely as knowledge and experience with nuclear plants has accumulated and been put into practice over the years. There have been none with consequences approaching serious in over thirty years of EBR-II operation. Nevertheless their impact, real or perhaps pictured only, is frightening. Radioactivity is the key, naturally. There must be *no* radioactive release under any circumstances. For potentially serious accident situations in the IFR the central points are these:

1. For initiating events due to failures of equipment in large commercial size reactors of the IFR type, the passive reactivity feedback characteristics will shutdown the reactor without damage, if called upon. They will only be needed in the event of simultaneous failure of the control and safety systems. All this is extremely unlikely, given the backups to the backups to the backups. However, the two principal such events have both happened in nuclear power history: they are loss of coolant flow (Chernobyl) and loss of heat sink (TMI-2 and Fukushima Daiichi).

2. If, for any reason, the feedback is insufficient to control the event safely and fuel melting begins, it takes place at the top of the fuel column, and the melt is swept out from the core by flow, by contained fission gases, and by sodium vapor pressure in the fuel pin, Such loss of fuel reduces reactivity drastically, and shuts the system down once and for all. The point here is that for this potentially very serious event, the relatively low melting point of metal is a real advantage. It acts as a fuse—allowing limited disassembly before energy can build up and temperatures rise to the point where when

disassembly comes, it is with a real explosive force perhaps sufficient to threaten containment. The power the reactor reaches before such melting is considerable, not a few percent above operating, say, which would mean that the effects would be sensitive to precise control of power. Melting requires a power of several times operating power. Failure always occurred at about four times normal power in the TREAT (an Argonne reactor designed for such experiments) tests done explicitly to examine these effects in an IFR. Importantly, in contrast, the high melting point of oxide fuel, about 3,000°C, does not allow such a fuse effect. Concern arises that the explosive energy release in this case might breach containment.

3. Fuel can be made to melt by operator mistakes. The two events that come to mind are 1) at EBR-I, in 1955, the first liquid-metal-cooled reactor ever, tiny in size and in power, designed mainly to establish the physics of fast neutron reactors. The operators intentionally increased the reactivity very rapidly in an experiment to test fuel feedback. They planned to shut the excursion down before the danger level was reached manually, by verbal command. The command at the penultimate moment was misunderstood, scram came late, and a portion of the fuel melted. It moved away from the center of the core (important because that reduces reactivity), driven by sodium vapor, and the reactor shutdown with little energy release. (These kinds of experiments in pioneering days would not even have been contemplated in later years.) And 2) at the Fermi-1 reactor of Detroit Edison, in 1966, a pioneering sodium-cooled reactor of about 60 MWe, where questionable design led to a metallic piece coming loose in the core. It blocked a fuel channel, starving it of coolant. During startup then, the operators did not properly respond to signals indicating difficulty and continued to increase power until two subassemblies were partially melted and four were affected. The reactor control system scrammed the reactor and again there was no energy release. The reactor was brought back into service after a few-year clean-up period, and operated for a few more years. Both these accidents were in the early days. Neither reactor used IFR Zr-alloyed fuel, and EBR-I didn't use sodium as coolant; it used a eutectic alloy of sodium and potassium with a lower melting point. None of the later discoveries, of course, were known then.

4. For ultimate safety, then, if an accident starts, the best result would have the reactor shut itself down, without damage, and if with damage, to lead to no radioactive release. It always will shut down harmlessly if the control and safety systems are operational. If they are not, in the IFR, for the most feared initiating events, the passive response will shut it down without damage. If events somehow are such that the passive response is insufficient, the low temperature fuse effect of IFR fuel, combined with the fact that the hottest region is at the top of the fuel, causes fuel first to elongate and introduce some further negative reactivity; and then, if that is insufficient to terminate

the event, to melt in that region and be swept out of the core by several different effects, taking the reactor a long way sub-critical and decidedly shut down.

5. These effects are not theoretical or subject to informed challenge. They have been proven by full-scale experiments in the assemblage of fast reactor test facilities in Idaho by Argonne National Laboratory.

References

1. Nuclear Regulatory Commission, "Regulation of Advanced Nuclear Power Plants; Statement of Policy," 59 FR 35461, July 12, 1994.
2. H. A. Bethe and J. H. Tait, "An Estimate of the Order of Magnitude of the Explosion When the Core of a Fast Reactor Collapses," U.S.-U.K. Reactor Hazard Meeting, RHM(56)/113, 1956.
3. J. E. Cahalan and T. T. C. Wei, "Modeling Developments for the SAS4A and SASSYS Computer Codes," *Proc. International Fast Reactor Safety Meeting*, Snowbird, Utah, August 12-16, 1990.
4. S. H. Fistedis, ed., *The Experimental Breeder Reactor-II Inherent Safety Demonstration*, North-Holland, 1987. (Reprinted from *Nuclear Engineering and Design*, 101[1], 1987.)
5. D. C. Wade and Y. I. Chang, "The Integral Fast Reactor Concept: Physics of Operation and Safety," *Nucl. Sci. and Eng.*, 100, 507, 1988.
6. C. E. Till and Y. I. Chang, "The Integral Fast Reactor," *Advances in Nuclear Science and Technology,* 20, 127, 1988.
7. R. A. Harris, et al., "FFTF Passive Safety Flow Transient Test," *Proc. International Topical Meeting on Safety of Next Generation Power Reactors*, 424-430, Seattle, WA, 1988.
8. "Domestic Licensing of Production and Utilization Facilities," *Code of Federal Regulations*, 10 CFR Part 50.
9. Nuclear Regulatory Commission, "Pre-application Safety Evaluation Report for the Power Reactor Innovative Small Module (PRISM) Reactor," NUREG-1368, February 1994.
10. Nuclear Regulatory Commission, "Framework for Development of a Risk-Informed, Performance-Based Alternative to 10 CFR Part 50," NUREG-1860, Working Draft, July 2006.
11. Nuclear Regulatory Commission, "Safety Goals for the Operations of Nuclear Power Plants; Policy Statement," 51 Federal Registrar 30028, August 21, 1986.
12. W. H. Zinn, "A Letter on EBR-I Fuel Meltdown," *Nucleonics*, 14, No. 6, 35, 1956.
13. J. H. Kittel, M. Novick, R. F. Buchnan, and W. B. Doe, "Disassembly and Metallurgical Evaluation of the Melted-down EBR-I Core," *Proc. International Conference on the Peaceful Uses of Atomic Energy*, Geneva, 7, 472, 1958.
14. R. O. Brittan, "Analysis of the EBR-I Core Meltdown," *Proc. International Conference on the Peaceful Uses of Atomic Energy*, Geneva, 12, 267, 1958.
15. D. W. Cissel, et al., "Guidelines for Sodium Fire Prevention, Detection, and Control," ANL-7691, 1970.

CHAPTER 8

THE PYROPROCESS

In the next three chapters we wish to provide an understanding of the electrochemical portion of the IFR spent fuel process, electrorefining—the heart of IFR pyroprocessing—what it can do, and importantly, what it can't do. In this chapter we will describe the processes, equipment, results, and status of IFR pyroprocessing. The following chapter will deal with the chemistry itself, and the final one will cover the application to LWR spent fuel. It shares some but not all the characteristics of electro refining. After that, the waste processes will be discussed in some detail.

It is fair to ask why we need any process at all. The answer is straightforward. In any reactor, for any fuel, there is a strict limit to the fuel "burnup" beyond which the fuel cladding will "fail." It may just be a pinhole which leaks radioactive fuel material into the coolant, or it may be worse; it may breach wide open, leaking quantities of radioactivity into the coolant. This contaminates the entire primary system, which is serious in that it makes personnel access difficult if the reactor design is such that routine access is required, or it may be of limited importance if such access is not normally required. Our cladding is steel. It is damaged gradually by exposure to neutrons, particularly those at the highest energies, and eventually it becomes brittle and subject to puncture or breach, either by contact with the fuel it contains, or by internal pressure from the buildup of fission gases. The fuel must come out of the reactor in a safe amount of time, well before these things can happen. In the IFR, as it is for most reactors, that time is three or four years. When it comes out it must be "reprocessed" so it can be "re-fabricated" into fresh fuel and returned to the reactor for another cycle.

The fuel must also be recycled for both economic and resource reasons. The higher fissile content of the IFR spent fuel, which has a fissile percentage about twenty times that of LWR, makes its recycle and reuse mandatory for good economics. And it is recycle, recharging the same fuel over and over again, that allows the huge extension of fuel resources of the IFR. Further, a substantial fraction of the new plutonium is bred in the depleted uranium blankets surrounding the core and processing is needed to recover it. And finally, processing allows removal of the very long-lived isotopes from the nuclear waste.

So the IFR spent fuel process has an impressive number of goals for so simple a process. There are "musts," "good ifs," "must nots," and "best if nots." The process must separate the fission products from the uranium and the man-made elements like plutonium and the other actinide elements. Each one—fission products, uranium, and actinide elements—must be recovered separately. The actinides and the uranium in the blanket must be recovered separately so the actinides can be used as enrichment in new fuel. It is best if all the actinides are recovered to go in new fuel, as all are good fuel for the IFR, but more importantly, their removal means that waste has a much shorter radiological lifetime. And the product must not be pure plutonium, suitable for use in weapons. Also, it would be best if the process has few steps only, so it is inexpensive to assemble, and thus its economics show promise.

8.1 Earliest Experience with Pyroprocessing: EBR-II in the 1960s

The term "pyroprocessing" now covers a variety of different processes that operate at high temperature but whose principles of operation depend on more than just heat itself. Electrorefining, the use of electricity to effect desired chemical reactions, is the best example. Electrorefining is the basis of the IFR pyroprocess. But the earliest process developed at Argonne, for the EBR-II in the 1960s, was a true pyrometallurgical process: heat alone was used. [1] The spent fuel was simply melted. Melting removed some of the fission products, but of more immediate importance the spent fuel could then be recast into new cladding and returned to the reactor. It was a crude process, but for its limited purposes it worked satisfactorily.

EBR-II was designed from its conception to have a complete fuel cycle. Although it would demonstrate recycling for the fast reactor, the pressing reason to recycle its fuel was the limited life of metal fuel in-reactor, as we have discussed in the previous chapters. The fuel had to be removed, refreshed, re-fabricated, and returned to the reactor if purchases of new fuel were not to become inordinately large. For the same reason, there was a premium on a short recycle time. So the simplest possible process was used: Melting the fuel released the fission product gases and allowed it to be re-melted in a casting furnace and recast into new fuel slugs.

"Melt-refining" it was called; it yielded a partially refined product, sufficient for reuse of the same fuel over and over in EBR-II operation. But it was always recognized that substantial improvements would be needed for recycle in commercial fast reactors. There was a few percent loss of product in each recycle, far too much to be sustained commercially. The fission products were only partly removed, and without the inadvertent withdrawal in the losses they would have eventually built up to unsustainable levels. But all in all it worked satisfactorily for

EBR-II purposes, and it established once and for all the feasibility of quick-turnaround recycle in a fast reactor.

The recycle operations were carried out in a Fuel Cycle Facility (FCF) adjacent to and connected to the reactor building. Shown in Figure 8-1, the facility consisted of a rectangular shielded cell for operations in a normal air atmosphere (where the fuel itself was not exposed) and an annular shielded cell where operations were carried out under argon gas (argon gas is “noble”; it does not react with exposed fuel).

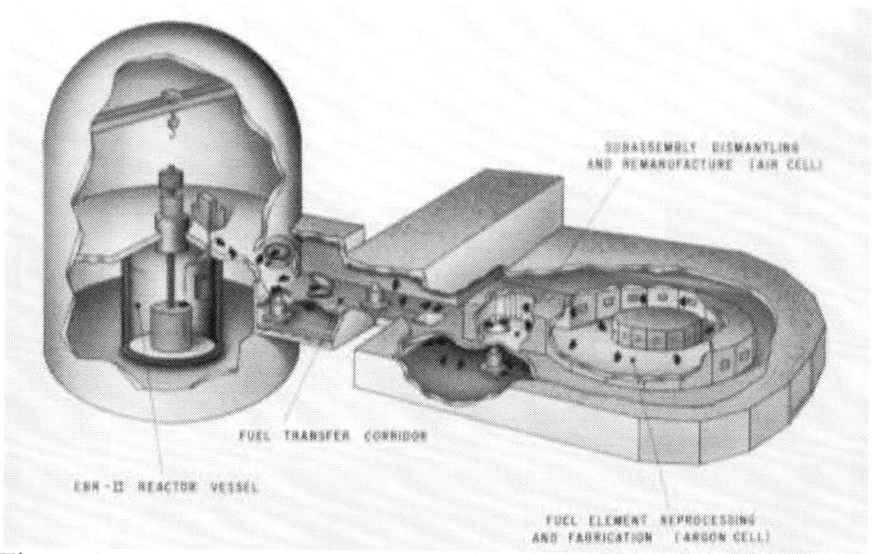

Figure 8-1. EBR-II and Fuel Cycle Facility showing reactor vessel, fuel transfer tunnel, air cell, and argon cell

After a cooling period of only fourteen days in a storage basket inside the reactor vessel, spent fuel assemblies were transferred into the air cell, where they were disassembled and individual fuel pins transferred to the argon cell. There the spent fuel pins were mechanically de-clad to remove stainless steel cladding, and the fuel slugs were chopped into small segments and loaded into a zirconia crucible for melting. When fuel was molten, fission product gases and fission products with high vapor pressures boiled off and all were collected in a fume trap. Other fission products reacted with the zirconia forming an oxide residue which remained in the crucible as a “skull” after the pouring operation. The noble metal fission products, like molybdenum, remained in the melt. The skull was 6 to 8 percent of the melt, the loss on each recycle. This drawing off of fuel in each recycle led to an equilibrium composition of fuel product, stable after a few recycles, with a substantial component of the noble metal fission products. These tended actually to improve uranium fuel performance, and later on when the fuel lifetime improved and fuel was no longer recycled, an approximation of the equilibrium composition from recycling became an intentional additive to EBR-II fuel in routine use. It was not used later for the IFR.

For fabrication of recycled fuel the melt-refined ingot was re-melted in a thoria-coated graphite crucible in a vacuum furnace. Above the crucible was a cluster of approximately one hundred Vycor (a form of glass) tubes used as molds. When the fuel charge was molten, the crucible was raised to immerse the lower end of the mold cluster into the melt. The furnace was then rapidly pressurized, driving the melt upward into the evacuated molds. After a few seconds, the melt froze, the Vycor molds were broken away from the castings, and the hundred or so pins were ready for cladding. The pins were loaded into stainless steel cladding along with a measured amount of sodium. The sodium was melted and the fuel pins settled by gravity. End plugs inserted into the claddings were welded to the cladding. The finished pins were reassembled in clusters called subassemblies and returned to the reactor.

An extraordinarily simple fuel cycle, the processing of irradiated fuel began in September 1964 and continued through January 1969. Approximately thirty thousand irradiated fuel pins, fifty-three hundred kg in total including scrap consolidation, were processed. The turnaround times from the receipt in FCF to return to the reactor averaged forty five days, with some fuel returned within thirty days. The average throughput rate was 100 kg a month, with the peak throughput reaching 245kg a month.

The melt-refining process was not suitable for scaleup, certainly not without substantial further development. It was acceptable to have the noble fission products remain in the product while metal fuel had very low burnup along with the substantial withdrawal due to the material left behind as skulls. The equilibrium concentration was low enough not to harm reactivity significantly. But with burnups up to 20 percent expected in future fast reactors, the noble metals must be removed by processing. Also, melt-refining doesn't selectively increase plutonium concentrations in any way; uranium and plutonium go right through the process in the same ratio as they come in. A satisfactory process must increase the plutonium to uranium ratio adequately to refresh core fuel. Finally, the amounts of actinides left in the "skulls" were simply unacceptable.

Without significant improvement, the melt-refining process for commercial fast reactor deployment was a dead end. For the IFR, some new process was going to be necessary. Electrorefining became the choice for the IFR, and we now turn to a description of the process and the equipment designed and built to carry out the process in the FCF.

8.2 Summary of Pyroprocessing

Electrorefining is commonly utilized in the minerals industry to purify metals, such as aluminum and zinc. In spent fuel processing, electrorefining allows the valuable fuel constituents, uranium and the actinides, to be recovered and the fission products to be removed. [2-5] A schematic of the IFR electrorefining process is shown in Figure 8-2. The electro-refiner is a meter in diameter and height; its "anode", the positive electrode, is a basket of chopped spent fuel; and two different kinds of "cathode," the negative electrode, are used. One cathode, a solid steel rod, collects uranium, and the other, of liquid cadmium, collects all the other actinides at the same time. The electrodes are immersed in a molten salt electrolyte. It floats on a liquid cadmium bottom layer, which serves a variety of purposes relating to collection of uranium and actinides that escape the cathodes during operation.

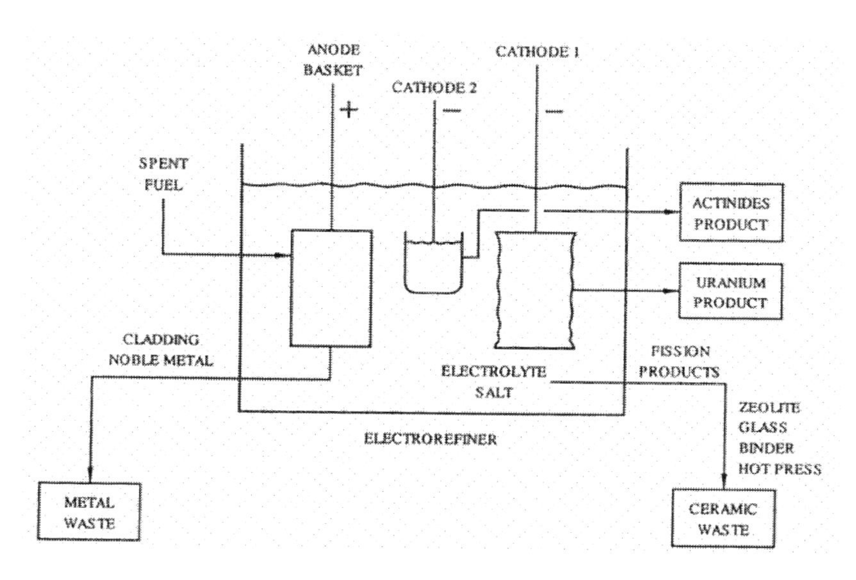

Figure 8-2. Schematic illustration of electrorefiner

The fuel pins are chopped into short lengths, loaded into perforated steel baskets, and introduced into the electrolyte. A low voltage (on the order of one volt) dissolves the fuel material in the electrolyte, transports the uranium and actinide ions to the appropriate cathode, reduces them (in electrical charge), and deposits them as electrically neutral metals at the relevant cathode. The bulk of uranium is collected first. When the desired amount of uranium has been removed by deposit on the solid cathode, the other actinides, along with some uranium, are collected in the molten cadmium metal cathode crucible. The electrorefiner, in place with the other components of the process, is shown in Figure 8-3.

The deposits then go to a cathode processor, also shown in Figure 8-3, which is basically a high-temperature vacuum furnace. It melts the uranium cathode deposit and evaporates the adhering salt impurities, leaving a pure metal product. For the liquid cadmium cathode product, in addition to evaporating the salt impurities, the liquid cadmium is also distilled off, leaving a pure metallic ingot of plutonium, uranium, and the other actinide elements. The lower portion of Figure 8-3 shows the components of the waste treatment processes that place the waste in appropriate forms for disposal, the subject of a later chapter.

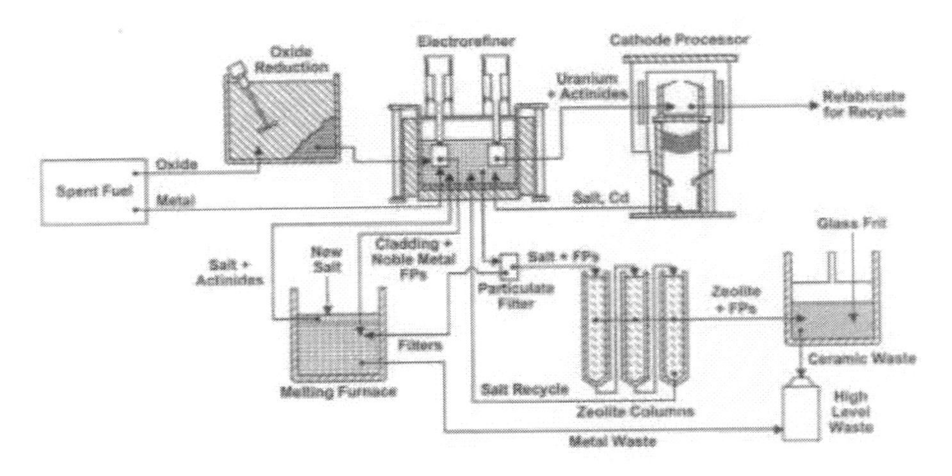

Figure 8-3. Schematic flowsheet of electrorefining based spent fuel treatment

We will want to go a little more deeply into the fundamentals of the electrorefining process in the next chapter. Without an understanding of the fundamentals it's difficult to see why the process works as it does. But for those who are interested only in a general description of the process, the next chapter can be skipped; the process equipment and operations will be described below.

8.3 The Fuel Conditioning Facility

We go into considerable detail in this descriptive section to give an idea of the two important things. First, the care that is taken in assuring safe, reliable operation in a remote environment served only by manipulators and other remotely operated equipment. And second, how compact and relatively simple the equipment is, and how few pieces of equipment there are, in recycling fuel by the IFR process.

8.3.1 The Facility Itself

It was recognized from the outset of IFR development that a full-scale demonstration of the entire fuel cycle closure using EBR-II as an IFR prototype would be required to convincingly establish the viability of the IFR concept. In the step-by-step progress of the IFR program, when the appropriate time arrived, the facility built for the original EBR-II pyroprocess, FCF, could probably be rehabilitated and put back into operation. But it would need considerable work to bring it up to the necessary modern standards.

In the years after 1969, the facility was used for examinations of irradiated fuels and materials and it had been renamed the Hot Fuel Examination Facility-South. (A more modern such facility, HFEF-North had also been built close by for such examinations.) Once in operation for the IFR program, the facility reverted to a form of its original name, as the Fuel Conditioning Facility (FCF). As noted earlier, the FCF consists of two shielded hot cells. The air atmosphere cell is a rectangle 4.6 m by 14.3 m. The argon atmosphere cell is an annulus about 5 m wide for process operations with an outer diameter of 19 m. All operations are conducted remotely. Master/slave manipulators are used at work station windows on the periphery of the annulus. Special viewing and limited-operation work stations are available in the center of the annulus as well, which is accessible from the basement. Both hot cells are heavily shielded (with approximately five feet of high-density concrete), with leaded-glass windows at the work stations. Heavier lifting and transport is done with overhead handling systems.

The FCF had been constructed to the U.S. Uniform Building Code criteria, and a detailed dynamic seismic analysis had shown that no significant structural damage would occur from a design basis earthquake. But the modern seismic analysis indicated that the existing hot repair area for contaminated equipment (which was added to the roof of the facility after original construction) would not withstand design basis wind or earthquakes without loss of confinement, so a new hot repair area was constructed inside the existing facility.

In remote operations like this, the ability to repair and service equipment is very important. New in-cell process equipment was designed to the extent possible to allow in-situ repair. When not possible, modules or components were brought to the new repair area for repair or disposal. It included provisions for remote transfer of equipment from the cells, remote decontamination, direct access to equipment by personnel in protective clothing, a special pit area for crane trolley and electromechanical manipulator carriage repair, and the capability for bag-transfer of equipment and low-level contaminated waste into or out of the area.

Confinement improvements were made both in physical barriers and in the air flow and/or filtration system. Utility penetrations and master/slave manipulators in the air-atmosphere cell were sealed, where in the past confinement had relied on maintaining pressure in the cell negative with respect to the operating floor. A new safety-grade exhaust system was installed, assuring no unfiltered release from the argon cell even in the unlikely event of an accident or earthquake causing a breach in the cell boundary. And a new seismic- and tornado-hardened building was constructed to house the two new safety-grade redundant diesel generators (375 kW each) for the emergency electrical power system.

The preparatory work on the FCF refurbishment project started in 1988 and refurbishment was completed in December 1994, ready for formal readiness review processes.

8.3.2 Process Equipment Development

EBR-II spent fuel assemblies are transferred into the air cell for temporary storage and dismantling. Individual fuel pins are removed from the hexagonal duct using remote dismantling equipment that was refurbished and improved. The fuel pins are then loaded into a cylindrical magazine and transferred into the argon cell through a small transfer lock.

Inside the argon cell, both refining and re-fabrication of fuel require only five discrete pieces of equipment. Their locations are depicted in Figure 8-4. The equipment systems were designed, fabricated, and qualified by a formal process designed to meet the highest quality assurance standards. After the normal disciplined engineering design and fabrication process, each piece of equipment was qualified in three phases: assembly, out-of-cell remote handling, and finally in-cell remote handling. During assembly qualification, the equipment was assembled and tested for functionality by the equipment design personnel. But the out-of-cell qualification was performed by the same operations technician who would use the equipment, and he verified that all assembly, operation, and maintenance could be done remotely. A mockup shop enabled testing of the equipment in the same configuration as in the cell. Finally, the in-cell qualification equipment was assembled in its final location and tested for operability. Operations were then started with depleted uranium, proven to be satisfactory, and finally the irradiated fuel was introduced.

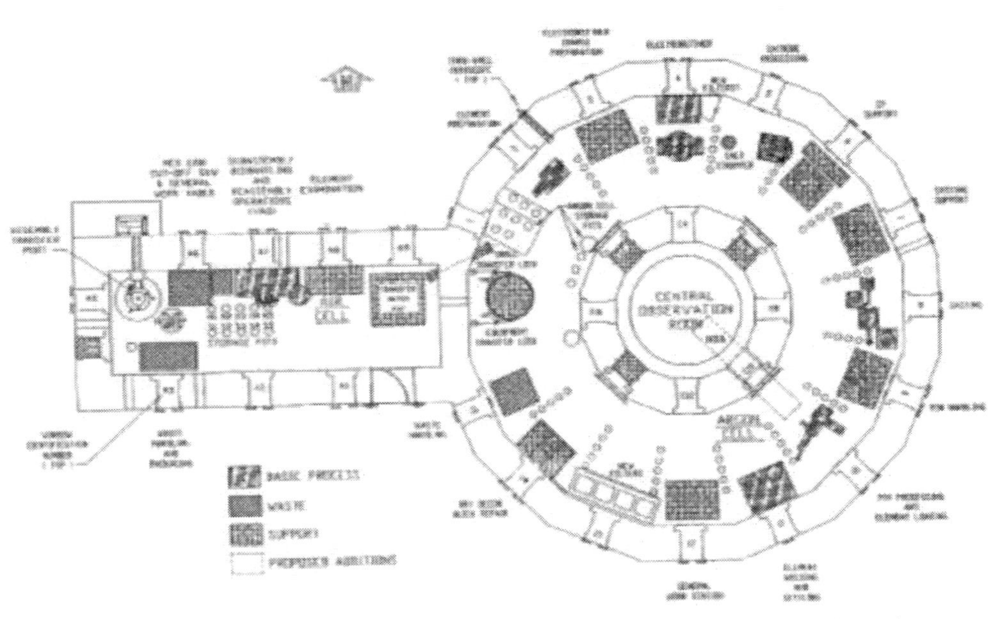

Figure 8-4. Air and argon cell proposed process equipment locations

The first operation is done in the pin chopper. Intact fuel pins are prepared for electrorefining by chopping them into short segments, exposing the fuel, and removing scrap such as the spacer wire, end caps, and plenum. The highly radioactive fuel is then introduced into the electrorefiner in perforated steel baskets.

Shown in Figure 8-5, the electrorefiner consists of the process vessel with associated heater/insulator assembly, cover, and support stand; electrode assemblies (anodes and cathodes) and their associated handling and placement mechanisms; ancillary systems for cover gas circulation, stirring, materials addition, sampling, and measurements; and an instrumentation and control system. A staging or support station services the electrorefiner. It attaches and removes anode and cathode parts, removes product from the cathodes, and empties the cladding hulls from the anode baskets. The design is based on the engineering-scale electrorefiner in service in the Chemical Technology Division at the Illinois site since 1987 and still ongoing.

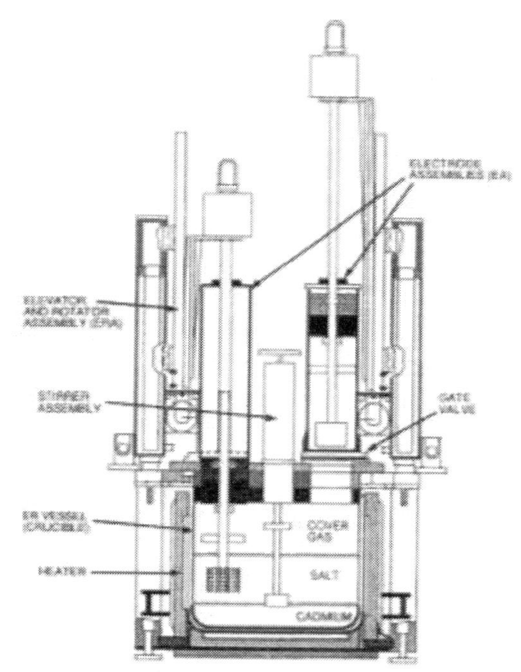

Figure 8-5. Schematic of electrorefiner

The electrorefiner operates at a temperature of 500°C. It has a layer of molten cadmium, ten centimeters thick under thirty-six centimeters of molten salt. It is heated by the thermal radiation from a furnace assembly of resistance heaters in low-density ceramic insulator blocks. Circular ports in the steel cover and thermal radiation baffles provide access for four electrodes, 28 cm in diameter, and a stirrer 20 cm in diameter. Other smaller ports allow material additions, sampling measurement, and cover gas control. The electrode assemblies have a long shaft with an adapter for an anode or a cathode bottom fixture, a rotation motor, rotating electrical contacts, a port cover, a containment housing, and various instrumentation components. The containment isolates the cover gas from the general cell atmosphere. The electrode assemblies are moved in and out of the process vessel by elevator/rotator mechanisms mounted on the support structure.

Baskets loaded with chopped spent fuel are attached to anode assemblies. The anodes and the cathode assemblies are then inserted into the electro-refiner. At the end of an electro-transport run, the electrodes are removed and transferred back to the support station. There the cladding hulls are removed from the anode baskets and the dendritic (like tendrils) solid cathode deposit or the actinide/cadmium ingot is removed from the cathode. (See Figure 8-6.) The cathode products are sent on for further processing, the cladding hulls are retained for later waste treatment, and the electrode assemblies are prepared for the next batch. Uranium is deposited on the solid cathode in run after run. When, after multiple runs, the actinides build up sufficiently in the electrolyte the uranium concentration is drawn down to allow processing of actinides. When the ratio of actinides to uranium reaches three or so, the liquid cadmium cathode is introduced and actinides are deposited.

Figure 8-6. Dendritic deposit of uranium on solid cathode

The next step is the cathode processor, schematically shown in Figure 8-7. Cathode products from the electro-refiner contain the heavy metals along with some salt and/or cadmium. The cathode product is loaded into a process crucible and heated under vacuum where, in sequence, cadmium itself, cadmium from the heavy metal intermetallic compounds, and finally the adhering salts, are evaporated. In heating, the distillate transports to the condenser region, condenses, and runs down into the receiver crucible. The heavy metal in the process crucible is then melted and consolidated into ingots in shaped recesses in the bottom of the process crucible.

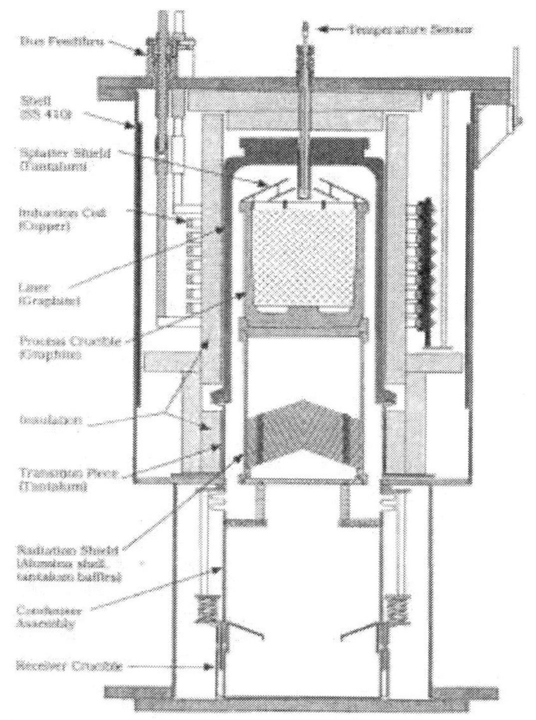

Figure 8-7. Schematic of Cathode Processor

The main features of the cathode processor are the vessel, the induction-heated furnace region inside the vessel at the top, and the condenser region inside the vessel at the bottom. The cathode processor is bottom-loaded and is therefore supported in an elevated position within the cell. An elevating mechanism raises and lowers the crucible assembly into position and the bottom flange seals the furnace prior to the heating cycle. When in the lowered position, a trolley moves these components to an adjacent window for crucible handling operations.

The third step is the injection casting furnace. There, the fuel ingots from the cathode processor are fabricated into new fuel slugs. Injection casting has been the primary metal fuel fabrication technique from the initial operation of EBR-II. The batch size is 10–25 kg for the FCF casting furnace, a limit set by criticality safety constraints.

Shown in Figure 8-8, the casting furnace vessel provides confinement for the casting operation. The furnace crucible is machined graphite. A center thermocouple well in the bottom of the graphite crucible allows melt temperature measurement. The furnace atmosphere is controlled by an external high-purity argon source to minimize pickup of argon cell atmosphere contaminants. Vacuum is drawn immediately before casting, and an accumulator with a fast-acting valve supplies the rapid and repeatable pressurization of the furnace necessary for casting repeatability. The fuel ingots are combined with the desired makeup material (zirconium and uranium or plutonium alloy) and recycle material (fuel pin casting heels and recycle fuel pin pieces) in a graphite crucible. The crucible is then placed in the induction coil of the furnace. All normal handling and loading operations are conducted through the "top hat." More extensive disassembly for cleanout or repair is also possible. Redundant, normally closed isolation valves on all argon supply and exhaust lines will isolate the furnace contents from the cell atmosphere in the event of any loss of cell integrity that might result from a design basis accident.

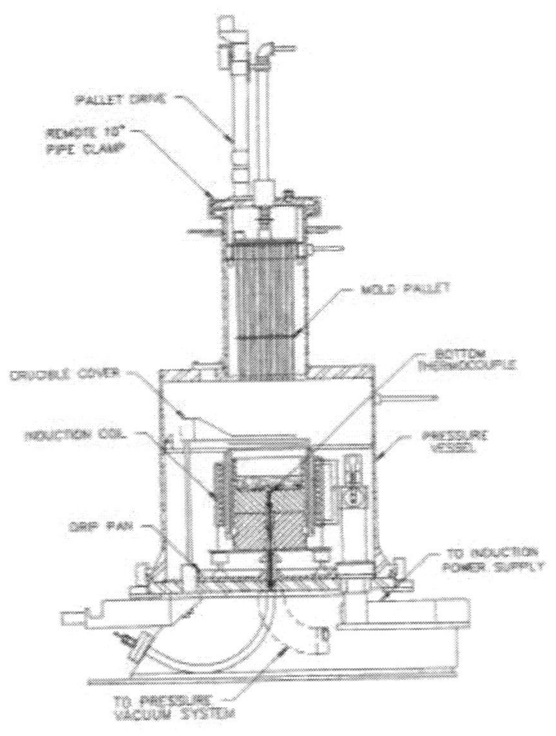

Figure 8-8. Schematic of injection casting furnace

The fourth step is pin processing. Fuel slugs are loaded into a magazine that feeds the fuel pin processing machine. The machine frees the fuel slug by breaking off the glass molds, shears the fuel slugs to the correct length, and conducts a number of physical inspections. Slugs which meet acceptance criteria are directly and automatically inserted into new cladding which has been loaded with sodium to provide the thermal bond. Rejected slugs are diverted to a special carrier for recycle. The top end cap is inserted and welded into the cladding, and the fuel slug is settled into the bond sodium. Automated equipment inserts the end caps and welds them. Following welding and settling, the pin ends and the magazine are decontaminated and the pins are transferred into the air cell.

In the final step, the pins are arranged in subassemblies for return to the reactor. In the air cell, the magazine is unloaded and the pins are inspected in an automated inspection system. The pins are then installed in the subassembly hardware on the assembler/dismantler. An existing machine that has been extensively refurbished assembles the newly fabricated fuel pins into assembly hardware for transfer to EBR-II.

A process control and accountability system gives the desired process information. It consists of logic controllers, operator control stations, central data logging, and mass tracking. An overview of the information which will be shared among the different areas and the general system configuration is shown in Figure 8-9. The logic controllers provide the interface between the process equipment and the control system. Operating technicians control and monitor all important process parameters through the operator control station (OCS).

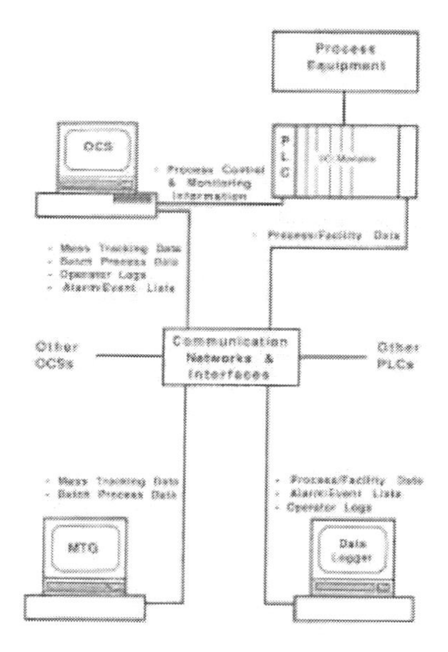

Figure 8-9. Process control and accountability system layout

The mass-tracking computer system tracks the movement, location, and elemental and isotopic composition of all nuclear material inside the air and argon cells in near real time. In addition, process materials such as chemical reagents and fission products can be tracked. Tracking is by discrete item, such as pin, assembly, storage container, crucible, or piece of processing equipment. The data comes from the programmable logic controllers, operator control stations, and analytical laboratory measurements.

The mass-tracking system assists operation and operations support personnel with materials control and accountability, compliance with criticality safety specifications, compliance with facility operating limits, assistance in process control, and simulation of fuel processing operations. At some process steps, hold points in the operating sequence require the operator to request approval and receive guidance from the mass tracking system. The operator cannot proceed until the mass tracking system determines that the actions to be completed meet the operating and criticality safety limits.

Accountability of fissile material is done by making the entire fuel cycle facility a single material balance area. The mass of incoming nuclear material is established from the initial fuel pin masses, with accuracies increased by detailed reactor burnup calculations. The calculated corrections are verified by chemical analysis of at least one sample from each incoming assembly. The accuracy of the calculation methods is verified in separate experiments with detailed sampling and analysis of representative assemblies. The initial fuel mass is established by weighing each newly fabricated fuel pin and by chemical analysis of at least one slug from each casting batch.

After the cathodes have been processed in the cathode processor, the change in the electrorefiner inventory is checked for the balance between the input to the chopper as given above, the change in electrorefiner inventory, and the output from the cathode processor. This balance provides the timely assessment of the chemical separation steps for both accountability and process control purposes. As part of the balance data, samples are taken for chemical analysis of the electrorefiner salt and the cadmium pool where there are a significant amount of in-process materials. Finally, samples are also taken from product ingots. This then provides the data to follow the fissile materials through the process from start to finish.

8.4 EBR-II Spent Fuel Treatment

When the DOE decision came to terminate the IFR Program, EBR-II was shut down on September 30, 1994 after thirty years of very successful operation. The reactor could have operated for many additional years. But the irradiated fuel, both in-reactor and spent fuel from years past, had now to be disposed of. EBR-II driver fuel contains highly enriched uranium even at discharge (53 to 75% U-235), which is enough to raise concern for the *in situ* criticality in any repository. EBR-II spent fuel also contains reactive metal, the bond sodium, which because of its chemical activity is prohibited from repository disposal. So the EBR-II spent fuel had to be treated in some way to remove the high fissile content and reactive materials before permanent disposal in a repository would be possible.

At the time of the IFR program termination, FCF refurbishment was essentially complete and the process equipment installation and in-cell qualification were in their final phase. It was decided that electrorefining in the FCF was the most cost-effective and technically sound approach to treat the EBR-II spent fuel for disposal. That is what it was designed for. The operational readiness reviews by Argonne and DOE were completed during 1995. The environmental permits also had to be completed, including the air quality permit modification by the State of Idaho as well as an environmental assessment by DOE. The latter was completed with a finding of no significant impact in May 1996. [6]

The EBR-II spent fuel treatment was to be pursued only after successful demonstration of its technical feasibility with up to one hundred driver assemblies and twenty-five blanket assemblies. This three-year feasibility demonstration was to be monitored and assessed by an independent panel established by the Board on Chemical Sciences and Technology, National Research Council of the National Academies. The Special Committee of the National Research Council issued its final report in 2000 [7]. It concluded that "The committee finds that ANL has met all of the criteria developed for judging the success of its electrometallurgical demonstration project." and that "The committee finds no technical barriers to the use of electrometallurgical technology to process the remainder of the EBR-II fuel."

The Department of Energy then completed its Final Environmental Impact Statement for the treatment and management of sodium-bonded spent nuclear fuel in July 2000, and the EBR-II spent fuel treatment began. [8] As shown in Figure 8-10, a total of 830 kg of driver fuel and 3,620 kg of blanket had been processed in the FCF as of the end of FY2010. The FCF equipment systems are capable of processing at far higher throughputs, but budget constraints and the low priority placed on the treatment operation have limited the amounts processed to date.

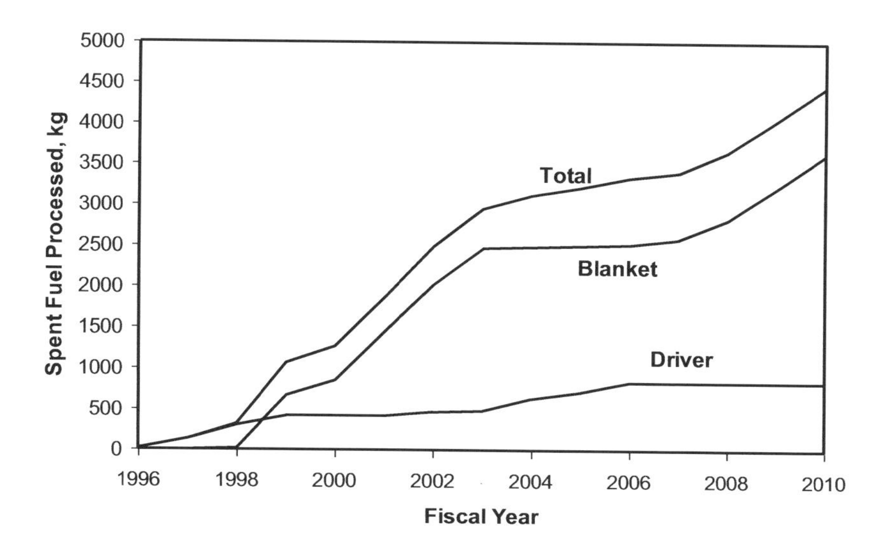

Figure 8-10. Cumulative amount of EBR-II spent fuel processed in the FCF

8.5 Waste Streams in Pyroprocessing

Electrorefining differs in kind from conventional PUREX reprocessing. It's much smaller and can be deployed much more cheaply than the conventional PUREX technologies deployed in Europe and Japan. But it also offers substantial improvements in waste management and proliferation resistance, as well as economic potential. Electrorefining was developed for metallic fuel from a fast reactor —the IFR. However, adding the oxide-to-metal reduction step discussed in Chapter 10 allows the process to treat the current inventories of oxide spent fuel from commercial light-water reactors as well.

As we have seen, the actinides are separated from both the uranium in the spent fuel and the fission product waste. The actinides have long half-lives and hence long-lived radiological toxicity. The fission products have much shorter half-lives; for a few decades they are intensely radioactive, but they gradually decay away to the point where in a few hundred years their toxicity is at the level of the original ore. The actinides should be removed from the waste. For one thing, they should be recycled as fuel in a fast reactor where they are beneficially destroyed by fission. But their removal from the waste also removes almost all the long-lived toxicity of nuclear waste. The fission products must be isolated, preferably in a repository, for the few hundred years necessary to have their activity die away. They should remain in a repository, as there are one or two very long-lived fission products that continue to have very weak activities, but for a very long time.

This is the best that can be done with spent fuel waste. It truly provides a solution to what has been an intractable problem. Dangerous waste toxicity dies away; it doesn't last hundreds of thousands of years. Nobody can predict, or even picture, storage for unimaginably long times like those. But spans of a few hundred years are well within human experience, and most certainly they do not challenge modern engineering practice. And we all have seen churches that have lasted a lot longer than that, so we don't have to be experts; our common sense tells us that storage in a repository, properly designed, for a few hundred years should not be difficult. Some form of a repository is necessary, but regulatory standards can be far more easily met without the actinides. And the amount of waste that can be stored can be increased. The amount is limited by the very long-term heat source, and as that comes from the actinides, without them the source is eliminated. Thus electrorefining does not eliminate the need for a repository, but it will allow the technical performance requirements to be met more easily and reduce the burden of long-term stewardship.

A fuel cycle based on electrorefining provides proliferation resistance, not because of what it can do, but because of what it cannot do. It is not capable of separating pure plutonium from other actinides and all fission products. As noted, the product is a mixture of uranium, plutonium, other actinides, and some fission products, highly radioactive and self-protecting against diversion. The electrorefining of spent fuel and fabrication of new fuel must be carried out remotely behind heavy shielding, with mechanical manipulators and possibly robots, but always with sophisticated and specialized equipment. You can't just handle it casually. In addition, the compactness of electrorefining makes it possible to locate the facility on the same site as the reactor. There is no need to transport fuel to the reactor; the uranium and plutonium of the spent fuel is recycled back to the reactor; and the waste has been reduced to a fraction of the volume of the spent fuel it came from.

The pyroprocessing waste products can be summarized in this way. The anode basket that contains the fuel cladding hulls and chemically noble fission products not dissolved during electrorefining is simply melted to form an exceptionally corrosion resistant alloy. The metal waste form consists of primarily stainless steel, with 15% zirconium as the base matrix. Rare earth and active fission products and adhering salt are stabilized in the ceramic waste, a glass-bonded sodalite waste form. The actinides are in chloride form in the electrorefiner, so to minimize their loss to waste, they are removed from the salt in a "drawdown" operation prior to salt purification. Molten alloys containing a reducing agent such as lithium are contacted with salt to reduce the actinides from the salt phase to the metal phase. After actinide removal, the salt is contacted with zeolite to remove fission products. The salt-loaded zeolite is eventually combined with additional zeolite and 25% glass and processed thermally into a monolithic waste form. In this process, the zeolite is converted to sodalite, a stable naturally occurring mineral. The process is reversed to charge the next salt batch and recycled back into the electrorefiner.

In electrorefining, all reagents such as electrolyte salts are recycled and there are no large volume low-level process waste streams. Secondary waste from electrorefining includes operational wastes such as failed equipment, rags, packaging materials, dross, mold-scrap from fuel fabrication, and other miscellaneous items. These waste streams are categorized using existing orders and regulations and disposed of as standard practice. Items unique to electrorefining, like dross and mold-scrap from fuel casting undergo further treatment to recover actinides for recycle.

8.5.1 Metal Waste Form

After electrorefining operations, the anode basket contains the stainless steel cladding hulls, fuel matrix alloy zirconium, noble metal fission products (molybdenum, technetium, ruthenium, rhodium, palladium, niobium, tellurium, and so on), and adhering electrolyte salt. The anode basket contents are heated in the metal waste furnace to distill off the adhering salt, and then heated at a higher temperature to consolidate the metal waste form. The base alloy for the metal waste form is stainless steel with the nominal 15% zirconium concentration, but the allowable range is 5–20%. The Fe-Zr phase diagram, Figure 8-11, shows that the alloy with 13% Zr is the low-temperature melting eutectic. The presence of zirconium also allows formation of a durable zirconium-iron intermetallic. Any minute quantities of actinides carried into the metal waste stream tend to be incorporated in this phase, and have excellent retention. The stainless steel to zirconium ratio for typical metal fuel pins would be about 85:15, but when the plenum sections are added to the metal waste form, some additional zirconium is added before the metal waste consolidation.

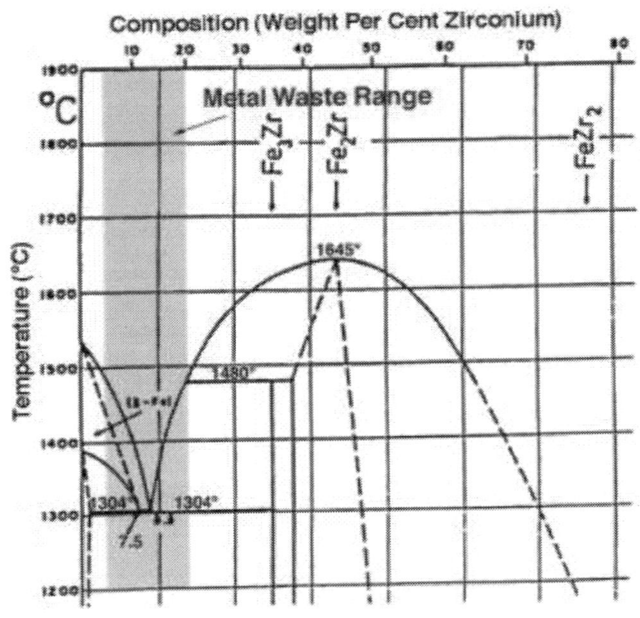

Figure 8-11. Iron-zirconium phase diagram [9]

As will be discussed later, for the LWR spent fuel processing, where the cladding hulls are based on zirconium alloys, the metal waste matrix will be zirconium with about 15% iron, which is another low-temperature eutectic on the other side of the Fe-Zr phase diagram.

The initial metal waste samples were produced by distilling adhering salt in the cathode processor and consolidating the final product in the casting furnace with appropriate zirconium addition. Later a dedicated large scale metal waste form furnace, which can produce up to 90 kg metal ingots, was installed in the Hot Fuel Examination Facility (HFEF). Its schematic is presented in Figure 8-12.

As shown in Figure 8-12, the innermost crucible assembly is 47 cm in diameter and 71 cm in height with a total usable volume of approximately 75 liters. The crucible assembly is composed of two parts—the crucible ring and crucible bottom connected with tongue-and-groove joints. The bottom of the crucible ring also has its own bayonet-style partial internal flange which allows the crucible ring to be rotated and locked onto the crucible bottom. This allows the crucible assembly to be transferred into and out of the furnace as a single piece, but separated for ingot dumping. The crucible ring and crucible bottom are constructed of graphite lined with aluminum oxide refractory, which is non-wetting to molten metal and resistant to damage from reactive molten chlorides.

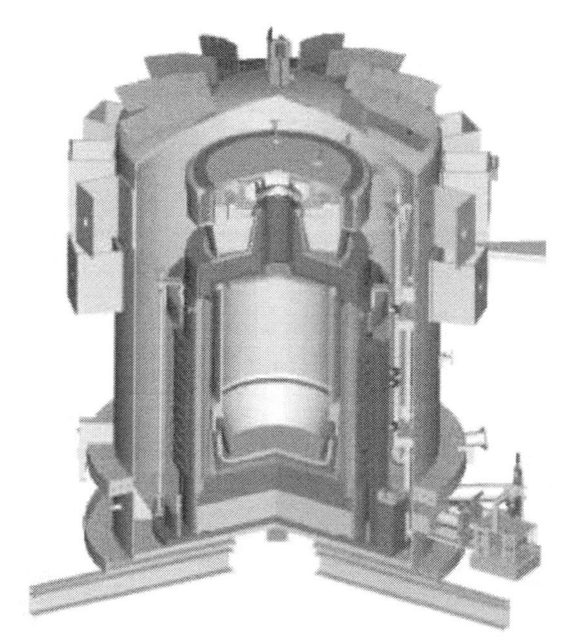

Figure 8-12. Metal Waste Form Furnace

The cladding hulls with the appropriate amount of zirconium added are loaded into a thin-walled stainless steel container and placed inside the crucible assembly. The crucible is then electromagnetically heated by induction coil to 1,350°C and held at temperature for ninety minutes to distill off the adhering salts. The distilled salt vapor rises through a 10 cm diameter throat and condenses inside the cool condenser volume above the crucible assembly. Condensation of the salt vapor in the cool condenser region creates a pressure drop which draws more vapor. During this distilling operation, the steel container prevents the reactive chlorides from contacting the crucible wall, after which it is sacrificed as part of the consolidated waste form. For the consolidation, the temperature is increased to 1,630°C and held there for three hours.

In the top condenser assembly, the condensed salt is frozen into individual wedges formed by removable stainless steel fins which fit the annular space. After several runs, the condenser assembly is opened to remove the salt wedges, which are recycled back into the electrorefiner.

8.5.2 Ceramic Waste Form

Most of the fission products other than noble metals accumulate in the salt phase, including alkaline metal (Cs and Rb), alkaline earth (Sr and Ba), halide (Br and I), and rare earth (Y, Sm and Eu). These fission products are immobilized in zeolite, a material that adsorbs the fission products in a manner that immobilizes them. The fission product cations in the salt are adsorbed onto the zeolite by ion exchange and a portion of the salt is occluded—that is, it enters molecular cages that trap it in the zeolite structure. Both the adsorbed ions and occluded salt are very resistant to leaching by water. To further improve the leaching resistance, the salt-loaded zeolite is consolidated into a monolithic form by combining it with borosilicate glass as binder and sintering it at high temperature. At high temperature, the zeolite is converted to sodalite, a stable naturally occurring mineral. The conversion from zeolite to sodalite can be avoided if the temperature and the time at temperature can be carefully controlled; however this would be difficult for large-scale waste processing operations. In any case, sodalite is preferred for salt retention. Zeolite has larger molecular cages to occlude salt molecule. Sodalite's smaller size cages provide better containment of the salt as they inhibit release of the occluded salt molecules. Figure 8-13 illustrates the molecular structures of zeolite and sodalite.

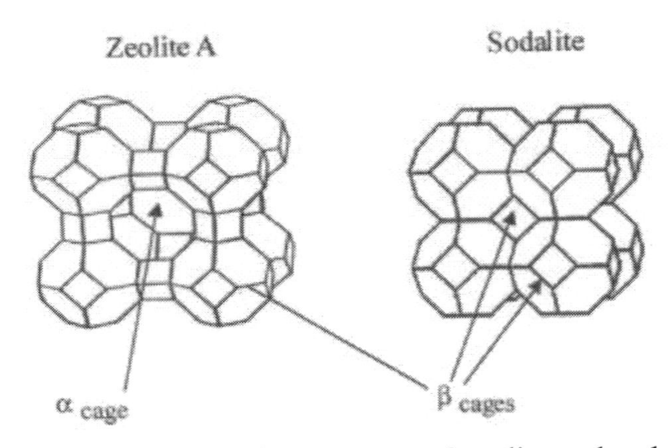

Figure 8-13 Molecular structure of zeolite and sodalite

In the EBR-II spent fuel treatment, the bond sodium is oxidized into NaCl in the electrolyte. Buildup of NaCl in the electrolyte salt will raise the melting temperature of the salt, so its buildup is limited to approximately 6 wt%. At this point portions of the salt are then treated for disposal. Commercially available zeolite powder contains as much as 22 wt% moisture, which is dried by heating to less than 1 wt% moisture. The disposed salt is crushed to appropriate particle size and then combined with zeolite powder in a V-mixer, heated to 500°C for approximately 15 hours to allow salt occlusion into the zeolite structure and mixed with 25 wt% glass.

If the bond sodium along with the salt waste gives too much ceramic waste volume or it is desirable to minimize the volume of the ceramic waste, the bond sodium can be distilled off as a front-end step before the pin-chopping and electrorefining. Then the waste salt can be passed through a zeolite column, fission products will be ion-exchanged into the zeolite, and the salt can be recycled back into the electrorefiner. It is expected that this approach will be adopted for future processing of the sodium-bonded metal fuel.

The final step of consolidating the fission product loaded zeolite was initially done in a hot isostatic press (850-900°C and 14,500-25,000 psi). Later, pressureless sintering was developed as a preferred approach since the latter can be more readily scaled up and is more amenable to remote operation. The reference processing conditions for the sintering kiln are 850°C for approximately four hours at 1 atmospheric pressure of argon gas.

8.6 Summary

Processes for each step of the IFR spent fuel recycle and waste disposal have been developed and put in place at EBR-II in the modernized Fuel Conditioning Facility (FCF) The necessary equipment has been designed, fabricated, tested, and installed, and remote operation is routine. Only five pieces of equipment are needed for recycle. All five are fully operable and some few tons of EBR-II assemblies have been processed.

The waste processes have been developed as well, and full-scale equipment has also been designed, fabricated, tested and installed. There are two principal products. The noble metal fission products and the structural materials are incorporated in a steel-matrix with approximately 15% zirconium, which is resistant to leaching by water. The active fission products from the electrolyte are immobilized in a glass-bonded zeolite, sintered, to form a monolithic product of sodalite, also resistant to leaching by water. Both waste products are suitable for repository disposal.

References

1. Charles E. Stevenson, *The EBR-II Fuel Cycle Story*, American Nuclear Society, 1987.
2. R. D. Pierce and L. Burris, "Pyroprocessing of Reactor Fuels," *Reactor Technology-Selected Reviews*, TID-8540, p.411, 1964.
3. L. Burris, R. K. Steunenberg, and W. E. Miller, "The Application of Electrorefining fpr Recovery and Purification of Fuel from the Integral Fast Reactor," *Proc. Annual AIChE Meeting*, Miami, Florida, November 2-7, 1986.
4. J. P. Ackerman, "Chemical Basis for Pyroprocessing of Nuclear Fuel," *Industrial & Engineering Chemistry Research,* 30-1, 141, 1991.

5. R. D. Pierce, et al., "Progress in the Pyrochemical Processing of Spent Nuclear Fuels," *J. Metals*, 45, 40, 1993.
6. DOE/EA-1148, "Environmental Assessment: Electrometallurgical Treatment Research and Demonstration Project in the Fuel Conditioning Facility at Argonne National Laboratory-West," May 15, 1995.
7. National Research Council, *Electrometallurgical Techniques for DOE Spent Fuel Treatment: Final Report*, National Academy Press, 2000.
8. DOE/EIS-0306, "Final Environmental Impact Statement for the Treatment and Management of the Sodium-Bonded Spent Nuclear Fuel," U.S. Department of Energy, July 2000.
9. T. B. Massalski, *Binary Alloy Phase Diagrams, Second Edition*, American Society of Metals, 3:1799, 1990.

CHAPTER 9

THE BASIS OF THE ELECTROREFINING PROCESS

In this chapter our aim is to provide a physical understanding of the electrochemical basis of electrorefining—the heart of IFR pyroprocessing. We are indebted to John Ackerman, Argonne's fine electrochemist, for his generous advice and help and in the editing of this chapter. Our goal also is to give as simple and straightforward a presentation as we can manage. In seeking simplicity there may be some error in detail, but the major themes of this chapter are as they are understood at the present time. Responsibility for the former lies with the authors; for assurance of the latter we are in debt to Dr. Ackerman.

The chapter is an abbreviated treatment of the material given in Appendix A. The appendix treats the electrochemistry of the process in detail from the fundamental particles on up—through the concepts used, the thermodynamics, a little of the kinetics of reactions, and the details of the calculations of the important properties and products. Those who wish a fuller understanding of the subject are referred to Appendix A.

9.1 Electrorefining Spent Fuel

The inexpensive compactness of the IFR fuel cycle, the effectiveness of the separations of product from waste, and in fact the very ability to recycle spent fuel at all, depend on one piece of equipment: the IFR electrorefiner. Its operation is based on the principles of electrochemistry, a science at the intersection of electricity and chemistry. One specific class of chemical reactions gives rise to the electrical phenomena that are the basis for all of electrochemistry. When an electrode (a conductor of electrons) is immersed in an electrolyte (which contains electrically charged ions and can conduct electricity) it forms an "electrochemical cell." Chemical reactions then occur naturally at the interface between electrode and electrolyte which convert chemical energy to electrical energy spontaneously and naturally, as a battery does. However, if the object of a process is to have a particular chemical reaction occur, the reverse can be done: a voltage can be imposed that alters the electrical phenomena so as to cause the desired chemical reaction. Current flows, electrical energy is converted to chemical energy, and chemical bonds are formed or broken as needed to form the products desired. This is what happens in electroplating, for example—and this is what happens in the IFR process.

The IFR process electrolyte is a molten mixture of lithium chloride and potassium chloride salts. Chopped-up fuel pieces are made the anode of the electrochemical cell; a metallic cathode will collect the product. Voltage is applied and the spent fuel gradually dissolves into the electrolyte. Here the first and very important separation takes place. The most chemically active (the most driven to react) of the fission products, which are also responsible for much of the radioactivity, react immediately with the ionic compound uranium chloride, UCl_3, in the electrolyte. It is present as a seed from initial operation and maintained by electrorefining operations. The active fission products displace the uranium and form their own chlorides. At actinide-refining voltages, the chlorides of the active fission products, once formed, are very stable, and they remain in the salt until they are removed as waste in a later operation. The positively charged uranium and higher actinide ions diffuse through the electrolyte toward the cathode, and only they deposit in quantity on the cathode because the higher stability of both the chlorides of the electrolyte materials and the chlorides of the dissolved active metal chlorides prevents them from also "reducing" to metals and depositing on the cathode at the voltages used.

9.2 Energy Transfer: The Thermodynamics of the Process

The fundamental bases of the process are actions at the molecular level of atoms, ions, and electrons. Chemical reactions are just these very small particles interacting with each other. Tiny energy changes occur in these interactions, and these energy changes determine what happens in the process. Classical thermodynamics, which deals precisely with energy relationships (e.g. energy cannot be created or destroyed), gives us the means to predict what chemical reactions are possible. Only if the energy content of the products of the reaction is less than the sum of the energy contents of the reactants going into it is a chemical reaction possible. The fraction of the energy that isn't dissipated in the reaction and is available to drive the reaction is called the free energy.

The free energy is the maximum energy available from a reaction for conversion to other forms of energy. It is a potential energy, energy stored and actually available to "flow downhill" and do useful work. The magnitude of the free energy in the reactions of various elements forming their chloride ionic compounds is available in tabulations in the literature and for the most relevant elements is given in Table 9-1 below. [1] Our process is based on the differences in free energies of chloride formation of the various elements and compounds in the electrorefiner.

Whether or not a given reaction can occur, and if it can, how completely, depends on the magnitude and sign of these differences.

Table 9-1. Free Energies of Chloride Formation at 500°C, - kcal/g-eq*

Elements that remain in salt (very stable chlorides)		Elements efficiently electro transported		Elements that remain as metals (less stable chlorides)	
$BaCl_2$	87.9	$CmCl_3$	64.0	$ZrCl_2$	46.6
CsCl	87.8	$PuCl_3$	62.4	$CdCl_2$	32.3
RbCl	87.0	$AmCl_3$	62.1	$FeCl_2$	29.2
KCl	86.7	$NpCl_3$	58.1	$NbCl_5$	26.7
SrC12	84.7	UCl_3	55.2	$MoCl_4$	16.8
LiCl	82.5			$TcCl_4$	11.0
NaCl	81.2			$RhCl_3$	10.0
$CaCl_2$	80.7			$PdCl_2$	9.0
$LaCl_3$	70.2			$RuCl_4$	6.0
$PrCl_3$	69.0				
$CeCl_3$	68.6				
$NdCl_3$	67.9				
YCl_3	65.1				

*The term kcal/g-eq is to be read as kilocalories per mass in grams of the material interacting with one mole of electrons. (For elements with a valence of one the mass is just the atomic weight in grams; for trivalent substances, uranium for example, the mass is one third of the atomic weight; and so on.) The sign of the numbers is understood to be negative.

Three groups of chlorides can be identified, separated in free energies of formation. Each acts differently. The first group is the active metals, which stay as stable chlorides in the electrolyte until they are stripped out in later waste processing. The second are the uranium and transuranics, which electro-transport to the cathode of the electrochemical cell, the only elements that are actually "electrorefined." They are the product. The third are the metals with still less stable chlorides, iron and the noble metals particularly, which do not form stable chlorides in the presence of more active elements; they collect as metals in the cadmium pool below the electrolyte or remain as hulls in the anode basket.

Thus the first big separation of product from waste comes directly from chemical reactions of the spent fuel ions with less stable chlorides—uranium chloride principally—in the molten salt. For the elements left in the anode, their non-reaction serves a similar purpose. Most of these separations, important as they are, have little to do with the imposed voltage. They are due solely to the energy relationships basic to the elements themselves. They cause reactions which leave most of the troublesome fission products in the salt, others in the liquid cadmium below the salt, and still others, non-reacted, in the anode basket. The imposed voltage is for the electrorefining process itself, drawing the product element ions to the cathode and there reducing them to metals as a stable product.

Uranium is deposited on a steel cathode, quite pure, with some adhering salt. In an electrolyte that contains significant amounts of uranium chloride, as ours does, plutonium and the other transuranics will not deposit in that way. Their stability as chlorides is greater than that of uranium chloride—that is, their free energies are greater, more negative than uranium. So when they reduce to metal at the steel cathode in the electrorefining process, instead of depositing as metal they immediately react with the uranium chloride and form their more stable chlorides once again. Stated plainly, they just exchange right back into the electrolyte again.

In the presence of ample uranium chloride, the normal case, plutonium and other transuranic elements therefore cannot be collected this way. However, by altering the free energies through formation of an intervening compound and reducing uranium chloride concentrations in the electrolyte, plutonium can be collected at a cathode. The alteration of the free energy relationships is done by use of a liquid cadmium cathode. The higher actinides, including plutonium, but importantly not uranium, form cadmium compounds—compounds of two metals, or "intermetallic compounds," whose effect is to lower the free energies of formation of their chlorides by the amount of free energy used up in forming the intermetallic. The result is that the free energy of the plutonium in its intermetallic compound then almost matches that of uranium—it is higher, but only slightly so. The intermetallic compounds thus stabilize it and the higher actinide elements in the cadmium. In this way, using the two different cathode types—metal for uranium, liquid cadmium for the higher actinides—it is possible to adequately, but not perfectly, separate transuranics from uranium.

However, the free energy of the reaction of plutonium or $PuCd_6$ in cadmium with the uranium chloride in the electrolyte still favors exchanging plutonium for uranium in uranium chloride. Not all the difference in free energies of formation of plutonium and uranium chloride can be eliminated by the formation of plutonium inter-metallic. Ninety percent of the driving force for the exchange reaction of plutonium for uranium in uranium chloride is removed, but some remains in the direction of removal of the plutonium metal by exchange with uranium in uranium chloride. (As opposed to the "back reaction," $PuCl_3$ in the presence of U breaking up to form Pu metal and UCl_3.) So something more has to be done. That something is to draw down the uranium chloride concentration in the electrolyte until the ratio of plutonium chloride to uranium chloride in the electrolyte is brought up to the point where the back reaction of $PuCl_3$ to UCl_3 is significant.

Calculation of the ratio of $PuCl_3$ to UCl_3 in the electrolyte necessary to make the "back reaction" sufficiently dominant for adequate plutonium depositions requires us to introduce briefly the kinetics of molecular reactions.

9.3 Kinetics of the Reactions

Chemical reactions between molecules result from collisions, where the more energetic molecules undergo reactions. Molecular kinetic energies cover a wide range, increasing with temperature. The fraction of molecules with kinetic energies sufficient to cause a reaction is given by a Maxwell-Boltzmann distribution. Differing slightly from the bell-shaped "normal distribution" seen in all kinds of phenomena, the Maxwell-Boltzmann distribution has a lop-sided bell shape with a high energy "tail" trailing out from the maximum. The distribution of energies broadens as temperature is increased and the maximum moves to higher energies.

The fraction of molecules with energies sufficient to cause reactions is proportional to an exponential of the form $\exp((E_f\text{-}E_b)/RT)$, where $(E_f\text{-}E_b)$ is the difference between the free energies of forward and back reactions and RT is the energy corresponding to temperature T. This expression allows us to actually calculate the balance between the forward and backward reactions. In addition to the principal reaction in the "forward" direction, there is always some "back" reaction; it may be small or very small if the free energy differences are substantial, but always there is some. Note that the smaller the exponential, the smaller the difference between the two free energies, and the closer the rates of forward and back reactions, and thus the back reaction fraction gets larger. In our case, because the back reaction—Pu metal forming from $PuCl_3$ in the presence of UCl_3—is what we want, the smaller $(E_f\text{-}E_b)$ can be made (again in our case by use of a liquid cadmium cathode), the easier it is to deposit plutonium metal.

9.4 The Power of Equilibria

The $PuCl_3/UCl_3$ ratio in the electrolyte must be high enough to give a back reaction sufficient for a useful rate of Pu metal deposition. The $PuCl_3/UCl_3$ ratios necessary for useful depositions can be calculated from equilibrium considerations. In fact, what goes where, and how much, and in what form is extremely important, and it is possible to calculate it from the simple principles of equilibrium in the rates of the reactions of each of the elements. The principles of equilibrium are simple, very general and very powerful. They state that in any reaction the forward rate must eventually equal the backward rate. Eventually the amount of product will build up enough that the backward rate—and there will always be a backward rate—equals the forward rate of product formation. The backward rate—the basic rate at which the product of a reaction dissociates into the original components of the reaction—may be small; it may in fact be microscopically small, and present only due to highly unlikely statistical fluctuations. But equality comes when the concentrations of reactants have so decreased and the amount of product has so increased that the two rates overall are equal. In those reactions where the backward rate is extremely small, present only because of statistical fluctuations, reactions go to almost perfect completion. For a solid cathode, without the effect of the plutonium-cadmium intermetallic compound on the free energies for plutonium to deposit the necessary ratio of plutonium chloride to uranium chloride in the salt is calculated to be over a million—1.2 x 10^6 to be exact. For our case, with liquid cadmium as the cathode, the $PuCl_3/UCl_3$ ratio in the electrolyte for equilibrium, and thus the $PuCl_3/UCl_3$ ratio for significant deposition of plutonium, is calculated to be 4.69, a ratio that is realistic to achieve.

The reaction we are concerned with is Pu metal (just created by electrorefining) reacting with the UCl_3 in the electrolyte to form U metal and $PuCl_3$ (returning the plutonium once more to the electrolyte). The reaction at equilibrium is written

$$Pu + UCl_3 <-> U + PuCl_3.$$

It expresses the fact that at equilibrium the rates of the two reactions—the forward reaction, Pu + UCl_3, and the back reaction, U + $PuCl_3$, will be equal. The concentrations given by the back reaction must increase until this is so.

The exponential derived from the kinetic considerations above enables us to put numbers to the ratio of forward to back reactions, as R is the well-known Gas Constant, T the temperature, and the free energies relevant to the reactions (E_f-E_b) are known and tabulated. The concentration ratios must distribute according to the equality below:

$$\exp((E_f - E_b)/RT) = (a_{PuCl3}/a_{UCl3})/(a_{Pu}/a_U).$$

The reactions rates are directly proportional to the "activities" denoted by the symbol "a." They are approximately proportional to concentrations. Where they are not, they are corrected by altering the concentrations by an amount given by an "activity coefficient," which is essentially a fudge factor, known and tabulated for our reactions of interest. The concentrations corrected in this way are called "activities" or "chemical activities." Activities, therefore, are the effective concentrations entering into the chemical reactions. They are the quantities that actually determine the reaction rates—concentrations are an approximation thereto.

Where the activity coefficients are constant with concentration over the concentrations of interest, concentrations can be used directly. And this appears to be the case for much of the IFR process. [2] Constant activity coefficients are a good enough assumption as long as the reactants are present in concentrations where they remain dissolved—that is, for unsaturated conditions. This comes up in a case important to the cadmium cathode. Calculations are simplified by the assumption of constant activity coefficients, as activities are then directly proportional to concentration from zero to the value at saturation in cadmium, where precipitation of the intermetallic compound begins.

The exponential $\exp((E_f-E_b)/RT)$ is termed the equilibrium constant K_{eq}, a very important number in our understandings of the process. (For the solid cathode it is the $1.2x10^6$ noted previously, and for the liquid cadmium cathode it is 4.69, also previously mentioned.)

The activity coefficients of the actinide chlorides are 6.62 for $PuCl_3$ and 5.79 for UCl_3 (see Appendix A) and their ratio is 6.62/5.79 or 1.14, thus the equilibrium ratio of the $PuCl_3/UCl_3$ concentrations in the salt is 4.69/1.14 = 4.1. The number 4.1 is a very significant number.

It is the concentration ratio of $PuCl_3$ to UCl_3 for equilibrium when both the uranium and plutonium cadmium intermetallic are saturated in the cadmium and in contact with the salt. And it provides the criterion for assessing the effect of any particular plutonium chloride to uranium chloride ratio in the salt on the ratio of plutonium to uranium in the product. The equilibrium constant therefore is

$$K_{eq} = (a_{PuCl3}/a_{UCl3})/(a_{Pu}/a_U).$$

And from this simple expression, the ratio of the concentrations of plutonium to uranium in the product can be calculated for any ratio of the plutonium and uranium chlorides in the electrolyte at equilibrium.

Knowing the equilibrium ratio of $PuCl_3/UCl_3$ in the electrolyte allows us not only to calculate the Pu/U ratio in the product at equilibrium, but also to estimate the Pu/U ratio in off-equilibrium conditions. Again, at equilibrium, with U and Pu both saturated in cadmium and in contact with a salt containing actinide chlorides, the $PuCl_3/UCl_3$ ratio in the salt is 4.1.

We know activities at saturation of U and Pu and we know their activity coefficients; from these, substituting the values in the above equation allows us to calculate the ratio of plutonium to uranium in the product for this equilibrium situation. (The details of the calculation are given in Section A.9 of Appendix A.) This important ratio is 1.55 to 1 plutonium to uranium.

It is the composition when both uranium and plutonium are saturated, with the actinide chloride ratio at its equilibrium value. It is not the cathode composition, unless by happy coincidence. It is the composition of the actinide product at the precise point of saturation of both uranium and plutonium before any solid phases have formed. As such, it gives a practical feel for the cathode behavior to be expected. But it surely isn't all we want to know. The total amount in all phases of U, Pu, and Cd in the cathode is the totality of our product. It is the total amounts of each that are important to us.

At equilibrium, with a concentration ratio of $PuCl_3$ to UCl_3 of 4.1 in the electrolyte, the actinide ratio Pu/U in the cadmium is about 1.55. But if the ratio of actinide chlorides in the salt differs substantially from 4.1, the ratio of the metals in cadmium will differ greatly from this. With one actinide saturated—plutonium, say—the plutonium to uranium ratio will be greater in the cadmium at a concentration ratio of $PuCl_3$ to UCl_3 greater than 4.1, and less at a lesser ratio. If both are saturated, the $PuCl_3/UCl_3$ equilibrium ratio can only be 4.1. The increase or decrease of the Pu/U ratio in the cathode toward equilibrium value is slow if both are unsaturated, rapid if just one is saturated. But by designing to saturate plutonium only, the ratio of the amounts of plutonium to uranium can be increased to achieve adequate plutonium enrichments at reasonable actinide chloride ratios in the electrolyte.

A related effect is how well the fission products are separated from the actinide product. The degree of separation of the waste from the product—the higher the better—is quantified by defining a "separation factor" which quantifies the success of this portion of the process. Stating matters as simply as we can, for this is important, the separation factor is how much of the waste element—cesium, say—is in chloride form divided by how much is in metal form, divided by the same ratio for uranium. The separation factors tell us how clean the separations are likely to be. Their basis, once again, is the difference in free energies of formation of the chlorides of the various elements relative to that of uranium.

Measured values of the separation factors are listed in Appendix A, taken from Ackerman and Johnson. [3] The values vary from separation factors of 43.1 to 1.6×10^9 for fission products, and from 1.88 to 3.52 for plutonium and the higher actinides. These are perfectly adequate fission product separations for IFR fuel, and for the actinides, small separations between plutonium and both uranium and the actinide above plutonium, assuring the desirable (for safeguarding purposes) imperfect actinide separations.

9.5 Actinide Saturation in Liquid Cadmium: Adequate Plutonium Depositions

In the operation of the liquid cadmium cathode, saturation effects in the liquid cadmium play a role. With saturation comes the ability to control, within limits, the possible composition and amount of the actinide product. [4] The concentrations of plutonium and uranium going into the cathode change markedly from the state when neither element is saturated, uranium or plutonium, to the state when one element, but not both, is saturated. The third state, when both are saturated, has still different characteristics, but practical difficulties generally rule out operation in that regime.

The electrorefining process brings actinides to the liquid cadmium cathode. At the start, they go into solution. But only up to a certain point, as each has a very well defined solubility limit. At concentrations in the range of 2 percent of the cadmium amount, their solubility limits are reached. There the mode of deposition changes dramatically, and it affects both the ratio of plutonium to uranium and the product amount.

When neither uranium nor plutonium is saturated, the actinide contents are low enough that their activities, their driving force to react, can be taken to be directly dependent on concentration. This can be expected to always be at least a little bit "wrong," but in these circumstances not significantly so.

The ratios of cathode actinide metal and electrolyte actinide chloride activities will satisfy the equilibrium constant expression and continue to do so as deposition progresses. This goes on smoothly with no dramatic changes until the solution becomes saturated with either uranium or plutonium.

When one or the other is saturated but not both, deposition changes markedly. Plutonium is the important case. No solid phase $PuCd_6$ forms until the Pu concentration reaches its solubility limit. When this limit is reached, no more of the saturating Pu metal enters in solution; it deposits preferentially as a solid phase, $PuCd_6$. When both U and $PuCd_6$ are saturated, they accumulate at the same ratio as ratios of activities $PuCl_3$ and UCl_3 in the salt and the ratio does not change until all the cadmium or actinide chloride is used up. These are the different possible operating regimes, but as mentioned above, the last regime is not very important as inconvenient things happen to the deposit in practice.

The principle here is that free energy change remains opposite in direction to the reaction we would like, but by loading up the ratio of $PuCl_3/UCl_3$ heavily in $PuCl_3$ an equilibrium is achieved that allows us a useful ratio of Pu/U in the cathode. The exact value of the equilibrium coefficient isn't important for purposes of understanding the process; the fact that it is an important criterion is. However, our calculated value of 4.1 is probably adequate, certainly for insight. Off equilibrium in actinide chloride ratios for saturation, the system tries to achieve saturation of both elements in cadmium and get back to the corresponding equilibrium ratio of actinide chlorides.

At an actinide chloride ratio of precisely 4.1, the uranium and the $PuCd_6$ activities will maintain the same ratio right from the beginning. Both saturate at the same time. All are nicely in equilibrium. The Pu/U ratio will be about 1.55. But let us vary the actinide chloride ratio and observe what happens. A consistent way of thinking about all this is provided by the equilibrium expression.

1. At an actinide chloride ratio of precisely 4.1, the uranium and the $PuCd_6$ activities will maintain the same ratio right from the beginning, to the extent that their activity coefficients are given by the "ideal approximation"—the saturation value multiplied by the ratio of the concentration to the saturation concentration. Both will saturate at the same time. All are nicely in equilibrium.
2. At an actinide chloride ratio of two, the activity ratio is half the saturated ratio; the uranium will saturate first while the PuC_6 is only half way to saturation. The UCl_3 concentration has to be decreased to get to the equilibrium value of 4.1, so uranium will preferentially deposit, trying to increase the actinide chloride ratio toward 4.1. In this regime the cathode is acting pretty much like a solid cathode and the resulting overall deposit will be largely U.
3. At the other extreme, at a ratio of eight, say, the opposite effects take place; the activity ratio is twice the saturated ratio, the $PuCd_6$ will saturate, and $PuCd_6$ will try to deposit as the system tries to evolve to decrease the $PuCl_3$ concentration and move the actinide chloride ratio toward 4.1. In this case the deposit will be plutonium rich, above the 1.5 Pu/U value for the both-saturated case 1, above.

Off equilibrium in actinide chloride ratios for saturation, the system tries to achieve saturation and get back to the saturated cadmium equilibrium ratio of actinide chlorides. Practical everyday considerations limit what can be done. Criticality considerations are important. The amount of fuel that the anode baskets can contain is to some degree limited. The cathode too almost certainly is limited in capacity. The actinide content of the salt is high relative to the amounts in the anode and cathode, making changes in its composition tedious. The composition of the anode feed may make higher $PuCl_3/UCl_3$ ratios difficult to achieve; they are still possible, but perhaps only after several batches have been processed. What is possible in practice is limited by other practical effects such as deposit growing out of the cathode.

Finally, it is of interest to note that with everything at equilibrium, the ratio of plutonium to uranium in the product of 1.55 is a factor of 2.65 less than the 4.1 ratio of plutonium chloride to uranium chloride in the electrolyte. This gives an indication of the difficulty of a pure product—whatever the $PuCl_3/UCl_3$ ratios are in the salt, the Pu/U product ratio will be degraded from this by a considerable factor.

9.6 Effect of Saturation on Chemical Activity

And now we come to the key point in the effect of saturation. The chemical activity of a solute in a solution is independent of the amount of solute once the solution is saturated in that solute. This is basic to understanding liquid cadmium cathode operation. Once the solution is saturated, the activity is fixed at the saturation value. And this provides a basis for gathering useful amounts of plutonium in the cathode. No solid-phase $PuCd_6$ will form until the Pu concentration reaches its solubility limit (nor will uranium metal form until its solubility limit is reached.) When the solution becomes saturated, either with uranium or with plutonium, no more of the saturating metal can enter solution; it must deposit as a separate phase. But its activity never changes from the saturation value. Activity of a pure solid or liquid is always unity, so a pure uranium phase has a uranium activity of one and $PuCd_6$ has an activity of one.

The meaning of activities may have remained somewhat puzzling at this point, so let us digress briefly to clarify the basis of chemical "activity." Activity is dependent upon, and its magnitude is defined by, the free energy that drives the tendency to react. It is proportional to the exponential, $\exp(\Delta G^o{}_f/RT)$, where ΔG is the free energy driving the reaction that the chemical activity quantifies. The exponential is a measure of the effect of the free energy change which determines the degree to which the reaction will proceed. Thus the free energy change, ΔG, is proportional to the natural logarithm of the activity (ln a) and if the free energy change is zero, as ln(0) is unity, the activity is exactly one. An activity of one means no net reaction at all. That, of course, is to be expected when a pure material simply adds to the pure material already there. But the point is that an activity of one is no arbitrary normalization, it very specifically means that there is no net reaction—no free energy change available to drive a reaction.

Taking plutonium as our example—the important case—once the plutonium saturates its activity does not change.

$$a_U = K_{eq}\, a_{Pu}/(a_{PuCl3}/a_{UCl3})$$

Knowing K_{eq} and a_{Pu} are constant, for any activity ratio of the chlorides (a_{PuCl3}/a_{UCl3}) we can calculate the corresponding activity of the uranium. From this and knowledge of the uranium amount at saturation we can calculate the amount of (not yet saturated) uranium in the product. We know the plutonium concentration at saturation, so we then have the calculated ratio of plutonium to uranium in the product. Each of the cases—plutonium saturated but not uranium; both saturated; uranium saturated but not plutonium—can be calculated in a similar manner.

This then gives the basis for estimating beforehand the ratio of plutonium to uranium in the product. The details of the calculation are laid out in Section A.11 of Appendix A.

9.7 The Plutonium Recovery Experiments

Three kilogram-scale extractions of plutonium from the electrolyte to a cadmium cathode have now been reported in the literature. [5] A fourth (Run 4) has also been done [6-7], which is very similar to Run 3. For our purposes it demonstrates repeatability and serves as a check on the previous run. All four are analyzed below. The ratio of $PuCl_3/UCl_3$ in the electrolyte varied from run to run, but all were in a range that from the considerations presented in the previous section would predict to give adequate plutonium deposition. (Runs 3 and 4 were only just in that range.) In all a deposit of a kilogram or more was sought; two of the four deposits were over a kilogram.

The plutonium recoveries were done as part of the uranium recovery processing program of spent fuel and blanket pins from EBR-II. With the many uranium recovery runs, a substantial amount of plutonium builds up in the electrorefiner electrolyte. The buildup is slow because the blanket elements contain only a percent or two of plutonium. However, the amount of plutonium that can be tolerated must be kept within limits, for criticality reasons if for no others, and the plutonium content of the electrorefiner was reduced somewhat in this way.

As expected, the lower the ratio of plutonium chloride to uranium chloride in the electrolyte, the less plutonium in the product, but all had considerable uranium deposited along with the plutonium. The highest ratio had an initial plutonium chloride concentration ten times the uranium chloride concentration. The product still contained 30 percent uranium. We will look at these very important results in some detail below. Their importance is clear. These are the only engineering-scale plutonium deposition results we are aware of to the present time. Further, the calculations we present are the only such calculations—aimed at predicting the results—that we are aware of to the present time.

The complete equilibrium calculations of the influence of plutonium to uranium concentration ratios in the electrolyte on plutonium deposition in the cadmium cathode are presented in Appendix A. The calculations of Appendix A were provided by Dr. John Ackerman. [4] Sufficient information is included for the calculations to be reproduced by those inclined. Here we will summarize the important data.

In Table 9.2 the first six entries are measured values; the next seven give the meaning of the measurements, converting concentrations to activities, listing the amounts of plutonium or uranium in the metal phase (not in cadmium solution), the amounts actually in solution, and where they are not saturated, the degree of saturation.

Table 9-2. Measured data and calculation of degree of saturation of cadmium

	Run 1	Run 2	Run 3	Run 4
Initial $PuCl_3/UCl_3$ ratio (meas.)	10.2	5.1	3.7	3.45
Final $PuCl_3/UCl_3$ ratio (meas.)	5.66	3.31	2.86	2.85
Final $PuCl_3$ concentration, wt% (meas.)	2.67	2.46	2.38	2.85
Final UCl_3 concentration, wt% (meas.)	0.47	0.74	0.83	0.84
Pu metal in cathode, kg (meas.)	1.02	1.08	0.49	0.55
U metal in cathode, kg (meas.)	0.34	0.66	0.82	0.77
Final activity ratio, $PuCl_3/UCl_3$	6.46	3.77	3.26	3.27
Pu in $PuCd_6$ phase, g	26.2	89.0	0.0	0.0
U in U metal phase, g	0.0	48.4	204.4	171.0
Pu in Cd solution, g	997.8	991.0	492.0	553.0
U in Cd solution, g	341.0	610.6	616.6	616.6
Pu saturation, %	100	100	49	56
U saturation, %	55	100	100	100

From the considerations of Section 9.6 we can calculate the expected result of each run—for its final $PuCl_3/UCl_3$ ratio—and compare the calculated results to the measured values in Table 9.2.

The important practical result is the richness of the plutonium deposit—the ratio of plutonium to uranium in the deposit. How well can this important ratio be estimated from our simple considerations? We know from Table 9.2 the actual degree of saturation of the two actinides when each run was terminated. The runs were to go until a kilogram of plutonium was deposited. How much uranium can we expect in each deposit as the $PuCl_3/UCl_3$ ratio varies in the three runs?

The final actinide chloride ratios in the salt ranged downward from 5.66 to 3.31 to 2.86 to 2.85 in the four runs, which are to be compared to 4.1, the calculated ratio $PuCl_3/UCl_3$ at equilibrium with both plutonium and uranium saturated in the cadmium. In Run 1, plutonium was saturated, uranium was not; in Run 2, both were saturated, and in Run 3 and Run 4, plutonium was not saturated, uranium was. Where amounts greater than saturation amounts were present in the product, the amounts in grams of the metallic phases are listed in Table 9.2.

The measured results are compared to calculation by our simple equilibrium expressions in Table 9.3 below.

The equilibrium constant for the Pu on UCl_3 reaction is fixed—it depends only on the free energy difference in the reaction. Each ratio of actinide chlorides will give the concentration ratio of uranium and plutonium in the cadmium relevant to it up to the point of saturation of one of the actinides. For a $PuCl_3/UCl_3$ concentration ratio of 4.1 in the salt everything is in equilibrium and the ratio of plutonium to uranium in the cathode will be 1.55. For actinide chloride concentration ratios above and below 4.1, the U/Pu ratio in the cadmium will follow the relationships given by the equilibrium coefficient expressions.

Table 9-3. Calculated vs. measured Pu/U product ratio

	Initial Actinide Chloride Ratio	Final Actinide Chloride Ratio	Pu/U Product ratio		
			Measured	Calculated in-solution	Calculated total
Run 1	10.2	5.7	3.0	2.2	2.25
Run 2	5.1	3.3	1.6	1.6	1.6
Run 3	3.7	2.9	0.6	1.1	0.75
Run 4	3.5	2.9	0.7	1.1	0.78

The degree of agreement between the measurements and our calculations based on equilibrium is gratifying.

1. The $PuCl_3/UCl_3$ ratio of 5.7 is above the equilibrium ratio of 4.1, and the system will try to equilibrate by subtracting uranium from the cathode to increase the UCl_3 in the electrolyte and bring the $PuCl_3/UCl_3$ ratio down. And less uranium than calculated is found in the product based on the $PuCl_3/UCl_3$ ratio of 5.7.

2. The $PuCl_3/UCl_3$ ratio of 3.32 is somewhat below the equilibrium ratio of 4.1. The amounts are both are slightly in excess of saturation. The actinide ratio in the salt now is less than the equilibrium ratio of 4.1, and some of the deposited plutonium would be expected to migrate back into the salt as the system tries to equilibrate. However, the ratio of plutonium to uranium in the deposit agrees with that calculated based on the $PuCl_3/UCl_3$ ratio of 3.32.

3. The $PuCl_3/UCl_3$ ratio of 2.86 is further below the equilibrium ratio. The system will try to add plutonium from the cathode to add $PuCl_3$ to the electrolyte, reducing the amount of plutonium in the cathode. Here less plutonium is found in the cathode that is calculated based on the ratio 2.87.

4. The $PuCl_3/UCl_3$ ratio of 2.85 is the same as Run 3 for all intents and purposes. The results are the same as well, to the accuracy of calculation and measurement. The run provides assurance of the consistency of results, but provides no other new information.

All four measurements show that even with high plutonium to uranium ratios in the salt, the plutonium to uranium ratio in the product will be much lower, by a factor varying but in the range of the 2.65 previously calculated for lowering of the product Pu/U ratio from the equilibrium $PuCl_3/UCl_3$ ratio of 4.1 in the salt. In other words, getting close to a pure plutonium product without accompanying uranium is extremely difficult, and the presence of actinides above plutonium makes a pure plutonium product impossible.

The measurements show that a perfectly adequate plutonium enrichment of the product for reactor fuel recycle is possible with reasonable $PuCl_3/UCl_3$ ratios in the salt. Neither the calculations nor the experimental numbers should be taken as highly accurate. But the systematics displayed by the calculations do show that the equilibrium calculations predict the results found in these very significant experiments surprisingly well.

In summary, we now have the means to estimate the ratio of Pu/U in the product as a function of the $PuCl_3/UCl_3$ ratio in the electrolyte salt. At equilibrium, with both uranium and plutonium saturated in the cadmium, the ratio of Pu/U product is 1.55. The $PuCl_3/UCl_3$ ratio of Run 2 (3.32) is nearest to the calculated equilibrium value of 4.1. It gave both a measured and a calculated value of the Pu/U ratio of 1.6, agreement that is probably fortuitous, but gratifying in any event. The conclusion to be drawn overall is that the important Pu/U ratio can be predicted with useful accuracy from these simple considerations and very simple expressions.

9.8 Summary

The inexpensive compactness of the IFR fuel cycle, the effectiveness of the separations of product from waste, and in fact the very ability to recycle spent fuel at all, depend on one piece of equipment—the IFR electrorefiner. Its operation is based on the principles of electrochemistry, a science at the intersection of electricity and chemistry. One specific class of chemical reactions gives rise to the electrical phenomena that are the basis for all of electrochemistry. When an electrode (a conductor of electrons) is immersed in an electrolyte (which contains electrically charged ions and can conduct electricity) it forms an "electrochemical cell." Chemical reactions then occur naturally at the interface between electrode and electrolyte which convert chemical energy to electrical energy spontaneously and naturally, as a battery does. However, if the object of a process is to have a particular chemical reaction occur, the reverse can be done: a voltage can be imposed that alters the electrical phenomena so as to cause the desired chemical reaction. Current flows, electrical energy is converted to chemical energy, chemical bonds are formed or are broken as needed to form the products desired. This is what happens in electroplating, for example—and this is what happens in the IFR process.

The most chemically active (the most driven to react) of the fission products, which are also responsible for much of the radioactivity, react immediately with the ionic compound, uranium chloride, in the electrolyte. At the actinide refining voltages, the chlorides of the active fission products once formed are very stable, and they remain in the salt until they are removed as a waste in a later operation. The positively charged uranium and higher actinide ions diffuse through the electrolyte toward the cathode, and only they deposit in quantity on the cathode. The higher stability of both the chlorides of the electrolyte materials and the chlorides of the dissolved active metal chlorides prevents them from also "reducing" to metals and depositing on the cathode at the voltages we use.

Classical thermodynamics gives us the means to predict what chemical reactions are possible. Only if the energy content of the products of a reaction is less than the total of the energy contents of the reactants going into it is a chemical reaction possible. The energy available to drive the reaction is called the free energy. Our process is based on the differences in free energies of chloride formation of the various elements and compounds. Whether or not a given reaction can occur, and if it can, how completely, depends on the magnitude and sign of these differences.

Three groups of chlorides can be identified, separated in free energies of formation. Each acts differently. The first group is the active metals, which stay as stable chlorides in the electrolyte until they are stripped out in later waste processing. The second are the uranium and transuranics that electro-transport to the cathode of the electrochemical cell, the only elements that are actually "electrorefined." They are the product. The third are the metals with still less stable chlorides, iron and the noble metals particularly, which do not form stable chlorides in the presence of more active elements; they collect as metals in the cadmium pool below the electrolyte or remain as hulls in the anode basket.

Uranium is deposited—on a steel cathode, quite pure, with some adhering salt. In an electrolyte that contains significant amounts of uranium chloride, as ours does, plutonium and the other transuranics will not deposit in that way. Their stability as chlorides is greater than uranium chloride—that is, their free energies are greater, more negative than uranium. However, by altering the free energies through formation of an intervening compound and reducing uranium chloride concentrations in the electrolyte, plutonium and the other higher actinides can be collected at a cathode. The alteration of the free energy relationships is done by use of a liquid cadmium cathode. "Intermetallic compounds" are formed with cadmium which lower the free energies of formation of the plutonium and higher actinide chlorides sufficiently to almost match uranium's—still higher, but only slightly so. In this way, using the two different cathode types—metal for uranium, liquid cadmium for the higher actinides—it is possible to adequately, but not perfectly, separate transuranics from uranium.

The free energy of the reaction of plutonium or $PuCd_6$ in cadmium with the uranium chloride in the electrolyte still favors exchanging plutonium for uranium in uranium chloride. Not all the difference in free energies of formation of plutonium and uranium chloride can be eliminated by the formation of plutonium intermetallic. Some difference remains in the direction of removal of the plutonium metal by exchange with uranium in uranium chloride. Drawing down the uranium chloride concentration in the electrolyte well below the plutonium chloride concentration allows useful concentrations of plutonium to be deposited, always along with higher actinides and substantial amounts of uranium.

Chemical reactions result from collisions between molecules, and the more energetic molecules undergo reactions. The fraction of molecules with kinetic energies sufficient to cause a reaction is proportional to an exponential of the form $\exp((E_f\text{-}E_b)/RT)$, where $(E_f\text{-}E_b)$ is the difference between the free energies of forward and back reactions and RT is the energy corresponding to temperature T. In addition to the principal reaction in the "forward" direction, there is always some "back" reaction; it may be small or very small if the free energy differences are substantial, but always there is some. The exponential allows calculation of the balance between the forward and backward reactions. In our case, because the back reaction—Pu metal forming from $PuCl_3$ in the presence of UCl_3—is what we want, the smaller $(E_f\text{-}E_b)$ can be made (again in our case by use of a liquid cadmium cathode), the easier it is to deposit plutonium metal.

The $PuCl_3/UCl_3$ ratio in the electrolyte must be high enough to give a back reaction sufficient for a useful rate of Pu metal deposition. The $PuCl_3/UCl_3$ ratios necessary for useful depositions can be calculated from equilibrium considerations. In fact, what goes where, and how much, and in what form is extremely important, and it is possible to calculate it from the simple principles of equilibrium in the rates of the reactions of each of the elements.

Pu metal (just created by electrorefining) reacts with UCl_3 in the electrolyte to form U metal and $PuCl_3$ (returning the plutonium once more to the electrolyte.) At equilibrium the rates of the two reactions—the forward reaction Pu + UCl_3, and the back reaction U + $PuCl_3$—are equal. The concentrations given by the back reaction must increase until this is so.

$Pu + UCl_3 <\!\!-\!\!> U + PuCl_3$.

The reaction rates are directly proportional to the "activities" denoted by the symbol "a" The activities in turn are approximately proportional to concentrations. Where they are not, they are corrected by altering the concentrations by an amount given by an "activity coefficient," essentially a fudge factor, known and tabulated for our reactions of interest. The concentrations corrected in this way are the "activities" or "chemical activities" that determine the reaction rates.

The concentrations at equilibrium distribute according to the equation

$\exp((E_f-E_b)/RT) = (a_{PuCl3}/a_{UCl3})/(a_{Pu}/a_U) = K_{eq}$, the equilibrium constant.

The equilibrium constant K_{eq}, a very important number in our understandings of the process, from the exponential above, is calculated to be 4.69. The ratio of activity coefficients of the actinide chlorides is 1.14, and the equilibrium ratio of the $PuCl_3/UCl_3$ concentrations in the salt is 4.69/ 1.14 = 4.1.

The number 4.1 is a very significant number. It is the concentration ratio of $PuCl_3/UCl_3$ for equilibrium when both the uranium and plutonium cadmium intermetallic are saturated in the cadmium and in contact with the salt. And it provides the criterion for assessing the effect of any particular plutonium chloride to uranium chloride ratio in the salt on the ratio of plutonium to uranium in the product.

Knowing the equilibrium ratio of $PuCl_3/UCl_3$ in the electrolyte allows us not only to calculate the Pu/U ratio in the product at equilibrium but also to estimate the Pu/U ratio in off-equilibrium conditions as well. Again, at equilibrium, with U and Pu both saturated in cadmium and in contact with a salt containing actinide chlorides, the $PuCl_3/UCl_3$ ratio in the salt is 4.1.

From these considerations the ratio of plutonium to uranium in the product with everything at equilibrium can be calculated. This important ratio is 1.55 to 1 plutonium to uranium.

At equilibrium, with a concentration ratio of $PuCl_3/UCl_3$ of 4.1 in the electrolyte, the actinide ratio Pu/U in the cadmium is 1.55. But if the ratio of actinide chlorides in the salt differs substantially from 4.1, the ratio of the metals in cadmium will differ greatly from this. With one actinide saturated—plutonium, say—the plutonium to uranium ratio will be higher in the cadmium at a concentration ratio of $PuCl_3/UCl_3$ greater than 4.1, and less at a lesser ratio. If both are saturated, the $PuCl_3/UCl_3$ equilibrium ratio can only be 4.1.

In an electrorefining run with an actinide chloride ratio of precisely 4.1, the uranium and the $PuCd_6$ activities will maintain the same ratio right from the beginning. Both saturate at the same time. All are nicely in equilibrium. The Pu/U ratio will be about 1.55. But off this equilibrium actinide chloride ratio, the deposit will behave in the following way.

1. At an actinide chloride ratio of precisely 4.1 the uranium and the $PuCd_6$ activities will maintain the same ratio right from the beginning. Both will saturate at the same time. All are nicely in equilibrium.

2. At an actinide chloride ratio of two, say, the activity ratio is half the saturated ratio, the uranium will saturate first while the PuC_6 is only half way to saturation. The UCl_3 concentration has to be decreased to get to the equilibrium value of 4.1, so uranium will preferentially deposit to increase the actinide chloride ratio toward 4.1. In this regime, the cathode acts pretty much like a solid cathode and the resulting overall deposit will be largely uranium.

3. At the other extreme, at a ratio of eight, say, the opposite effects take place, the activity ratio is twice the saturated ratio; the $PuCd_6$ will saturate and $PuCd_6$ will try to deposit as the system tries to decrease the $PuCl_3$ concentration and move the actinide chloride ratio toward 4.1. In this case the deposit will be plutonium rich, above the 1.5 Pu/U value for the both-saturated case 1, above.

Finally, it is of interest to note that with everything at equilibrium, the ratio of plutonium to uranium in the product of 1.55 is a factor of 2.65 less than the 4.1 ratio of plutonium chloride to uranium chloride in the electrolyte. This gives an indication of the difficulty of a pure product—whatever the relevant chloride ratios are in the salt the product ratio will be degraded from this by a considerable factor.

The place of saturation of uranium and plutonium is basic to understanding liquid cadmium cathode operation. Once the solution is saturated the activity is fixed at the saturation value. And this provides a basis for gathering useful amounts of plutonium in the cathode. No solid-phase $PuCd_6$ will form until the Pu concentration reaches its solubility limit (nor will uranium metal form until its solubility limit is reached). When the solution becomes saturated, either with uranium or with plutonium, no more of the saturating metal can enter solution; it must deposit as a separate phase. But its activity never changes from the saturation value. Activity of a pure solid or liquid is unity—a pure uranium phase has a uranium activity of one and a $PuCd_6$ phase has an activity of one. When either plutonium or uranium saturates, the degree to which the other is saturated can be determined from the equilibrium constant expression, and from that in turn we can calculate the ratio of plutonium to uranium in the product.

Four measurements of plutonium-extraction runs in the kilogram range have been reported. The agreement with calculation is gratifying, providing convincing proof that the richness of the product in plutonium can be estimated quite accurately. The relatively simple considerations and expressions relying on equilibrium are shown to give satisfactory results. Both measurement and calculation show that reasonably high but achievable plutonium to uranium ratios in the salt are necessary for an adequate product. The corollary is that the plutonium to uranium ratio in the product will be considerably much lower than that in the salt by a factor that varies but lies in the approximate range of the precise 2.65 calculated for the degradation of the product Pu/U ratio from the well-defined equilibrium situation where the $PuCl_3/UCl_3$ ratio is 4.1 in the salt. In other words, getting close to a pure plutonium product without accompanying uranium is extremely difficult, and the presence of actinides above plutonium makes a pure plutonium product impossible.

Thus the agreement with measurement demonstrates that we have the means to estimate the ratio of Pu/U in the product as a function of the $PuCl_3/UCl_3$ ratio in the electrolyte salt. It is predictable with useful accuracy from simple considerations and very simple expressions.

Finally, for a much more complete treatment of our subject the reader is directed to Appendix A.

References

1. W. H. Hannum, Ed., "The Technology of the Integral Fast Reactor and its Associated Fuel Cycle," *Progress in Nuclear Energy*, 31, nos. 1/2, Special Issue, 1997.
2. J. P. Ackerman and T. R. Johnson, "Partition of Actinides and Fission Products between Metal and Molten Salt Phases: Theory, Measurement and Application to Pyroprocess Development." *Proc. Actinides-93 International Conference*, Santa Fe, New Mexico, September 19-24, 1993.
3. J. P. Ackerman and T. R. Johnson, "New High-Level Waste Management Technology for IFR Pyroprocessing Wastes," *Global '93 International Conference on Future Nuclear Systems, Emerging Fuel Cycles and Waste Disposal Options*, Seattle, Washington, September 12-17. 1993.
4. J. P. Ackerman, private communication.
5. D. Vaden, S. X. Li, B. R. Westphal, and K. B. Davies, "Engineering-Scale Liquid Cadmium Cathode Experiments." *Nuclear Technology*, 162, May 2008.
6. D. Vaden, private communication.
7. C. Pope, private communication.

CHAPTER 10

APPLICATION OF PYROPROCESSING TO LWR SPENT FUEL

The pyroprocesses described in Chapters 8 and 9 were developed for recycling the IFR metal fuel. Can the same technology be applied to process light water reactor (LWR) spent fuel? That is the question that will be explored in this chapter. A front-end step is needed to convert the LWR oxide fuel to a metal so that electrorefining by the IFR process is possible. Furthermore, because of the greater bulk of LWR fuel, the IFR electrorefining process will need scaleup to handle the higher throughput rates.

10.1 Background

Present day reactor fuel from commercial pressurized and boiling water reactors —generically light water reactor (LWR) fuel—is an oxide. There is real incentive to process the spent fuel from existing reactors. It makes the waste easier to deal with and it allows re-use of the useful portion, by far the bulk of the spent fuel. Use in fast reactors like the IFR greatly extends fuel resources (more than a hundredfold over once-through fuel use.) At the time of writing (2011) the Obama administration has stopped funding for the Yucca Mountain waste repository, apparently removing any possibility of bringing this badly-needed waste disposal facility into operation. As a result, there is literally no plan or process in place now in the U.S. to deal with LWR spent fuel. It will continue to build up in spent fuel pools and temporary storage locations at the hundred or so plants around the country. It is unwise and unnecessary to leave spent fuel indefinitely at the plant sites. It remains a potential hazard in circumstances that may not be foreseen. The recent events in Japan underline this—spent fuel was a source of radioactive release.

Here we show how the IFR process could be adapted to treat such fuel. The radioactive fission products could be put in impermeable long-lasting containment, with the very long-lived actinide elements in a small volume suitable for later reactor recycle and the uranium separated for later reuse in reactors as well. This would provide the basis for a tractable and cost-effective nuclear waste policy.

When pyroprocessing was being developed as the centerpiece of the IFR Program, Japanese utilities began to show interest in its potential for economic and proliferation-resistant spent fuel reprocessing. A joint program with Argonne was established with the Central Research Institute of Electric Power Industry (CRIEPI) of Japan in 1989. The original scope was limited to the research on pyroprocessing. Two years later, the contract was expanded to include the Japan Atomic Power Company as a representative of all other utilities, and the scope was expanded to include the planned fuel cycle demonstration in the refurbished fuel conditioning facility (FCF) at Argonne-West as well. A separate contract was also signed with the Tokyo, Kansai, and Chubu Electric Power Companies to investigate the feasibility of applying pyroprocessing to LWR spent fuel. Another new contract with the Power Reactor and Nuclear Cycle Development Corporation (PNC), a predecessor of the Japan Atomic Energy Agency, was also agreed to just before the IFR program was canceled in 1994. Altogether, these agreements represented over $100 million contribution from Japan. They were terminated when the program was canceled.

Although the pyroprocessing development efforts were focused on its IFR application, the contract with the three Japanese utility companies gave practical impetus to feasibility work on applying pyroprocessing to the LWR spent fuel. A large glove box facility to handle twenty kilograms of uranium oxide fuel was constructed as part of this joint program. The initial work on the oxide-to-metal conversion was based on use of lithium metal for the reduction (from oxide to metal) process. [1] For uranium oxide it worked well, but the process wasn't satisfactory for plutonium oxide. Therefore, an alternative process based on electrolytic reduction was developed and turned out to be far more satisfactory. This process will be discussed in detail in Section 10.2.

As mentioned above, scalability of the electrorefining process is obviously a must as well. The main difference between the LWR and the fast reactor spent fuel is in the actinide (in this context, all elements higher on the periodic table than uranium) contents. LWR spent fuel has 1–2% actinides, depending on discharge burnup, and fast reactor spent fuel has about a 20–30% actinide content. This factor of twenty between the two means that batch sizes can be much larger for LWR spent fuel without violating the criticality constraints necessary in processing fast reactor fuel. The perception that electrorefining is advantageous for the smaller scale fast reactor application but is not suitable for the LWR spent fuel isn't necessarily true. In industry, zinc and copper, for example, are routinely electrorefined in large quantities. The magnitude issue will be discussed in some detail in Section 10.3.

10.2 Electrolytic Reduction Step

10.2.1 The Process

There are three closely related processes for the electrolytic reduction step, all based on lithium metal acting on the oxides to reduce them to metals. Development work continues both at ANL-E where the earlier was largely done, [2-4] and at ANL-W (now INL) where later work used actual spent fuel. [5-6]

A ceramic LWR oxide fuel is hard and tough, like the ceramic in bathroom fixtures. There is no possibility of its direct use in IFR-type electrorefining, as it is an electrical insulator. It must be converted to metal, and this can be done. Development work has been underway at a relatively low level for several years at Argonne, at both its sites earlier, and now at Argonne-Illinois and at the Idaho National Laboratory in the part that was formerly Argonne-West. The development work has had considerable success. Other nations have picked it up as well. Republic of Korea, in particular, has made excellent progress too, and Japan has seen useful results as well.

Uranium oxide is the sole constituent of fresh fuel, and after irradiation it remains the principal constituent. But other actinides will have built up as well, and they and many of the fission products are present as oxides too. After the reduction process they will be metallic as well.

The first process used was direct pyrochemical reduction by lithium metal in LiCl molten salt at 650°C. There was no electrical input. Lithium oxide, Li_2O, is produced in the reduction of the targeted oxides. It dissolves in the LiCl, and its amount must be controlled or the oxides in the spent fuel, other than uranium—plutonium notably—will not reduce effectively. In this process, then, the salt was regenerated in a second vessel by removal of the Li_2O. The process worked, but it was improved in introducing electricity and by combining the two-vessel operation into one. However, much of the thermodynamic understanding of all the processes came from this early work.

The successor process is electrolytic reduction. In this process, the molten LiCl electrolyte is seeded with a small amount of Li_2O, which ionizes, providing initial current-carrying capability. The anode is platinum and the cathode is the oxide fuel. The metallic oxides are reduced at the cathode, and the oxygen ions so produced convert to oxygen gas at the anode and removed. In this way the Li_2O concentration is controlled. The necessary reactions are produced with a voltage between the electrodes selected to convert the $Li+$ ions of the dissociated Li_2O to Li metal at the cathode. The Li metal produced in this way gives the same reaction as the pyrochemical reduction process described above, but the source of it now is reduced Li_2O.

The most recent process is a variation on this wherein the voltage is set for direct electrolytic reduction of the uranium oxide. The voltage is high enough to reduce the UO_2 but not to touch the Li_2O. An advantage is that metallic lithium droplets are not formed. But as the process goes along under controlled current conditions, the voltage increases from a number of causes, and after a few hours both direct and lithium-metal-assisted reduction results.

10.2.2 The Reactions

All the oxide reduction reactions boil down to electrons being supplied to the metal oxide in question, resulting in a product consisting of the metal itself and negative oxygen ions. Depending on the particular process, the oxygen ions may then form another compound—in our case Li_2O, which dissolves in the electrolyte—or have the electrons that give them their negative charge stripped from them at an anode. The oxygen gas formed is then swept away. In either case we are left with the desired metallic product.

We'll use uranium oxide as our example, but all the actinide oxides act similarly. The end result of the necessary series of reactions is simply:

$UO_2 \rightarrow U + O_2$ (gas).

The basic reduction reaction then is:

$$UO_2 + 4e^- \rightarrow U + 2O^-. \qquad (1)$$

And this is exactly the reaction of the direct electrolytic reduction process. The cell potential (voltage) must be set at the correct value to give this reaction. Its magnitude must be enough to initiate this reaction, but not so high that it disassociates any Li_2O present or the electrolyte, LiCl, itself. (All voltages are negative.) At the anode the reaction is:

$$2O^{--} \rightarrow O2 + 4e^-. \qquad (2)$$

Now turning to the electrochemical process where lithium is used, the electrolyte contains 1 w% Li_2O in the LiCl. Here the reaction is:

$$UO_2 + 4Li \rightarrow U + 2Li_2O. \qquad (3)$$

The Li metal is generated at the cathode by setting the voltage somewhat more negative than for direct reduction, at a value that disassociates the Li_2O, but again not so high that it disassociates the electrolyte itself. The reaction at the cathode is:

$$4Li^+ + 4e^- \rightarrow 4Li \text{ metal}. \qquad (4)$$

The pyrochemical process simply supplied Li metal directly, with no electrical supply. Equation 3 above then applied as it stands. The relevant potentials are: [5]

$$
\begin{aligned}
&UO_2 \rightarrow U + O_2 && 2.40 \text{ volts} \\
&Li_2O \rightarrow 2Li + \tfrac{1}{2}O_2 && 2.47 \text{ volts} \qquad (5) \\
&LiCl \rightarrow Li + \tfrac{1}{2}Cl_2 && 3.46 \text{ volts}
\end{aligned}
$$

And for the lithium metal reaction on UO_2, reaction (3) above, the driving force is given by a negative Gibbs free energy change of -6.5 kcal/mol at the 650°C operating temperature. This is an adequate driving force, but not as high as for some other choices possible—calcium, for example. However, lithium fits well with the electro refining step to follow, and was the choice in the fairly extensive development that followed the early survey work.

A variation from the basics above was used by Hermann et al. [5-6] to control the metallic lithium created at the cathode so it stayed in the fuel material of the cathode and carried out the desired reduction, but did not diffuse through the salt to attack and damage the platinum anode. A second circuit with a separate power supply oxidized to Li^+ any Li metal that reached the wall of the basket (the basket wall forming the anode) and then reduced it to Li metal in the spent fuel bulk of the cathode, where it continues to add to the reduction process.

10.2.3 The Energy Relationships

We discussed electrochemical energy relationships in terms of Gibbs free energy changes in the previous chapter. The cell potential, or voltage, is directly related to the Gibbs free energy change by a simple mathematical formula:

$$\Delta G = -nFE,$$

where ΔG is the change in Gibbs free energy in joules (1 joule=0.239 calories), n is the moles of electrons transferred, F is Faraday's constant (98,485 amp seconds/ mole of electrons) and E is cell potential in volts.

So, for example, if n=1, then the free energy change resulting in a one volt cell potential, would be

$$\Delta G = -nFE = -(1 \text{ mol } e^-)(96485 \text{ C/mol } e^-)(1.00 \text{ V}) = -96.5 \text{ kilojoules.}$$

As we have discussed previously, ΔG must be negative for the process to proceed spontaneously, that is, without electrical help. The lithium metal process with its -6.5 kcal/mol ΔG will do so. But the direct reduction process and the creation of lithium metal from dissolved Li_2O require imposed voltages in the 2.5 volt range shown in the equations (5) above.

For the other actinides, and plutonium in particular, there is an added constraint. Although the action of lithium metal on the actinide oxides like PuO_2 has free energies of formation of Li_2O of similar magnitude to UO_2 (and also negative), and thus would be expected to reduce completely to their metals, they may not do so. If the concentration of Li_2O dissolved in the electrolyte is too high, an intermediate oxide compound will form, Pu_2O_3, instead of the reduction going right on to metal. This compound has a positive ΔG for Li_2O formation (about +4.5 kcal/mol), and thus the action of lithium metal on it will not reduce it further. The way around this problem is to avoid forming the compound in the first place. By maintaining the Li_2O concentration in the LiCl salt to less than 3.3 w% throughout the reduction process, the intermediate compound is avoided and the reduction goes directly to the metal, as desired.

It may be noted that the oxides of several of the rare earth fission products have limits on the Li_2O as well, some well below the concentrations necessary for the actinides. To reduce rare earths, the Li_2O concentration needs to be lowered into the 0.1% range. The decomposition potential of Li_2O rises as the concentration is lowered, and these concentrations bring the Li_2O decomposition potential close to those necessary for LiCl decomposition, which could result in undesirable chlorine gas evolution. Where these fission product oxides are not reduced, they remain as oxides in the waste. Fortunately, most of the rare earth fission products are formed as metal in the UO_2 matrix, and therefore do not need to be reduced.

10.2.4. The Equipment

The principle of electrolytic reduction equipment is schematically illustrated in Figure 10-1. At the anode oxygen gas is swept from the cell. The cathode process yields metallic product that will then be electrorefined. In fact, the cathode basket in the reduction step can be directly transferred to the electrorefiner as the anode basket for the next step.

Engineering-scale reduction cells were designed and fabricated, and they are currently used to understand the parameters for process scaleup. In particular the effect of cathode bed thickness and oxygen removal/handling on the extent and rate of reduction are being examined. These cells are operated at a one-kilogram scale using depleted UO_2 feed. The engineering-scale tests will provide electrochemical engineering data needed for design and modeling of pilot-scale cells. So far, the tests have shown significant improvements, ~40% to ~65% in current efficiency, while producing a high quality metallic product. Although most of the cell studies have used a platinum anode, several inert ceramic materials are being evaluated as alternatives to costly platinum as well.

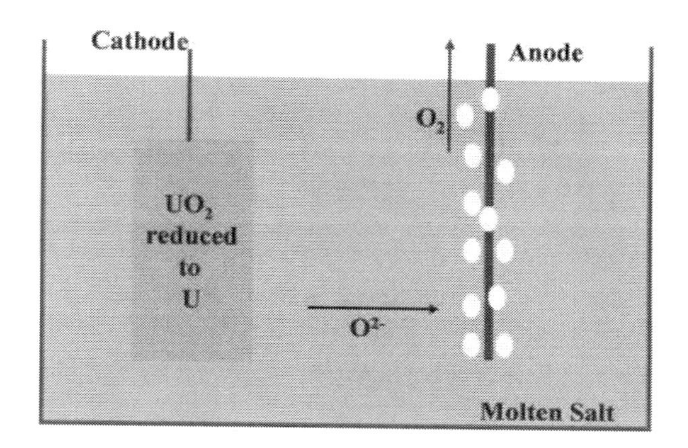

Figure 10-1. Electrolytic reduction process

The process parameters that require additional development have been identified. One is the incorporation of forced flow in the cells, which becomes increasingly important as the scale of cell increases from engineering- to pilot- to plant-scale. It is important to experimentally demonstrate and understand the beneficial effects of forced flow. We want flow through the oxide bed to help move the oxide ions to the anode, but we do not want the flow to entrain oxygen bubbles in the electrolyte which could back-react with product metal. Improvements of this kind are not issues of feasibility. A prototype reduction vessel to handle 250–500 kg has been designed for eventual testing of all such engineering features.

10.2.5 Experience

Electrolytic reduction has been successfully demonstrated with UO_2 and UO_2-5wt% PuO_2 at Argonne-East and with spent light water reactor oxide fuel at Argonne-West. The initial experiments at ANL-E at laboratory-scale used about 20–50 g of heavy metal to demonstrate the extent and rate of conversion. Experimental results indicate complete conversion (i.e. >99.95%) of uranium, plutonium, and americium oxide to metal and the rate is satisfactory as well. The process appeares to be robust and scalable.

Turning now to the results of two series of experiments that established the process, and are representative samples of the whole, we will see that they give a good practical sense of the various features and idiosyncrasies of the process.

The first set is representative of the early stages of development, experiments done at ANL-E aimed at establishing the important principles of direct electrochemical reduction by Gourishankar et al. [3]. The experiments were done with about 15 g of UO_2 chips, from 45 microns to 5 mm in size in most cases. The electrolyte was LiCl–1%Li_2O at 650°C, based on previous experience. The cells were operated at one ampere constant current, a level that maximized oxygen evolution at the anode and prevented chlorine evolution at the cathode. As operation began, there was an initial gradual drop in anode potential. After four hours there was a sharp drop in the cathode potential of 300 to 500 mV, observed in all experiments, as it shifted from direct reduction of UO_2 to the lithium reduction potential.

The combined effects of a metallic surface layer building up on the fuel particles, the surface area of the oxide particles decreasing, and the buildup of oxygen ions in the immediate area of the cathode, causes the voltage to drop down to the lithium deposition potential when constant current is maintained. In this circumstance, lithium metal is generated, and its deposition contributes significantly to the cell current and to the reduction mechanism itself. The balance between the two processes, direct reduction and lithium chemical reduction, depends on factors such as electrical isolation of particles and on the stage of the process. The current must be adjusted to avoid generation of excess lithium at the cathode while maximizing the reduction rates. Lithium generation should therefore match its consumption by the chemical reaction. Designing to allow stirring in the area of the cathode may well be necessary.

The product retained its shape generally, some fines were generated, and a significant amount of electrolyte was entrained. Reduction proceeded from the outer surface inward and a metallic layer of uranium formed on the particle surface, decreasing the process rate. Particle size, it was concluded, will be important.

There are three important factors in operation. Anode potential has to be maintained below the level where chlorine gas is evolved, the cathode potential must avoid lithium droplets and films, and the oxide ion concentrations have opposing important effects on both anode and cathode reactions and must be optimized. As they can be controlled, such optimization is possible.

A second set of experiments with actual spent fuel, representative of later stages in the development, were performed at Idaho National Laboratory. [5] Here again the electrolyte was 650°C LiCl with 1w% Li_2O. A series of ten experiments were done with LWR spent fuel, long out of the reactor. The fuel was from the Belgian reactor BR-3, irradiated in 1979. The fuel was crushed into particles, with batch sizes of about 50 grams. Two power sources were used, with the cathode lead in the center of the cathode fuel mass acing as the negative electrode for both. The voltage of the platinum anode in the primary circuit was kept below the value which would cause platinum to dissolve, and the primary current was controlled to maintain the center lead voltage below lithium formation potential. The secondary circuit anode was the basket wall, and this circuit was activated when the potential on the basket wall indicated lithium formation. In effect, the metallic lithium was pumped from the wall to the center of the fuel mass, where it aids in reduction, and cannot cause other troubles.

The distribution of fuel constituents between the salt and metal phases was largely as expected. The soluble fission products, notably cesium and strontium, which are radioactive fission products both plentiful and radioactively penetrating, clearly accumulated in the salt phase as desired. Rare earths and zirconium were partially reduced; they and the noble metals remained in the fuel basket. The reduction of the actinide oxides was, again, largely as expected. Reduction efficiencies were high; 99.7% for uranium, >97.8% for plutonium, 98.8% for neptunium, and 90.2 % for americium. All in all, the tests showed very positive results, and further tests will be done to explore the separation of the actinides from fission products.

In conclusion, given this work, the principal current need is proof of the scalability of the process, and the necessary equipment is now being assembled at ANL- E.

10.3 Electrorefining Scaleup

The electrorefiner for LWR spent fuel application must handle a much higher throughput rate or a larger batch size. The evolution of the electrorefiner designs for the EBR-II spent fuel treatment indicates that such scaleup in fact is reasonably straightforward. The first electrorefiner installed in FCF, Mark-IV, was patterned after the Mark-III electrorefiner in operation for many years at ANL-East. Mark-III has a 28 in. vessel diameter and two electrode ports, each with an 8 in. diameter. Mark-IV has a 40 in.-diameter vessel and four electrode ports, each with a 10 in. diameter. Mark-IV was intended to operate with two anode baskets and two cathodes in parallel. Each anode basket can contain about 10 kg of chopped pins. This batch size was limited by the criticality constraints because of high enrichment (above 60%) of U-235 in the EBR-II fuel. For the purpose of the demonstration of IFR fuel cycle closure this batch size was adequate although the throughput rate of the Mark-IV electrorefiner was limited to amounts on the order of 10 kg/day.

But to process the large inventory of the EBR-II blanket assemblies, this throughput rate was not adequate, and the Mark-V electrorefiner was designed, constructed and installed in the FCF. It utilized concentric anode-cathode modules in a spare electrorefiner vessel identical to Mark-IV electrorefiner. The anode-cathode module was designed to increase the throughput rate with higher currents. The current cannot exceed the mass transport limits. The mass-transport-limiting current for the anode and cathode reactions is proportional to the electrode's wetted surface area. Shortening the distance between the anode and cathode reduces the cell resistance and improves the throughput rate.

The concentric anode-cathode module illustrated in Figure 10-2 accomplishes most of this. The heavy metal loading in the anode baskets is 100 kg total for four electrode modules, the surface areas of both anode and cathode have been increased, the anode-cathode distance has been reduced, and as a result, the Mark-V throughput has increased to the order of 50 kg/day—a factor of 5 improvement in the same size electrorefiner vessel and with the same size and number of electrode ports. In the Mark-V design, two rings of segmented annular anode baskets rotate between three concentric stationary cathode rings. The cathode deposits are scraped off by beryllia scraper blades mounted on the leading edge of the anode baskets. The scraped deposits fall to the bottom and collected in the product collection basket, emptied periodically.

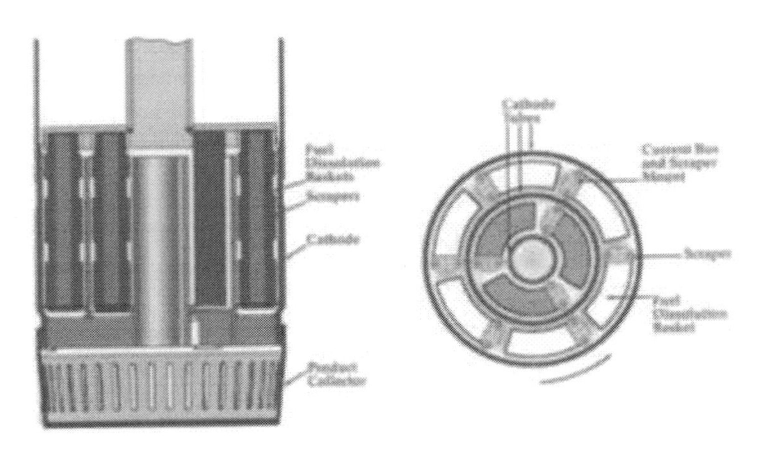

Figure 10-2. Schematic arrangement of Mark-V electrode

For LWR spent fuel processing, where the criticality constraint is considerably lessened, a much higher throughput rate is possible, and indeed is required for commercial viability. The annular anode-cathode module can be scaled up in radius by adding additional rows of concentric rings as well as increasing the height. However, a new approach has been developed that appears more amenable to large batches and higher throughput rates. Parallel planar design incorporates thin rectangular anode baskets stacked vertically, sandwiched with cathode plates or multiple rods. It is essentially an uncoiled concentric anode-cathode module, but allows simplified basket geometry and simplifies scaleup. Cathodes are scraped intermittently. The scraper assembly, motor-driven, is placed above the salt pool during electrorefining. The electrorefining process is halted during the scraping to prevent shorting. The scraped material falls onto a sloping surface inside the salt pool and then slides or is pushed down into a trough located at the bottom of the vessel.

A prototype of the planar electrode concept has been put in operation. It has greater than 90% scraping efficiency under various operating conditions. A high current density deposit in the cathode promotes granular-type deposits which are more readily scraped. With parallel planar electrode arrangement, it is expected that an electrorefiner with a 500–1,000 kg batch size and throughput rate of 250–500 kg/day is possible, which would imply good economics even in the batch operated mode. But it may also be possible to develop continuous processing if that is best for commercial applications.

10.4 Pre-Conceptual Design of Pyroprocessing Facility for LWR Spent Fuel

Before we discuss a pre-conceptual design of a pyroprocessing facility, we should point out that internationally there have been two different pyroprocesses developed for application to the LWR oxide spent fuel. One based on oxide electrorefining is being developed at Research Institute of Atomic Reactors in Dimitrovgrad, Russia. This technology originates from an earlier application of electrorefining UO_2-PuO_2 as a front-end step in fabricating high-density vibrocompacted fuel in the remotely operated automated fuel fabrication facility for the small reactor BOR-60. [7]

In this electrorefining application, the UO_2-PuO_2 feed is dissolved in the NaCl-CsCl electrolyte salt at 650°C by reaction with chlorine in the presence of oxygen. The plutonium valence must be regulated throughout electrorefining to obtain the desired UO_2-PuO_2 ratio in the product by maintaining a controlled chlorine-oxygen composition in the cover gas. The UO_2-PuO_2 is deposited on a pyrographite cathode and falls off readily when the cathode is cooled. The deposit is then washed with water before being crushed and sized for vibrocompacting. This technique results in a high-density fuel, actually denser than pelletized fuel, and its irradiation performance has been excellent.

Although this application as the front-end step in vibropac fabrication has been demonstrated successfully since the late 1970s, the experiments of spent fuel processing application began only in the late 1980s and progress has been slower. The kinetics and thermodynamics of the oxychloride system composed of the spent fuel constituents are far more complicated than in a simple binary system of UO_2-PuO_2. Nevertheless, because of its apparent compatibility with MOX fuel, Japan Atomic Energy Agency (JAEA) selected this process as an option to evaluate in Phase II of their "Feasibility Study on Commercialized Fast Reactor Cycle Systems." [8]

The other approach is ours, of course: converting the oxide spent fuel to metallic form and then taking advantage of the metal electrorefining process successfully demonstrated at engineering-scale as part of the EBR-II spent fuel treatment project. Electrolytic reduction operates with a potential in the range of three volts, compared to less than one volt in electrorefining, and the throughput rate in the reduction step is expected to be much higher than that of electrorefining itself. Once electrolytic reduction is developed fully, the oxide reduction step is not expected to add much to the oxide processing cost over metal fuel processing. Electrorefining is fundamentally more amenable to batch processing than continuous processing, but this does not necessarily mean that scaling up involves multiple units. As discussed above, a single electrorefiner, incorporating a planar electrode arrangement, can be designed for a 500 kg or even 1,000 kg batch size.

A pre-conceptual design for a pilot-scale pyroprocessing facility (100 ton/yr throughput) for LWR spent fuel has been developed at Argonne. [9] Assuming a 500 kg/day throughput rate and 200 days of operation leads to an annual throughput rate of 100 metric tons, the basis of the pre-conceptual design. A bird's-eye view of the process equipment layout is presented in Figure 10-3, and the floor plan in Figure 10-4. Two electrorefiners with a 500-kg batch capability each were provided, assuming the process time will be more than 24 hours, to be conservative. Only one electrolytic reduction vessel is assumed because of its higher throughput rate than that of an electrorefiner. The cathode baskets of the electrolytic reduction vessel are designed to be transferred, with their reduced metallic spent fuel contents, directly into the electrorefiner to be used as its anode baskets.

Figure 10-3. Bird's eye view of a 100 ton/yr pyroprocessing facility for LWR spent fuel

A more detailed engineering plan is required for a truly reliable cost estimate. However, a preliminary analysis indicates that even a pilot-scale pyroprocessing facility for LWR spent fuel should be economically viable. Further scale up can be done by duplicating the process equipment systems, and some economies of scale will come in. It is plausible that a pyroprocessing facility with, say an 800 ton/yr throughput, could be constructed far below the capital cost of an equivalent-size aqueous reprocessing plant. The economics potential of pyroprocessing facilities will be discussed in more detail in our later chapter on economics.

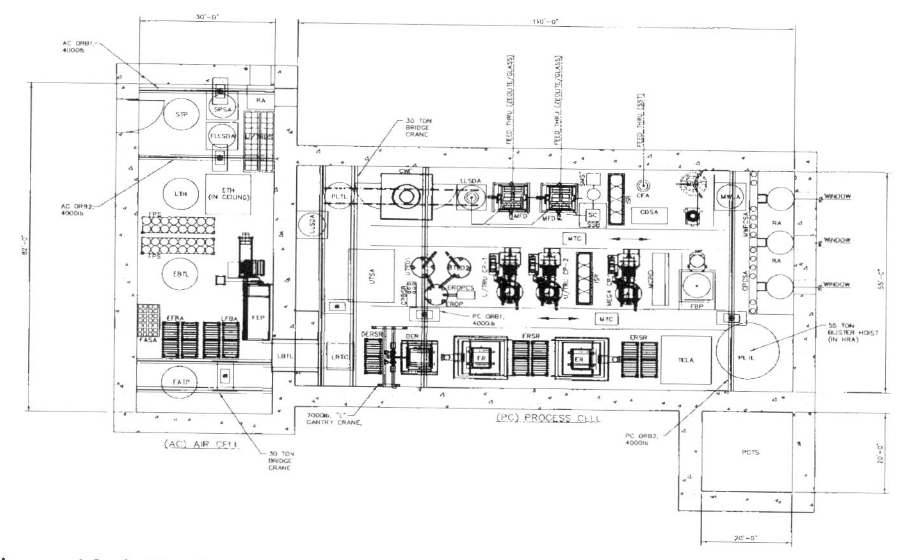

Figure 10-4. Equipment layout for a 100 ton/yr pyroprocessing facility for LWR spent fuel

10.5 Pyroprocessing Activities in Other Countries

We have mentioned the different type of pyroprocessing activity being carried out in Russia, developed originally as a front-end step of the vibropac fabrication technique. Whether oxide electrorefining through oxychloride can be successfully developed for the recovery of actinides from oxide spent fuel has not yet been demonstrated. But success will be a valuable contribution to pyroprocessing processes and their understanding.

France is committed fully to the development of advanced aqueous reprocessing technologies as the next generation process to recover all actinides, and only a small effort is allocated to basic chemistry work on pyroprocessing.

In Japan, due to early interests in pyroprocessing by the utility industry, CRIEPI still maintains an expert cadre and a laboratory infrastructure for pyroprocessing. However, the main thrust of the national development activities is on advanced aqueous process development activities and oxide fuel, in a somewhat different direction than the French.

India and China have started showing interests in pyroprocessing and modest R&D programs have been launched in those countries.

Republic of Korea has by far the most ambitious program for pyroprocessing development. Republic of Korea now has twenty-one reactors in operation, an additional seven reactors under construction, and their current energy plan calls for a 39 GWe nuclear capacity by 2040 to contribute 60% of the nation's electricity. Most of the currently operating reactors will have their spent fuel storage capacity filled within the next several years, and as a long-term solution, they are considering implementation of pyroprocessing, followed by sodium-cooled fast reactors with metal fuel. [10]

The Korea Atomic Energy Research Institute (KAERI) is the site of Korea's pyroprocessing technology development. It is based on the Argonne processes, but they have also introduced additional innovations such as a graphite cathode, more amenable to continuous electrorefining, and a crystallization method for removal of Cs and Sr in waste treatment. [11-12] Their current activities are focused around an engineering-scale integrated-process demonstration facility, called PRIDE (PyRoprocess Integrated inactive DEmonstration facility) which has a capacity of ten tons of uranium per year using inactive simulated fuel, to be constructed by 2016. [13]

10.6 Summary

Present-day reactor fuel from commercial LWRs is an oxide. The IFR process could be adapted to treat such fuel. The radioactive fission products could be put in impermeable long-lasting containment, with the very long-lived actinide elements in a small volume suitable for later reactor recycle and the uranium separated for later reuse in reactors as well. This would provide the basis for a tractable and cost-effective nuclear waste policy.

There have been three closely related processes based on lithium metal acting on the oxides to reduce them to metals. All the oxide reduction reactions boil down to electrons being supplied to the metal oxide in question, resulting in a product consisting of the metal itself and negative oxygen ions.

The first process was direct pyrochemical reduction by lithium metal in LiCl molten salt at 650°C, with no electrical input and removal of the problematical Li_2O so formed in a second vessel. The process worked, but it was improved in introducing electricity and by combining the two-vessel operation into one.

In the successor process, the molten LiCl electrolyte is seeded with a small amount of Li_2O, which ionizes to give initial current carrying capability. The metallic oxides are reduced at the cathode, and the oxygen ions so produced convert to oxygen gas at the anode and are removed. In this way the Li_2O concentration is controlled. The voltage is selected to convert the Li^+ ions of the dissociated Li_2O to

Li metal at the cathode. Lithium metal produced in this way gives the same reaction as the pyrochemical reduction process described above, but the source of it now is reduced Li_2O.

The most recent process variation sets the voltage high enough for direct electrolytic reduction of the uranium oxide but low enough not to touch the Li_2O. As the process goes along under controlled current conditions, the voltage increases from a number of causes, and after a few hours both direct and lithium-metal-assisted reduction results.

All these oxide-reduction reactions boil down to electrons being supplied to the metal oxide, resulting in a product consisting of the metal itself and negative oxygen ions. Depending on the particular process, the oxygen ions may then form another compound, in our case Li_2O, which dissolves in the electrolyte, or else the electrons that give the ions their negative charge are stripped from them at an anode. Oxygen gas is evolved that is then swept away. In either case we are left with the desired metallic product.

Electrolytic reduction has been demonstrated successfully with UO_2 and UO_2-5wt% PuO_2 at Argonne-East and with spent light water reactor oxide fuel at Argonne-West. The initial experiments at ANL-E at laboratory-scale used about 20–50 g of heavy metal to demonstrate the extent and rate of conversion. Experimental results indicate complete conversion (i.e. >99.95%) of uranium, plutonium, and americium oxide to metal, at a satisfactory rate. The process appeares to be robust and scalable. The principal need now is proof of the scalability of the process, and the necessary equipment for this is now being assembled at ANL- E.

For LWR spent fuel processing, where the lower fissile content makes criticality considerations much less constraining, a much higher electrorefining throughput rate is possible and is required for commercial viability. The existing annular anode-cathode module can be scaled up in radius by adding additional rows of concentric rings as well as increasing the height. However, a new approach more amenable to large batches and higher throughput rates has been developed. It is based on a parallel planar design with thin rectangular anode baskets stacked horizontally, sandwiched with cathode plates or multiple rods. It is essentially an uncoiled concentric anode-cathode module, but allows simplified basket geometry and simplifies scale up. Cathodes are scraped intermittently. The motor-driven scraper assembly is situated above the salt pool during electrorefining. For scraping, the electrorefining process is halted to prevent shorting. The scraped material falls onto a sloping surface inside the salt pool and then slides or is pushed down into a trough located at the bottom of the vessel.

A pre-conceptual design for a pilot-scale pyroprocessing facility (100 ton/yr throughput) for LWR spent fuel was developed at Argonne. Assuming a 500 kg/day throughput rate and 200 days of operation leads to an annual throughput rate of 100 metric tons, which was the basis of the pre-conceptual design. Two electrorefiners, with a 500-kg batch capability each, were provided, assuming the process time will be more than 24 hours (to be conservative). Only one electrolytic reduction vessel is assumed because of a higher throughput rate than that of an electrorefiner. The cathode baskets of the electrolytic reduction vessel containing reduced metallic spent fuel are designed to be transferred directly into the electrorefiner as its anode baskets.

A more detailed engineering demonstration and operation is required for a really solid basis for the cost estimate. However, a preliminary analysis indicates that even a pilot-scale pyroprocessing facility for LWR spent fuel would be economically viable. Any further scale up can be achieved by duplicating the process equipment systems and some economies of scale, where they come in. It is plausible that a pyroprocessing facility with, say an 800 ton/yr throughput, could be constructed at far below the capital cost of an equivalent size aqueous reprocessing plant.

The Argonne pyroprocessing program has piqued some interest in several countries, where modest research programs have been initiated. Republic of Korea has the most ambitious such program and has made significant progress in development.

There is real incentive to process the spent fuel from existing reactors to make it much easier to deal with, as well as to reuse the useful portion, and in so doing, to extend fuel resources by more than a hundredfold over the once-through-and-throw-away fuel cycle. With the Obama administration cancelling funding for the Yucca Mountain waste repository in 2010, there is literally no plan or process in place now in the U.S. to deal with LWR spent fuel. It will continue to build up at nuclear plants around the country, not wise as recent earthquake events in Japan may indicate.

References

1. K. V. Gourishankar and E. J. Karell, "Application of Lithium in Molten-Salt Reduction Processes," *Light Metals 1999*, ed., C. Edward Eckert, 1123-1128, The Minerals, Metals & Materials Society, 1999.
2. K. V. Gourishanker, L. Redey and M Williamson, *Light Metals 2002*, ed. Wolfgang Schneider, The Minerals, Metals & Materials Society, 1075-1082, 2002.
3. E. J. Karell, K. V. Gourishankar, J. L. Smith, L. S. Chow, and L. Redey, "Separation of Actinides from LWR Fuel using Molten Salt Based Electrochemical Processes," *Nuclear Technology*, 136, 342-353, 2001.

4. D. W. Dees and J. P. Ackerman, "Three-Electrode Metal Oxide Reduction Cell," U.S. Patent No. 6911134B2, June 28, 2005.
5. S. D. Herrmann, S. X. Li, M. F. Simpson, and S. Phongikaroon, "Electrolytic Reduction of Spent Nuclear Oxide Fuel as Part of an Integral Process to Separate and Recover Actinides from Fission Products," *Separation Science and Technology*, 41, 1965-1983, 2006.
6. S. D. Herrmann, S. X. Li, D. A. Sell, and B. P. Westfall, "Electrolytic Reduction of Spent Nuclear Oxide Fuel—Effects of Fuel Form and Cathode Containment Materials on Bench-Scale Operations," *Proc. Global 2007, Advanced Nuclear Fuel Cycles and Systems*, Boise, Idaho 2007.
7. V. Skiba, et al, "Development and Operation Experience of the Pilot Plant for Fuel Pin and Assembly Production based on Vibropac Uranium-Plutonium Oxide Fuel," *International Conference on Fast Reactors and Related Fuel Cycles*, Kyoto, Japan, October 28-November 1, 1991.
8. Y. Sagayama, "Feasibility Study on Commercialized Fast Reactor Cycle Systems: Current Status of the Phase II Study," *Proc. GLOBAL-2005*, Tsukuba, Japan, October 9-13, 2005.
9. A. Frigo, D. R. Wahlquist and J. L. Willit, "A Conceptual Advanced Pyroprocess Recycle Facility," *Proc. GLOBAL 2003*, New Orleans, November 2003.
10. S. W. Park, "Why South Korea Needs Pyroprocessing," *Bulletin of the Atomic Scientists*, October 26, 2009. http://thebulletin.org/node/7982.
11. H. Lee, et al., "Development of Pyroprocessing Technology at KAERI," *Proc. Global 2009*, 908-911, Paris, France, September 6-11, 2009.
12. E. H. Kim, et al., "A New Approach to Minimize Pyroprocessing Waste Salts through a series of Fission Product Removal Process," *Nucl. Technol.*, 162, 208, 2008.
13. K. C. Song, et al., "Status of Pyroprocessing Technology Development in Korea," *Nucl. Eng. and Technol.*, 42, 131, 2010.

CHAPTER 11

IMPLICATIONS OF THE IFR PYROPROCESS ON WASTE MANAGEMENT

It is a commonplace that radioactive waste management is an issue of first importance in the acceptance of nuclear power. The "back end" of the fuel cycle was identified early by the organized anti-nuclear community as being vulnerable to attack through extended legal wrangling, demonstrations, and finally legislation. A repository—of some kind—is necessary; spent fuel must not be allowed to keep on piling up at reactor sites around the country forever. Indeed, the very purpose of the anti-nuclear attacks on waste disposal is to choke off nuclear power by making it increasingly difficult to deal sensibly with the waste products. Yet the need for nuclear power is becoming more obvious day by day, as explained in Chapter 4. Nuclear waste has to be dealt with, and it should be possible to do it in a sensible way.

For fuel sustainability, the fast reactor is mandatory if nuclear is to play the role it must in the future energy mix. Waste management advantages, such as a waste lifetime drastically reduced from millions of years to a few hundred years, and much enhanced repository capacity utilization, are attendant bonuses of fast reactor introduction. If nuclear capacity expands to replace the current role of fossil fuels, fast reactors will be needed primarily to conserve the uranium resource, but will solve the waste disposal problem simultaneously. Reprocessing of LWR spent fuel becomes economic simply due to the recovery of valuable actinides. The long-lived waste component is eliminated automatically. One thing follows from the other. But the present situation is very different.

Reflecting the very different goals of proponents and opponents, there is a long and tangled legislative history of repository licensing. It has given standards for radiological release, legally binding, but still under litigation. Meeting the standards makes the licensing of Yucca Mountain, or indeed any repository, very uncertain indeed. The action of the present (Obama) administration in withdrawing the license application of the Yucca Mountain repository, of course, very much compounds the uncertainty surrounding the issue.

Our intent for the IFR technology was to remove the substances from the waste that are the only significant contributors to possible radioactive release in the long term. The goal: to extract them from the waste, recycle them back into the reactor, and destroy them. The IFR process does this naturally, without extra expense or increase in process complexity. In so doing, the recycled elements—the actinides—all fission and themselves produce power. Without these long-lived elements in the waste, the standards that are now difficult to meet no longer are. A path opens to a satisfactory, licensable repository, in a reasonable time.

11.1 Legislative Background

The Nuclear Waste Policy Act (NWPA) of 1982, amended in 1987, made the development of repositories for the disposal of high-level radioactive waste and spent nuclear fuel a federal responsibility. [1] This followed directly from a U.S. policy, laid down by the Carter administration, that spent fuel would be disposed of whole, without treatment of any kind.

In the NWPA there was provision for Monitored Retrieval Storage (MRS), but no MRS could be constructed until the NRC had issued a license for the construction of the permanent repository. This was certain to be a very long time in coming. A further provision prohibited any MRS from receiving spent fuel if the permanent repository license was revoked or its construction ceased. Intended or not, the effect was clear: spent fuel would simply accumulate at the operating nuclear plants. The events at Fukushima, where the spent fuel pools became a major source of radioactivity release at the site, made real the fecklessness of the implementation of this legislation.

The NWPA was received enthusiastically by utilities. Utilities would pay a fee of one mill/kWh—a tenth of a cent per kWh—and for this, the entire responsibility for the spent fuel would be transferred to the federal government. As the corresponding costs for the back end of the fuel cycle even at that time were estimated to be three or four mills/kWh or more, the disposal fee was a great bargain.

The original legislation mandated parallel exploratory activities at three different potential repository sites: Yucca Mountain in Nevada, Hanford in Washington, and Deaf Smith County in Texas. A 1987 amendment mandated that evaluation be limited to only one site, Yucca Mountain; the other two would be dropped.

The Nuclear Waste Fund Contract in the original legislation stated, in part: "Following commencement of operation of a repository, the Secretary shall take title to the high-level radioactive waste or spent nuclear fuel involved as expeditiously as practical…; and in return for the payment of fees established by this section, the Secretary, beginning not later than January 31, 1998, will dispose of the high-level radioactive waste or spent nuclear fuel involved as provided in this subtitle."

When the legislation was formulated, the repository was estimated to open by January 31, 1998. The clause went unnoticed in the 1987 amendment. In the absence of a repository (even today, more than a decade after this planned date), the utility industry filed lawsuits holding the DOE to this date for transfer of the title to the spent fuel. Seventy-two different lawsuits were brought against the DOE for its failure to begin accepting spent fuel as stated in the Act.

One utility, Exelon Corp., has dropped the breach-of-contract litigation against the DOE, and settled for reimbursement of spent fuel storage costs it has incurred, and will incur, as a result of DOE inaction. [2] Eighty million dollars to cover past storage costs and additional annual reimbursements to cover future spent fuel storage costs, estimated at as much as $300 million, were agreed to. Other utilities began following suit. It was assumed that Yucca Mountain would open in 2010, as then planned. With Yucca Mountain now "off the table," in the words of the current administration, costs must keep accumulating.

Twenty-nine years after the NWPA was enacted, there is still no operating repository in the U.S. However, no other countries, including major nuclear energy electricity generating countries like France and Japan, have decided on their repository sites either, or even on their ultimate technology options for disposal. In retrospect, the NWPA in 1982, specifying as it did a firm course of action for dealing with spent fuel, may have been ahead of its time. One thing is sure, however: in implementing better technology for nuclear waste disposal, the NWPA's agreement of a one mill/kWh fee for transfer of spent fuel to DOE will be a stumbling block. Consensus will be difficult when utilities by law simply transfer their spent fuel to the government for this minimal fee. Any alteration in present policy will have to deal with this issue—and it will not go away if ignore it.

11.2 Repository Regulatory Background

The standards and regulations for the repository have been controversial as well. Responsibility was divided. The Environmental Protection Agency (EPA) was mandated to promulgate the standards for releases of radioactive material, and the Nuclear Regulatory Commission (NRC) to promulgate technical requirements and criteria for licensing consistent with the EPA standards.

The EPA issued standards in 40CFR Part 191, "Environmental Radiation Protection Standards for Management and Disposal of Spent Nuclear Fuel, High Level, and Transuranic Radioactive Wastes". [5] Under this standard, the yearly dose to any member of the public must be less than 15 mrem (0.15 mSv) for ten thousand years. Containment requirements are specified in terms of cumulative releases, also over ten thousand years. The limits are stated in terms of probability: less than one chance in ten of exceeding the limits and less than one in a thousand of exceeding ten times the limits.

But even before the EPA standards were issued, in 1983 the NRC issued 10CFR Part 60, Disposal of High-Level Radioactive Wastes in Geologic Repositories. It defines licensing requirements, siting criteria, performance objectives, and the performance requirements of engineered barrier systems. [6] The two most important performance requirements are: a) containment of high-level wastes in the waste packages will remain "substantially complete" for a period determined by NRC (not less than three hundred years nor more than a thousand years); and b) following the containment period, the release rate of any radionuclide from the engineered barrier system shall not exceed one part in 100,000 per year.

As these were inconsistent with the EPA standards, the NRC issued draft revisions of 10CFR Part 60 that were consistent with EPA standards, but eventually the revisions were dropped. The EPA standards themselves, originally issued in 1985, were litigated and ultimately reissued with some minor revisions in December 1993.

The Energy Policy Act of 1992 [7] mandated that the EPA promulgate standards applicable specifically to the Yucca Mountain repository, based upon and consistent with the findings and recommendations of the National Academy of Sciences. It also required the NRC to modify its technical requirements and criteria to be consistent with the new EPA standards. Basically, the Energy Policy Act of 1992 sent 40CFR Part 161 and 10CFR Part 60 back to be made applicable to the Yucca Mountain repository.

The National Academy of Sciences (NAS) special committee on the Technical Bases for Yucca Mountain Standards issued its final report in 1995. [8] The NAS study recommended a risk-based standard as opposed to the EPA's approach of release limits. However, the NAS study did not make any recommendation as to what level of risk is acceptable. Its basis was that this is not a question of science but one of public policy.

However, in response to a specific question in the Energy Policy Act of 1992, the NAS study recommended adopting the International Commission on Radiological Protection (ICRP) maximum annual effective dose of 100 mrem (1 mSv) from all anthropogenic sources other than medical exposures, typically allocating 10 to 30 mrem (0.1 to 0.3 mSv) to high-level waste disposal. Another significant recommendation of the NAS study was that the compliance with the standard be assessed at the time of peak risk, which is tens to hundreds of thousands of years—even further into the future than the EPA's ten thousand years.

Following these recommendations, the EPA promulgated standards for Yucca Mountain in 2001 as 40CFR Part 197 [9], including the following:

> "The DOE must demonstrate, using performance assessment, that there is a reasonable expectation that, for 10,000 years following disposal, the reasonably maximally exposed individual receives no more than an annual committed effective dose equivalent of 150 microsieverts (15 millirem) from releases from the undisturbed Yucca Mountain disposal system."

The NRC then followed, promulgating its regulation in 10CFR Part 63 [10], consistent with the new EPA standards. It also had the following additional provision to complement the individual protection standard:

> "DOE must calculate the peak dose of the reasonably maximally exposed individual that would occur after 10,000 years following disposal but within the period of geologic stability. No regulatory standard applies to the results of this analysis; however, DOE must include the results and their bases in the environmental impact statement for Yucca Mountain as an indicator of long-term disposal system performance."

Then, in response to a lawsuit filed by the State of Nevada, the U.S. Court of Appeals for the District of Columbia ruled on July 9, 2004 that the EPA violated federal law in issuing regulations that require the DOE to show that the facility can meet ground protection requirements for at least ten thousand years. The court said the agency was required by law to follow National Academy of Sciences recommendations calling for a much longer compliance period. [11]

Based on this court ruling, EPA amended 40CFR Part 173 in October 2008, incorporating compliance criteria applicable at two different time frames: 15 mrem (0.15 mSv) per year for the first ten thousand years and 100 mrem (1 mSv) per year at times after ten thousand years but within the period of geologic stability (up to a million years). Subsequently, NRC amended 10CFR Part 63 in March 2009 to be consistent with the EPA standards, adopting the criteria applicable at two different time frames.

All this illustrates the tortuous path for licensing of a repository—Yucca Mountain or any other. Now that the Yucca Mountain has been abandoned by the government, the future course for spent fuel disposition is further in jeopardy. Releases, tiny as they are, are difficult to specify with exactness many years into the future; the times we are dealing with are all longer than any recorded human history—incomprehensible, really. But one thing is both clear and very important: *if the source of the radiation is made small, such standards become moot.* And the

IFR process, as we shall see, does just that. It removes or very greatly reduces the source of radiation for times far in the future, and makes the prediction of the containment behavior of a repository meaningful: The time over which containment must be strictly maintained drops into the few-hundred-year range, well within the capability of engineering practice—and even within the capabilities of the practices of millennia ago, as the continued presence today of ancient temples, tombs and cathedrals can testify.

11.3 Radioactive Life of Spent Fuel

Light Water Reactor (LWR) spent fuel consists of fission products (elements created by neutrons captured by the uranium and plutonium nuclei that resulted in fission), actinides (elements created by neutron capture in uranium that did not cause fission), and uranium remaining as it was when it went into the reactor. Uranium is the largest fraction by far, representing well over 90 percent of the bulk of spent fuel.

Three or four percent of the heavy metal (uranium and plutonium, principally) will have fissioned. Of the hundreds of fission product isotopes that result, most decay rapidly to stable (non-radioactive) isotopes. Two very significant exceptions are Sr-90 and Cs-137, which have twenty-nine and thirty-year half-lives, respectively, and which decay by emitting very energetic gamma rays. They are generally the two fission products of most concern. There are very long-lived fission products also, Tc-99, I-129, and Cs-135, which decay with low energies by the expulsion of low energy beta particles (electrons or positrons), the least damaging of radioactive emissions. As a general guide, the longer the life of the radioactive fission product, the lower the energy is of its radioactive emission.

About 1 percent of the heavy metal will have captured a neutron, undergone subsequent decay, and thus have been transmuted into higher actinides (also called transuranic isotopes, or TRU) which will not have fissioned. Typically their half-lives are long, some stretching into hundreds of thousands of years. Their decay is often by alpha particle emission and they are an ingestion hazard, even in fairly small amounts. Alpha particles are helium nuclei (two protons and two neutrons), heavy as particles go on the atomic scale, and although their range is short and penetrative ability low, their considerable energy is harmful particularly to internal tissue (the skin is sufficient to block external alpha emitters).

Uranium, about 95% of the bulk, does not contribute much to the radiological toxicity of the spent fuel. Uranium isotopes found in nature have extremely long half-lives (4.5×10^9 years for U-238 and 7×10^8 years for U-235), and thus emit relatively little radioactivity.

The radiological toxicity of typical LWR spent fuel over the years is shown in Figure 11-1. Radiological toxicity here is a relative measure of the cancer risk if ingested or inhaled, which we have normalized to that of the natural uranium ore. As mined, the ore contains uranium along with daughter products that have accumulated by its slow decay over the millennia. That is the form of uranium in nature, as it is accumulated in deposits and distributed in the earth and water (uranium compounds are soluble) all over the globe.

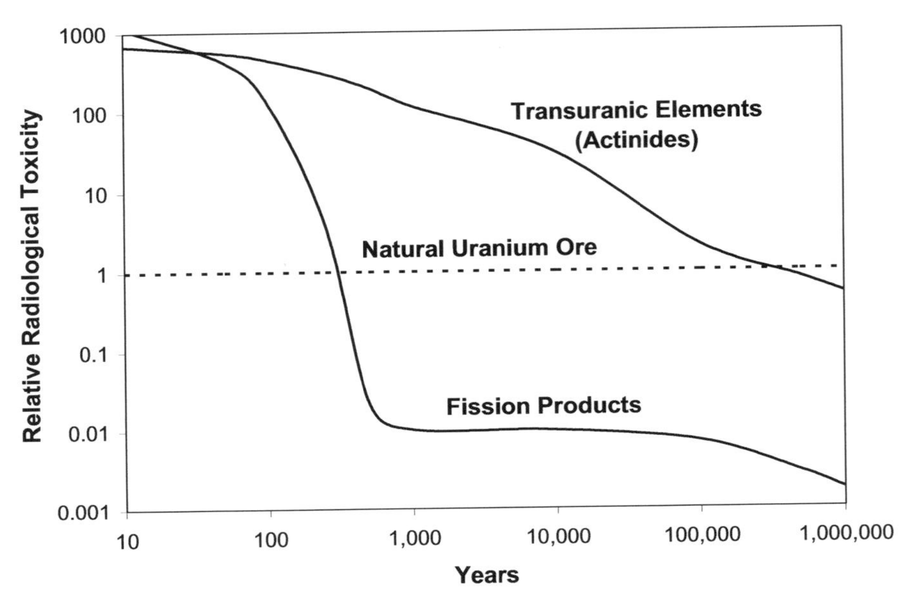

Figure 11-1. Relative radiological toxicity of spent fuel constituents

The normalization to the natural uranium ore from which the spent fuel originated is the standard we have chosen. If the radiological toxicity drops below the natural uranium ore level, radioactive nuclear waste presents no greater hazard than the ore had in nature. The radiological toxicity curve that crosses the natural uranium line can then be loosely defined as an effective lifetime of the waste components.

The radiological toxicity due to the fission product portion of the waste decays with the thirty-year half-life expected from the dominance of strontium and cesium. It drops below the natural uranium ore level in about three hundred years (ten half-lives), and becomes relatively harmless (by two orders of magnitude) in less than a thousand years. On the other hand, the toxicity level associated with the actinide portion stays far above that of natural uranium ore and remains at least three orders of magnitude greater than fission products for hundreds of thousands of years. If 99.9% of actinides are removed from the waste form (see previous chapter), then the radiological toxicity of the remaining 0.1% actinides stays below the natural uranium ore *at all times* and the effective lifetime of the waste is dictated by the fission products.

In today's commercial reprocessing based on PUREX, only uranium and plutonium are recovered. The minor actinides are disposed as waste along with other fission products. The recovered plutonium is recycled as plutonium-uranium mixed oxide (MOX) fuel only once, and its spent fuel is then stored for future recycling in fast reactors. Plutonium is not a particularly good fissile material in a thermal spectrum and its recycle value is limited, for reasons further explained in the next section. In single pass MOX recycle, about one third of plutonium can be fissioned in net, but the remaining plutonium (most originally are Pu-239) evolves into a mixture containing even higher mass plutonium isotopes and the heavier actinides such as americium, neptunium, and curium. As a result, the radiological toxicity is almost unaffected by MOX recycle. This is illustrated in Figure 11-2.

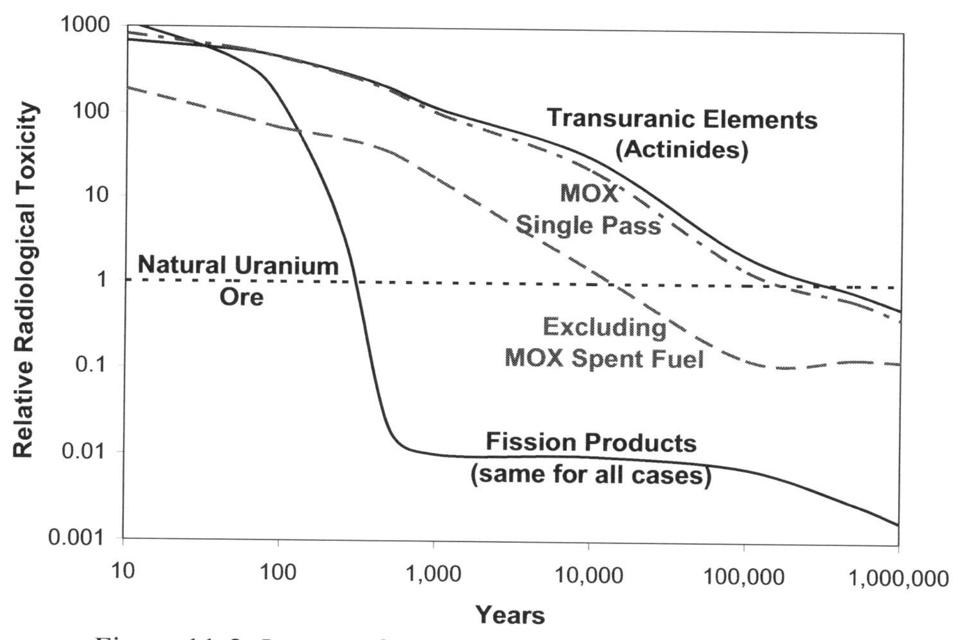

Figure 11-2. Impact of MOX recycle on radiological toxicity

The radiological toxicity of the original spent fuel is transferred whole to the MOX spent fuel. MOX recycle is often mistakenly claimed to reduce the radiological toxicity and effective radiological lifetime by a factor of ten. In making this claim, the actinides that were in the original spent fuel and are now contained in the MOX spent fuel are ignored. This case, where the toxicity of the spent fuel of the MOX is ignored, is also illustrated in Figure 11-2. The justification for doing so is that the MOX spent fuel need not be dealt with, as it will ultimately be recycled in fast reactors. The same claim could of course be made of the original spent fuel, with no MOX recycle. Nothing is added or subtracted by recycle. Irradiation in a thermal spectrum changes the amounts of particular actinides but doesn't affect the overall actinide content much.

The degree of dominance of the actinides in long-term radiological risk is vividly illustrated by normalizing the various components of the radioactive inventory to an easily understood release limit, and noting the degree to which each exceeds it. The cumulative release limits in the original 40CFR Part 191 no longer apply to Yucca Mountain, but they are easily understood and convenient to use for illustration. Typical LWR spent fuel radioactivity levels at three different time points are presented in Table 11-1 in terms of these EPA cumulative release limits. (That is to say, the components of the radioactive inventory in spent fuel are divided by the cumulative release limit.) The ten-year time point represents the state of the spent fuel ready to be packaged for disposal. The thousand-year time point represents the period after short half-life fission products have decayed away. The ten-thousand-year time point represents the time period of the EPA cumulative release limits.

In Table 11-1, all actinides, or transuranic isotopes, are listed together, and "Other Fission Products" are the long-lived isotopes not listed separately in the table, like Cs-135, Sn-126, Zr-93, Nb-93, Pd-107, and so on.

Table 11-1. LWR spent fuel radioactivity normalized to EPA cumulative release limits

Radio-nuclide	Activities at 10 yrs	Activities at 1,000 yrs	Activities at 10,000 yrs
Sr-90	60,000	0.0	0.0
Cs-137	90,000	0.0	0.0
I-129	0.3	0.3	0.3
Tc-99	1.4	1.4	1.4
Other F.P.	1,050	5.1	4.4
Actinides	76,000	19,000	4,000

The ten-year activities are dominated by Sr-90 and Cs-137. Decaying with their approximate thirty-year half-life, they are gone in a thousand years. At one thousand years and longer, only long-lived fission products and actinides are important. It is important to note that the inventories of Tc-99 and I-129, which are readily dissolvable in ground water and thus considered more likely to be released from the repository, are of the same magnitude as the cumulative release limits. If their entire inventory were released, the release limit (The cumulative release limits in the original 40CFR Part 191) could still be met. This is also demonstrated by the total system performance assessments for Yucca Mountain [12], which list the doses from Tc-99 and I-129 equilibrating around 2 mrem/yr, and 0.05 mrem/yr, respectively in the hundred-thousand-year time frame, well below the 15 mrem/yr limit specified in 40CFR Part 197.

It is obvious that the actinides completely dominate as the source of radioactivity at times even approaching a thousand years. Because the solubility of actinides in ground water is extremely low, they will not be readily released from the repository. Their toxicity is three orders of magnitude above that of other contributors shown in Table 11-1, and that tells the story. *If the actinides were removed from the spent fuel, the EPA standards, whether adopted from 40CFR Part 191 or 40CFR Part 197, or whether for the thousand years or millions of years, could be met on a priori basis.* Needless to say, this is an extraordinarily important conclusion. *And the actinides can be burned effectively only in fast reactors.* This will be discussed in detail in the next section.

As a repository design matter, strict limits have been placed on repository temperatures to assure the integrity of the barriers to radioactive release. Here again, the actinides are the dominant source, this time of heat. The decay heat in the spent fuel is plotted in Figure 11-3 for actinides and fission products.

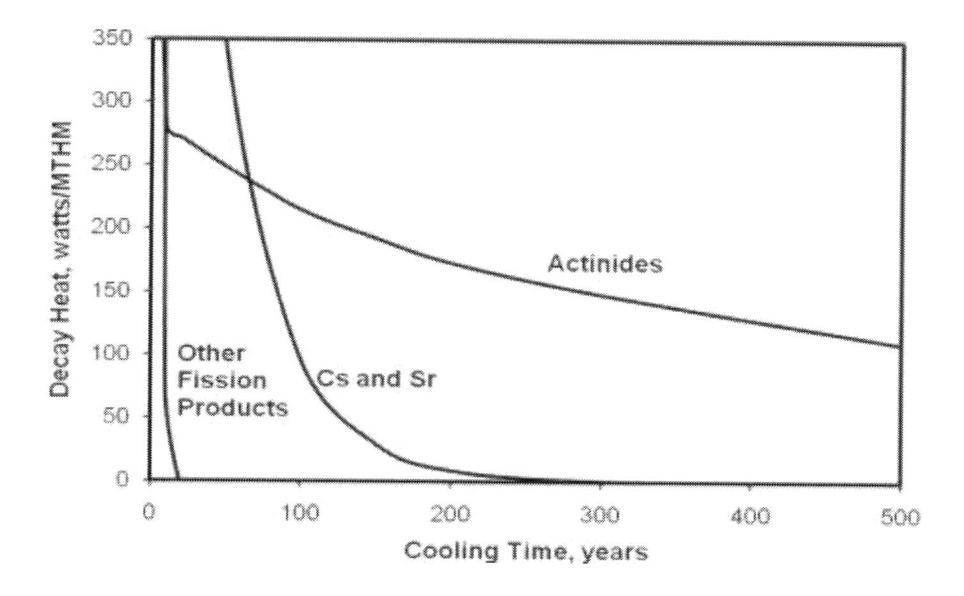

Figure 11-3. Components of spent fuel heat as years pass in watts per metric ton of heavy metal

Limits are placed on "near field" and "far field" temperatures. For "near field" —that is, the area within the drifts (tunnels)—the waste form centerline temperature is to be kept below the melting temperature. The waste package container temperature is to be kept low (say, below 100°C) because corrosion increases with increasing temperature. The drift wall temperature may have to be maintained below the boiling point of water to reduce uncertainties associated with thermal-hydrologic and thermal-mechanical processes at higher temperatures. [12] This would force a "cold repository design," which in turn would necessitate forced ventilation for an initial period up to three hundred years. In this mode, actinide removal does not reduce the heat loading by a large factor (less than factor of two, depending on the cooling time). However, actinide removal would lessen the required forced ventilation period and provide some flexibility managing the thermal loading.

"Far field" temperatures, such as at the midpoint between drifts, peak at times of several hundred to a thousand years. If their limits are the constraint on loading, removal of actinides will have a very significant impact on improving the repository capacity.

In summary, actinides present at the level they are in unprocessed spent fuel completely dominate the long-term radiological hazard. The IFR process removes the actinides from the waste, to a purity level at present a fraction of a percent. This completely eliminates the difficulty in meeting radioactive release standards at times beyond a few hundred years.

But now—turning back to the ninety-nine-plus percent of the actinides—they recycle back into the IFR. Exactly what happens to them there is now our next subject.

11.4 Actinide Transmutation

Actinide transmutation, from actinide to non-actinide, or "actinide burning," can be done by fission and by fission only. Neutron capture in the actinides without fission results only in their evolution to other actinides of ever higher mass, as a rule more and more radioactive. To burn actinides effectively, high-energy neutrons are needed. There is a huge difference between the transmutation possible from the low energy neutrons of thermal spectrum of an LWR and the high energies of the fast spectrum of an IFR. The transmutation probability, the percentage of neutrons absorbed that cause fission, of typical thermal and fast spectra for the actinide isotopes are compared in Table 11-2.

In a thermal spectrum, only a limited number of isotopes fission effectively. If fuel is recycled continuously, higher actinides will continue to build up until they approximately equal the amount of plutonium in the fuel. In a fast spectrum, all the isotopes fission substantially and the equilibrium composition is reached with relatively small, quite normal, amounts of higher actinides. The isotopic evolution in thermal recycle is presented in Figure 11-4. The isotopic evolution was accelerated assuming pure actinide fuel in inert matrix without uranium, which produces fresh actinides with irradiation. The fissile isotopes, such as Pu-239 and Pu-241, can be burned readily, but fertile isotopes, such as Pu-242 and various americium and curium isotopes, go on building up as burnup progresses. They have no reactivity value in a thermal spectrum and are useless there as fuel.

Table 11-2. Transmutation Probabilities (in %)

Isotope	Thermal Spectrum	Fast Spectrum
Np-237	3	27
Pu-238	7	70
Pu-239	63	85
Pu-240	1	55
PU-241	75	87
Pu-242	1	53
Am-241	1	21
Am-242m	75	94
Am-243	1	23
Cm-242	1	10
Cm-243	78	94
Cm-244	4	33

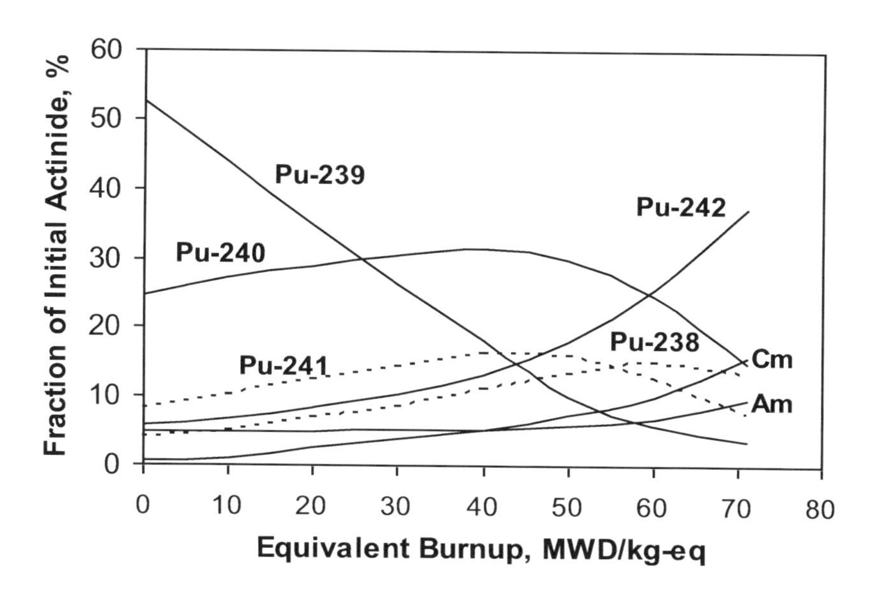

Figure 11-4. Isotopic evolution of actinides in thermal spectrum

Equilibrium compositions evolved through continuous recycle, starting from a natural uranium feed, are shown in Table 11-3. [13-14] For simplicity in calculation, they are compositions reached without regard to maintaining criticality or other operational characteristics. In the thermal spectrum, all the non-fissile isotopes (the ones with even numbers) are reactivity "poisons" (i.e., they absorb neutrons without releasing any new ones by fissioning) . Their reactivity constraints in a thermal spectrum are such that after a single three- to five-year cycle, the actinide compositions have no fissile value. At this point, only 30–50 % of actinides will have been transmuted. This is the ultimate limit for transmutation in thermal spectrum. Of course, additional transmutation is possible if the reactivity deficit is made up by more enriched uranium or fresh actinides in the form of first-generation discharge from uranium fuel. However, this merely stretches out the time of the evolution to several cycles. The reactivity limit of the actinides will still be reached with only the same fraction burned.

Table 11-3. Equilibrium Composition for Continuous Recycle

	Thermal Spectrum	Fast Spectrum
Np	5.5	0.8
Pu	51.1	97.0
Am	8.7	1.5
Cm	34.6	0.8
>Bk	0.2	$<10^{-4}$

Complete transmutation of actinides is possible only in fast spectrum reactors. Actinides are a valuable fuel for fast reactors. That's why recycle can be maintained until the uranium is all used, and in turn is the reason why uranium resource utilization is improved by a factor of a hundred over the thermal reactor systems. Recycle and actinide burning in the IFR accomplishes the two principal goals for reactors simultaneously: massive resource extension and elimination of long-term radiological toxicity of nuclear wastes.

11-5 The Long-Lived Low-Energy Radioactive Isotopes: Technetium and Iodine

Next, how much attention need be paid to the extremely long-lived non-actinide isotopes? The two such long-lived fission products are Tc-99 ($T_{1/2} = 2.13 \times 10^5$ yr) and I-129 ($T_{1/2} = 1.6 \times 10^7$ yr). They have been raised as potential health risks, worthy of special attention. Recall that the longer the half-life of a fission product, the less damaging its emissions will be. There have been proposals to isolate the long-lived fission products and transmute them as well. [15] In particular, Japan launched an ambitious research and development program (OMEGA Program) in 1988, carried on through 2000, evaluating various partitioning and transmutation technologies. [16-17]

Technetium-99, in particular, has been singled out. It has a long half-life, 213,000 years, but a plentiful yield; it represents 6 percent of all fission products. The health effects are small compared to other fission products and actinides, just over the permissible regulatory limit. In the ground, it's in oxide form; it dissolves readily in groundwater and migrates accordingly.

Although Iodine-129 has also been mentioned as a candidate for transmutation, it's hard to see the necessity for this, as its activities are only a fourth of those of technetium and are well within guidelines at all times.

The repository performance models that estimate the radionuclide releases are typically based on a dissolution-and-migration scenario whereby the radioactive nuclides that dissolve in groundwater are then released by groundwater transport. Both contribute to the early portion of the long-term dose rates in repository performance assessments.

Tc-99 and I-129 dominate the dose from the repository through the first twenty-five to fifty thousand years. However, as shown earlier, even if the entire inventory of these isotopes is released, the resulting dose is only the same order of magnitude as that allowed in the EPA standards. The actinides are quite a different case. They totally dominate radioactive release over very long time periods. Figure 11-5 from Reference 12 illustrates the point. It shows part of the analysis of the total system performance of a fully loaded Yucca Mountain repository. In the analysis a probability curve was assigned for each subcomponent or event, and repetitive Monte Carlo runs were done for thousands of cases, providing a simulation with error band estimates. The spread between 5% and 95% probabilities is shown in the figure. The results are strongly influenced by the assumptions on failure rates of the waste package and migration rates of the actinides. The probability of failure of the high-nickel container is essentially zero up to twenty thousand years, and the calculations reflect that assumption. No material will be calculated to escape before that time. After that, the calculations include the effect of release of iodine and technetium and then the slow release of actinides, governed by their solubility in water. The peak dose occurs in about two hundred thousand years. If waste package integrity is not assumed, the dose would peak earlier. The principal point, however, is that the actinides will be released in time and the magnitude of their effect is very substantial.

But what does this tell us? The important point from these data is simply this: If just the actinides are removed from the waste that enters the repository, any doses will be benign, close to the standards at all times, and the detailed model assumptions become irrelevant. To be exact, actinide contents should be made less by a factor of a thousand or so—only a 0.1 % loss in the recovery process would be acceptable if this statement is to be completely true. But with this, the EPA standards and the NRC dose limit can be met on *a priori* basis, regardless of the regulatory time period. It doesn't matter whether it's defined as ten thousand years or longer.

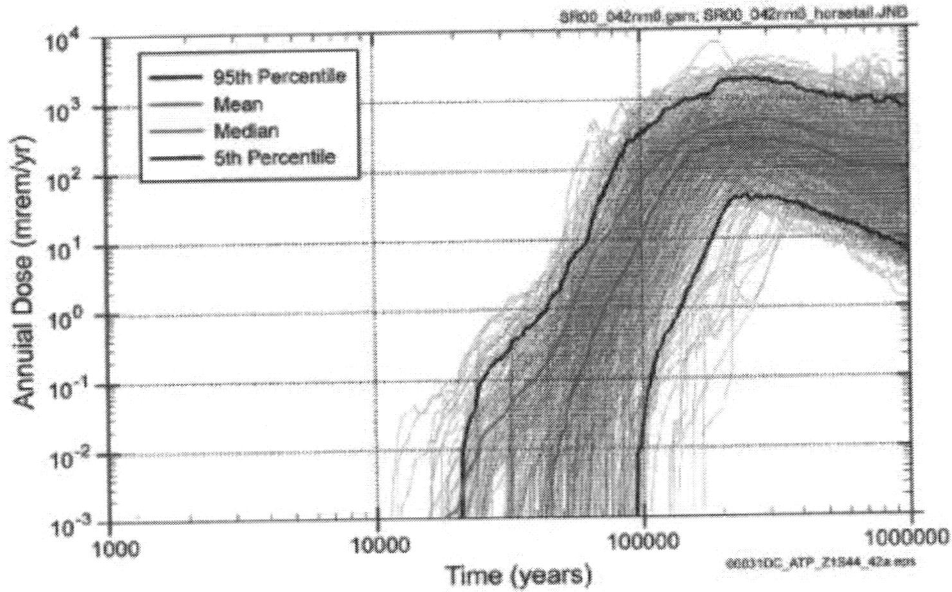

Figure 11-5. An example of long-term dose calculations to one million years (Source: Reference 12)

Another important point is that if Tc-99 were disposed of in a more durable waste form, release to the environment would be even further reduced. And this is precisely the case for the metal-waste form from electrorefining. In electrorefining, Tc-99 remains in the anode basket and along with other noble metal fission products is incorporated into the stable metallic, principally steel, waste, a form much more leach-resistant than a water-soluble oxide.

11-6 Highly Radioactive Medium-Term Fission Products: Cesium and Strontium

In the first few decades, and most after just a year or so, the highly radioactive short-lived fission products decay away to stable forms. The two thirty-year half-life fission products, Sr-90 and Cs-137, then contribute most of the radioactivity and most of the decay heat as well. There have been proposals in recent years to remove cesium and strontium from the waste stream and store them, to allow them to decay to the point where they could ultimately be disposed of as low-level waste. [18-19] Without Sr and Cs, and the actinides, the amount of the remaining isotopes would not be constrained by the heat loading. Space utilization would be hugely improved. For the specific temperature limits assumed for the repository, it has been shown that actinide removal alone can increase the repository space utilization by a factor of 5.7, but removal of Sr and Cs in addition to actinides would increase it by the huge factor of 225. [20]

Cesium and strontium control the near-field temperature limit at the inside wall of the repository tunnel, which is taken to be 200°C. This limit is imposed so that the uncertainties in the thermal-hydrologic response of the repository and the corrosion behavior of the waste packages are both reduced sufficiently to assure long-term performance characteristics. If the long-term performance is assured from the start, by the elimination of the actinides—the source of the long-term activities—the inside-wall thermal constraints may be less important.

There is no reason to question the correctness of the thermal analysis quoted above, showing the benefits of Sr and Cs removal. However, the very reason for a repository at all in the time scale of a few hundred years, over many human lifetimes, is to safely store Sr and Cs. Only in the very long term, ten thousand years and more, are the actinides the problem. That is why the original NRC regulations in 10CFR Part 60 required (1) substantially complete containment for a period not less than three hundred years or more than a thousand years, to deal with Sr and Cs, and (2) a restricted release rate beyond that period, to deal principally with actinides. In the absence of these two activities, dominant at different time scales (given that the remaining fission products as a practical matter decay away in a few years), there would simply be no need for a highly engineered repository.

Further, for a separated Sr and Cs waste form to decay below the low-level definition would take at least two hundred years. Even then, the waste package will contain some actinide contamination and minute quantities of long-lived Cs-135, in any case enough to prevent its near-surface land disposal as low level waste. Further, and most importantly, the low-level Class C definition in 10CFR Part 61 [24] was defined for land disposal of low-activity-level medical or industrial radioisotopes. It therefore would not be applicable to high-level wastes that have been stored for decay.

Strontium and cesium are high-level waste. They will eventually have to be disposed in a repository, even if it is engineered for a three-hundred-year time scale. There is little point in separating the more benign fission products and disposing of them in the repository. Even if the waste form were to be stored to reduce the heat load before emplacement in the repository, the conclusion is the same—store them together in this case as well.

11.7 Summary

The fast reactor is a must if nuclear is to play the role it must in the future energy mix. Waste management advantages, such as drastically reduced waste lifetime, from millions of years to a few hundred years, and much enhanced repository capacity utilization, are attendant bonuses. But the principal driving force is the need for fast reactors to conserve the uranium resource in the face of rapidly expanding need for nuclear power. Reprocessing of LWR spent fuel becomes economic simply due to the recovery of valuable actinides (further details in Chapter 13), and the long-lived toxic waste component is eliminated automatically. One thing follows from the other.

What the EPA standards and NRC regulations for the repository *should be* are really not important issues—whether they are 15 mrem (0.15 mSv) per year or some other number, for ten thousand years or longer, and so forth. The technical details necessary to demonstrate how the standards and regulations can be met are the important issue. But if actinides are recovered for use in fast reactors, separated from the waste sufficiently cleanly (e.g. 0.1 % of actinides remaining in the waste), the standards and performance criteria will be met on an *a priori* basis. Exhaustive proofs will not be necessary.

The actinides are the primary concern, and three orders of magnitude reduction of their amounts in the waste is possible. At that level, the residual risk is of the same order as the other fission products, and any further reduction isn't warranted. The risk from the long-lived fission products, Tc-99 and I-129, is similar—without reduction they are at or close to the release limits at worst, and reduction in their amounts is not warranted at all. If 99.9 % of actinides are removed, and Cs and Sr are contained for three hundred years or so, the public risk from the repository will be well below a small fraction of the natural background *at all times in the future*.

Removal of Cs and Sr to reduce the heat load in the repository does not stand scrutiny either. The separated Cs and Sr have to be safely stored for three hundred years, which is exactly why the repository is needed in the first place. If indeed heat load management is important, the entire waste package can be stored for an interim period to let it decay, but the repository itself is still the best place to store it while that is happening.

The properties of IFR waste—strict limits on actinides, provided by the process, containment of technetium to a considerable degree, and leach resistance equal to greater than present waste glasses—go a very long way toward a permanent solution to the problem of nuclear waste.

References

1. "Nuclear Waste Policy Act of 1982," Public Law 97-425, Enacted on January 7, 1983, and Amended as Public Law 100-203 on December 22, 1987.
2. "Exelon Settles with DOE Over Yucca Delays," *The Energy Daily*, August 11, 2004.
3. "Analysis of the Total System Life Cycle Cost of the Civilian Radioactive Waste Management Program," DOE/RW-0533, May 2001.
4. "Nuclear Waste Fund Fee Adequacy: An Assessment," DOR/RW-0534, 2001.
5. Code of Federal Regulations, 40 CFR Part 191, "Environmental Radiation Protection Standards for Management and Disposal of Spent Nuclear Fuel, High-Level, and Transuranic Radioactive Wastes."
6. "Disposal of High-Level Radioactive Wastes in Geologic Repositories," *Code of Federal Regulations*, 10CFR Part 60.
7. "Energy Policy Act of 1992," Public Law 102-486, Section 801.
8. National Research Council, *Technical Bases for Yucca Mountain Standards*, National Academy Press, 1995.
9. "Public Health and Environmental Radiation Protection Standards for Yucca Mountain, Nevada," *Code of Federal Regulations*, 40CFR Part 197.
10. "Disposal of High-Level Radioactive Wastes in a Geologic Repository at Yucca Mountain, Nevada," *Code of Federal Regulations*, 10CFR Part 63.
11. "Court Rejects Yucca Safety Standard in Big Win for Nevada," *The Energy Daily*, July 12, 2004.
12. "Yucca Mountain Science and Engineering Report," DOE/RW-0539 Rev. 1, February 2002.
13. Robert Avery, private communication.
14. Robert N. Hill, private communication.
15. National Research Council, *Nuclear Wastes: Technologies for Separations and Transmutation*, National Academy Press, 1996.
16. T. Mukaiyama, "Partitioning and Transmutation Research and Development Program (OMEGA) in Japan," IAEA-TECDOC-693, *Proc. IAEA Specialists Meeting on Use of Fast Reactors for Actinide Transmutation*, Obninsk, Russia, September 22-24, 1992.
17. T. Inoue, et al., "Development of Partitioning and Transmutation Technology for Long-Lived Nuclides," *Nuclear Technology*, 93, 206, 1991.
18. F. Vandergrift, et al., "Designing and Demonstration of the UREX+ Process Using Spent Nuclear Fuel," *Proc. ATLANTE 2004: Advances for Future Nuclear Fuel Cycles International Conference*, Nimes, France, June 21-24, 2004.
19. C. Pereira, et al., "Lab-Scale Demonstration of the UREX+2 Process Using Spent Fuel," *Proc. Waste Management Symposium*, Tucson, Arizona, February 27-March 3, 2005.
20. R. A. Wigeland, T. H. Bauer, T. H. Fanning, and E. E. Morris, "Separations and Transmutation Criteria to Improve Utilization of a Geologic Repository," *Nuclear Technology*, 154, 95, 2006.

CHAPTER 12

NONPROLIFERATION ASPECTS OF THE IFR

Although halting proliferation of nuclear weapons and unsecured weapons-suitable fissile material is of obvious importance, the contribution of civilian nuclear power to such proliferation is not obvious at all. In contrast to subjects that can be settled by agreement on technical facts, over the years opinion has been offered freely on all manner of proliferation matters, including a linkage to civil nuclear power. But really, how much has civilian nuclear power contributed to the proliferation of nuclear weapons and how much does it today? What can be said on the subject about the IFR specifically? The technical facts of the IFR relevant to proliferation are important. They need to be understood.

Our aim in developing the IFR technology was to go as far as we could in answering objections to nuclear power. But we were acutely aware that it was proliferation arguments that had become the weapons of choice for those who would stop the growth of civilian nuclear power. Further, nuclear fuel recycle systems with their spent fuel processing were under particular attack. Recycle is important. It allows expansion of nuclear power unlimited by fuel shortage. Recycle also requires processing of used fuel to reuse it. Neither expansion of nuclear power nor processing of spent fuel to allow its reuse is desired by the antis. During the Clinton administration they were successful in terminating IFR development based on just this issue. But what are the facts?

Our purpose in this chapter is to concentrate specifically on the attributes of the IFR relevant to safeguards and non-proliferation. But to provide context we will first touch briefly on the history of weapons development, the principles underlying efforts at different times to control the increase in the number of nations developing and possessing them, and the substance, or more accurately the lack thereof, of evidence to support a significant proliferation role for nuclear power in general. We then turn to our principal purpose—presenting the facts relevant specifically to the IFR.

12.1 Introduction

Assessment of a possible role for civilian reactor technology in the proliferation of nuclear weapons is to some degree a balance of risk against benefit. Nuclear power is here to stay. Some nations in special circumstances have found it expedient to announce their intentions to supply their energy in some other way and phase out nuclear power. Energy realities may catch up even with these nations when the paucity of real alternatives becomes plain and some serious rethinking may be in store down the road. But what is certain is that the nations with big populations—China and India particularly, but others as well—are increasing their amounts of nuclear power as quickly as they can. The U.S. is a notable exception, but, it can be expected, not for too long. The existence of nuclear power and its inevitable growth is a fact which can neither be denied nor wished away. Realistically the risk/benefit is not reactor technology or no, but rather a trade involving judgments—how much does civilian nuclear power really affect the possibility of further proliferation of nuclear weapons, balanced against the importance of the benefits of huge amounts of reliable, low-polluting electrical power? Therein has been enough and more than enough room for dispute.

Today, with sixty years of development and much specialized knowledge of the full range of nuclear science in every respectable scientific institution in the world, just how much risk does practical, working, civilian nuclear power add today? There can be no dispute about the importance of non-proliferation measures. The dispute comes in whether civilian nuclear power today plays *a* significant role, and really whether it plays *any* significant role in such proliferation. IFR technology had a principal aim of improving the present situation in safeguards against use of civilian technology for weapons purposes. Yet its opponents in the Senate debate on termination of IFR development—ignoring facts—focused their attacks on just this issue. [1] (For a spirited rebuttal, see, for example, Blees's "Prescription for the Planet." [2])

Relevant facts on the weapons themselves are not always easy to establish in the secrecy surrounding details of their development, but much can be established without probing much in these areas. Activist groups have put forth many writings and statements aimed at furthering their goals, in large measure political, which do little to illuminate and seem mostly meant to obscure. Their technical bases generally are incomplete and at worst deceptive. Still, enough is now known, and in the open literature, for an adequately factual picture of the principal points about weapons to emerge.

Where exactly does the risk lie? IFR technology poses the trade-off in this way—on one hand, unlimited energy production; on the other, technical attributes that minimize proliferation potential, but cannot eliminate it entirely. Risk can be said to be present always, with or without the IFR, but the real question is how much further risk does the IFR add, if any? In fact, in some ways does it not actually improve the situation?

Specialized knowledge about detail of weapons fabrication doesn't add as much to understanding as one might think. The "secret" of atomic weapons has been no secret for decades. Over the years, the politics, definitions, and nuances of non-proliferation have developed a language of their own, which doesn't excessively enlighten. But common sense goes a long way.

There is an important distinction to be made between the ability of nations to successfully develop nuclear weapons, with all the resources and advantages a state possesses, and the efforts of a sub-national group attempting to assemble a weapon in secret for whatever their purposes may be. A nation can establish a laboratory of highly technically skilled scientists and engineers, construct the kinds of reactors and enrichment facilities precisely suited to the purpose, and put in place all of the ancillary facilities needed to create and maintain an armory of deliverable nuclear weapons, storable for as long as desired. The clandestine group must rely on stealth and thievery. The state's need is for specialized knowledge, the clandestine group's for theft of a weapon, or at the very least the theft of the material for it. Both require suitable fissile material of the right isotope. A state can build the facilities to make it, or, if it so chooses, it could build facilities to transform material diverted from existing civilian facilities to material usable in a weapon. Unauthorized diversion from civilian reactors under safeguards would be necessary if this were the path to be taken, and a price would be paid. But construction of the relatively simple reactors adequate for weapons plutonium production has been the usual course, rather than involving their power production capacity, for the nations now possessing nuclear weapons. For a nation, considerations such as these and their decisions to move ahead are matters of political will, not of technological barriers. For practical purposes, the clandestine group is limited to purchase or theft.

Knowledgeable weapons designers have made the point that almost any composition of plutonium in theory could be used to make a weapon. [3] Nevertheless, as we shall see, apart from a single U.S. test in the early 1960s with plutonium whose isotopic composition was called "reactor-grade" at that time (as opposed to "weapons-grade," 7% or less Pu-240) but which may have been much lower in Pu-240 than plutonium that would be called "reactor-grade" today, the practicality hasn't been demonstrated successfully, although apparently there were two British trials in the 1950s. A former director of the U.K. Atomic Weapons Research Establishment at a conference in London on plutonium disposition in 1994 stated flatly (in the presence of one of the authors (Till)) "we tried reactor-grade plutonium a couple of times. We never will again." In fact, history has demonstrated that nations that have successfully developed weapons using plutonium invariably have used plutonium that was at least 93% pure Pu-239, which suggests good reasons for doing so.

History provides empirical evidence regarding probable future actions. The routes used by weapons states—specialized reactors and specialized facilities—exist today or can be built just as they were in the past, and for the very same reasons: walls of secrecy, optimum conditions for development, optimum fissile material, and perhaps even a general unwillingness to involve civilian nuclear power programs, with less provocative options so easily available. The only significant connection to nuclear power could be in the plutonium fissile material it could provide, and choosing simple reactors specialized to weapons plutonium (as the U.S. and other nations did) avoids much of the hassle in abrogating safeguards agreements and in accepting an imperfect plutonium product. Such simple reactor systems are certainly not unduly costly.

Nuclear power, in fact, has been largely irrelevant to proliferation. Not because it isn't part of a possible route to a weapon, but as decades of history and recent events show as well, the routes that have been used in the past are the likely ones to be used in the future. But IFR technology goes further. It poses the risk/benefit tradeoff in a particularly favorable form, minimizing proliferation potential while maximizing the energy production possible from the world's uranium resource.

12.2 History

U.S. policy on non-proliferation since the Carter administration has been strongly influenced by antis, and has put extraordinary emphasis on access to plutonium—as much as implying that access to any isotopic composition of plutonium is far along the path to developing a successful weapon, which is not a particularly easy position to defend considering the history of the actual isotopic composition of plutonium used in weapons. Access to fissile material obviously is necessary, but certainly not nearly sufficient. To proliferate (make a usable bomb) using plutonium in addition to plutonium from a suitable reactor, or such a reactor itself, chemistry is needed (PUREX, developed specifically for weapons purity plutonium), as is complex explosive design, triggers, delivery, and so on. Motivation, capital, and political will and/or cover all are necessary. And testing and the freedom to test are very important. (The U.S. has done hundreds of such tests.) Access to plutonium can play a necessary role, but with today's uranium enrichment capabilities, even that role can be questioned. Concentration on access to plutonium as though it is obviously *the* important requisite for proliferation is certainly outmoded—if indeed it was ever true. For uranium weapons, the necessary U-235 isotope is purified from natural uranium, now possible by the recently perfected high-speed centrifuge. Nations today, in fact, need have no reactor facilities at all for weapons production.

A glance at the situation with Iran or North Korea today makes the point. The recent statement by an ex-Los Alamos Laboratory Director Siegfried Hecker [4] that he had been shown a wholly new and very impressive uranium enrichment capability while in North Korea in November 2010 implies at least two things. The capability was assembled successfully in secrecy, illustrating once again that this is how nations proceed. North Korea's much-publicized weapons program had relied on the pure plutonium from a dedicated reactor (nominally 5MWe) processed by PUREX. They conducted two tests, one in late 2006 and the other in May 2009. The degree of success of the tests remains controversial and there has been no further testing. As uranium bomb fabrication is known to be far easier and requires less testing than the plutonium weapons, even down to none in the case of the first U.S. atom bomb (although it must be said it was assembled by the best nuclear physicists and engineers in the nation) the development of a uranium-235 capability (a big project indeed) suggests trouble with their plutonium weapons development. At the very least it underlines the need for more extensive testing—which hasn't taken place.

This need for testing is emphasized by Stephen M. Younger, a former nuclear weapons designer and head of nuclear weapons research at Los Alamos, in a wide-ranging, very readable recent book on the history of nuclear weapons, their current status, strategies and policies. [5] His chapter on nuclear proliferation is illuminating, and in particular his comments are significant on the difficulties facing a nation today in developing a weapon, coming as they do from someone very much in a position to know the relevant facts. Quoting,

> "The fact that nuclear weapons are *not* easy to make is demonstrated by the setbacks that all the nuclear weapons states have experienced in their well-funded (typically several billion dollars per year) and nationally supported programs."

Going on to describe the difficulties found in the early French program as an example, he then commented further,

> "Key to the success of all nuclear powers has been the ability to conduct one or more nuclear tests. Nuclear testing is more than a demonstration of success - it enables scientists to understand in detail the complex processes that occur during a nuclear detonation."

Further to the point,

> "... nuclear weapons development still requires the resources of a nation-state. To think that a terrorist group working in isolation with an unreliable supply of electricity (authors note: vital to centrifuge operation) and little access to tools or supplies could accomplish such a feat is far-fetched at best."

It is important to remember that in earlier administrations, prior to Carter, U.S. help with civil nuclear power was offered specifically as an incentive for nations to give up ambitions for weapons programs. Eisenhower's "Atoms for Peace" program offered nations help with nuclear programs for peaceful purposes in exchange for undertakings not to pursue indigenous weapons programs. The International Atomic Energy Agency emerged from this initiative. One of its principal roles was to monitor adherence to such agreements and their accompanying safeguards. With a very few exceptions, this international regime worked well. At the time, it was observed that as many as a dozen and perhaps up to twenty more nations already were pursuing such programs. In the fifty years that followed, only India, Pakistan, almost certainly Israel, and imperfectly, North Korea, became weapons states. India's nuclear device testing of 1974 was fresh in mind when the Carter administration adopted the new policy in large part directly opposite to the policies in place in the U.S. from the Eisenhower administration onward.

12.3 The International Nuclear Fuel Cycle Evaluation

The programs of the Carter administration were an attempt to deny the fissile materials themselves to non-nuclear weapons countries—to be implemented by banning reprocessing of spent fuel worldwide. The U.S. had maintained a dominant position in nuclear reactor development up to this time and other nations did tend to follow the U.S. example in development, although in actual commercial application this wasn't so true. The U.S. no longer enjoys such an advantage, in part at least due to the deleterious effect of these very policies on our own civilian nuclear reactor development and implementation. The principal actions of the Carter administration, with long lasting-effects, were principally on our own programs—to actively discourage reprocessing of irradiated uranium fuel (successfully imposed on our own program) and to cancel development of the Clinch River Breeder Reactor, the U.S. demonstration of breeder technology (also successful, but only after a six-year fight in Congress).

The principal diplomatic thrust was an "International Nuclear Fuel Cycle Evaluation," a two-year series of meetings in which the government hoped to convince other nations to limit or eliminate plutonium use in their civil nuclear power programs. [6] Accustomed to U.S. leadership in the field of nuclear power, even though the Carter administration brought many of the most prominent anti-nuclear people into key positions in the U.S. government, the nations with nuclear programs willingly involved themselves in this new initiative of the new U.S. administration.

The U.S. cancelled construction of its own civilian reprocessing plant at Barnwell in South Carolina, then nearing completion. It was to use the PUREX to process the spent fuel from the growing number of LWRs ordered and beginning construction. As it cancelled its capability for doing so, the U.S. abjured any intent to process spent fuel from its then rapidly increasing number of nuclear power plants. Disposal of nuclear waste became significantly more difficult.

The administration in 1977 also attempted to cancel immediately the Clinch River Breeder Reactor (CRBR), the U.S. demonstration plant for the oxide-fueled breeder reactor, being planned for construction in Tennessee. Congress resisted and it wasn't until November of 1983 the project was finally cancelled.

The stated intent of all of this was to set an example for others. As such it was successful—albeit in a very limited way. The INFCE report concluded that the two points in the fuel cycle that were "sensitive" were uranium enrichment facilities (which at this time were huge and extremely expensive plants based on diffusion processes, possessed by only by a very few nations) and plutonium separation facilities (at this time based exclusively on the aqueous PUREX process, designed specifically to produce pure plutonium). This conclusion occasioned no surprise—at either of these two points, pure fissile uranium or pure fissile plutonium is, or can be, produced and accumulated in a form that may be directly usable for weapons. Certainly this is so for uranium, and depending on the isotopic distributions and other factors we have mentioned, possibly so for plutonium. However, all nations with either capability, or both, were already "nuclear weapons states" and they proceeded with their programs of fuel enrichment for power production and nations with spent fuel reprocessing proceeded as well, with appropriate safeguards, as before. Only the U.S. turned away, and having cancelled its civil PUREX processing plant, announced a new policy to dispose of spent fuel whole, without processing.

All this had several effects, none desirable. In the U.S., stress was then put on the early need for a repository for the now much-increased tonnage of spent fuel when disposed of whole, creating a problem the nation is still dealing with, and, it may be said, completely unsuccessfully today. Cancellation of Barnwell civilian reprocessing plant, nearly complete, did not influence other nations' plans at all—Britain, France, Japan, and later, India, Pakistan, and others went ahead in any case—but it did create a real spent fuel problem for the U.S., a nation preeminently a nuclear weapons state with military PUREX plants in operation since WW-II. The U.S. repository at Yucca Mountain now being in limbo means that spent fuel continues to build up at the plant sites. The recent Japanese experience suggests that, although the quake and tsunami conditions there may be unlikely, this is not the best of ideas.

Repository construction has been held up by the same groups who championed the no-reprocessing decision in the first place. That example, at least, has been followed by the organized anti-nuclear groups internationally. Prior to their actions, the U.S. was on a path to take care of its spent fuel, reuse it where possible, and dispose of it safely in a single location where not. Today there is no such plan.

Other nations are recycling their separated plutonium in existing reactors, a usage that extracts only a little extra energy, and is certainly uneconomic. The U.S. is to some degree slowly following their lead in disposing of weapons plutonium, now surplus, in this unsatisfactory way. The word "disposal" isn't particularly apt, as the major effect is not to destroy the plutonium, but rather to evolve the plutonium isotopic composition to higher fractions of Pu-240 and the isotopes above it, taking the compositions out of "weapons grade," and thus denaturing it for weapons use. There is real irony here [7] in the obvious contradiction with earlier statements that "plutonium of any composition can be used for a weapon"—whose implication is that it's practical as well. The irony is made even more obvious in noting that, after all is said and done, it was just this statement that was used as the principal justification in the denigration of nuclear power of the Carter administration policies in the first place.

On the other hand the INFCE study did cause nations to think once again about decisions on reactor types and deployment that had been reached back in the early fifties and sixties. And in the U.S., it led directly to the Argonne initiative for development of the IFR.

12.4 Present Policies

U.S. policy today remains focused on denial of access, denial of the capabilities needed for weapons, and repetition of the suitability of fissile material from nuclear power reactors for weapons (in the face of the ironies mentioned above). Denial of access to the necessary fissile materials is undeniably simple in concept. It is unarguable that if the spent fuel is never processed—if say, it stays buried in a repository—there is no possibility of its use for weapons. On the other hand, if it does go to a repository it doesn't necessarily stay there. As years pass, radioactivity dies away, rapidly at first, then more slowly, but inexorably, and it leaves a product much easier to handle. It can be retrieved. As a stream of spent fuel is stored, year after year, in a repository, the stream becomes a lake—a massive amount of plutonium contained in the spent fuel in storage. The same, obviously, is true if the accumulated spent fuel of decades remains at the hundred or so reactor sites around the U.S. If reactor spent fuel is considered to be useful for weapons, its use is made easier as it becomes progressively more accessible as the years pass and the shorter-lived isotopes decay away. And its energy content, huge as it is, and accessible using fast reactors, is foregone for no obvious gain.

The plutonium in typical reactor spent fuel from a commercial LWR power plant contains a large proportion of the higher isotopes of plutonium. Although all are fissionable in the high energy neutron spectrum of a weapon, they are not desirable for weapons because they produce heat deleterious to weapon stability, emit spontaneous neutrons deleterious to timing of detonation in simpler designs at least, and they decay to other elements that increase radioactivity deleterious to handling in fabrication and in access to the weapons. The stated U.S. position is that they are undesirable in weapons because they cause "increased complexity in designing, fabricating and handling them." [8] The secrecy which surrounds weapons work, completely appropriate in the main, does serve to blur and weaken absolute statements of the possible and impossible. But what is clear is that impurities that spontaneously produce neutrons, generate heat, and increase radioactivity, all characteristics of spent fuel, are at best undesirable. It is the first-time acquisition of nuclear weapons by irresponsible states and the clandestine acquisition of weapons-usable material by groups seeking to cause terror that are of concern. Plutonium that would require very significant sophistication in weapons fabrication, storage, and detonation isn't a likely choice for weapons by a neophyte. The radioactivity of reactor-grade plutonium makes it an unlikely choice for the hands-on work that is necessary in any case.

12.5 The Subject of Plutonium

The existence of plutonium is not a matter for debate. That is a settled issue. The world inventory of plutonium is on the order of two thousand tons, the bulk of which now comes from civil nuclear power. In the main, it is contained in the spent fuel rods from present nuclear power plants, largely from light water reactors. The amount of separated plutonium in the world has increased in the last two decades. The inventory of separated civil plutonium was about 230 metric tons in 2004. [9,10] Plutonium fabricated into mixed plutonium-uranium oxide (MOX) fuel for recycle in LWRs and the plutonium declared as excess weapons materials in the U.S. and Russia is not included in this estimate. At the present time the commercial reprocessing plants, La Hague in France, THORP in the U.K., and Rokkasho in Japan, recover about thirty tons of plutonium per year. Not all this plutonium is recycled back into LWRs, so the inventory of separated plutonium grows.

All nuclear power plants produce plutonium. Plutonium production is not a matter of choice—it is produced as a matter of course in uranium-fueled reactors by transformation of the most common isotope of uranium, U-238, first to Np-239 and then in a few days by radioactive decay to Pu-239. The annual amounts produced and left in the spent fuel of an LWR equal, and as a rule more probably exceed, those produced in an IFR of equal power. The IFR, of course, produces more plutonium, but it burns much more in place. Its fuel, after all, is plutonium. By contrast, the LWR is fueled with fissile uranium and creates plutonium, and while it burns a considerable amount of it in place, it also leaves a lot unconsumed in the spent fuel.

There has been sporadic experimentation with thorium fuel reactors going all the way back to the early years of reactor development. (In recent decades the thorium uranium-233 cycle has been half-heartedly put forward by the more-or-less anti-nuclear administrations when they come to power in the U.S., which, intentionally or not, does serve to divert effort from more promising alternatives.) In earlier times, one or two prototypes were built in the U.S. The basic fact is that thorium is non-fissile and cannot be used as a fuel as such—it must breed its fissile uranium-233 after initial fueling with uranium-235 or plutonium. Thorium means reprocessing. No ifs, ands, or buts; the actual fuel in this cycle is fissile uranium-235 or plutonium initially, and fissile uranium-233 afterward. Once the uranium-233 is bred in thorium, the result can be reprocessed to make the U-233 available for recycle. But by fairly rapid radioactive decay, the latter produces U-232, which has very penetrating gamma radiation, making it extremely difficult to handle after a relatively short time in storage. It is fair to say thorium has its own problems and as a result has not been a significant factor in any nuclear power program in the world, with the limited exception of India.

It has been said that "plutonium has been demonized." All kinds of easily refutable statements are made about this valuable substance. In an otherwise interesting book on plutonium, the final paragraph contains the astonishing statement that apart from weapons, "it (plutonium) has almost no other use." [11] This was said about the material that right now generates at least 40 percent of the world's nuclear electricity in various types of water-moderated reactors—PWRs, BWRs and HWRs. Probably the fraction of power produced in plutonium is even more in the most modern versions where fuel burnup is longer, and plutonium stands to generate most of the world's electricity in the future.

Like uranium and thorium, plutonium is quite similar to lead in weight and color. It is radioactive like radium, an element widely distributed in the earth's crust. It is toxic like other heavy metals, also common, such as arsenic (found in groundwater), cadmium (used in solar photovoltaic cells), or mercury (released into the atmosphere when burning coal for electricity). Handled with ordinary care, it presents no toxicity problem greater than many other substances in common use. It is its use as the fissile material for weapons that makes it dangerous. However, with modern advances in isotope separation, uranium itself can be classed as at least as dangerous. Uranium is common in the earth's crust. The fact that it now represents at least an equal danger for illicit weapons development seems to have been ignored by those who continue to decry the very existence of plutonium. The simple point here is that plutonium cannot with honesty be singled out as uniquely dangerous, as has been said repetitively by antis. It most certainly is not. And in it lies the principal capability for massive energy production in the future.

Early in the atomic age, in the U.S. plutonium was considered safe from weapons use due to the technical sophistication needed for its use in a weapon, but that changed when other nations after experimentation and testing demonstrated their own capability to produce plutonium-fueled weapons. Controls were then put in place. In them, and in the international diplomatic structure centered on the International Atomic Energy Agency (IAEA) from the Eisenhower administration, there has been a considerable, but not complete, success in slowing and limiting nuclear-weapons proliferation.

12.6 Plutonium and the IFR

12.6.1 Practicalities of the Fast Reactor Neutron Energy Spectrum

The fast reactor, because of its fast-neutron spectrum, has a high tolerance for neutron-absorbing impurities—the fission products, the higher isotopes of plutonium, and the higher actinides elements created in reactor operation. It can burn mixtures of actinides that are completely unpalatable to an LWR, or indeed to any thermal-spectrum reactor. All actinides are fissionable in a fast spectrum. In fact, all are excellent fast reactor fuel for the IFR. And as we have seen in previous chapters, in contrast to thermal-spectrum reactors, these higher-actinide isotopes do not need to build up and go into the waste. They burn, and by so doing they can be kept limited in quantity.

The fast reactor maintains a low-capture cross section for Pu-239, so absorption of a neutron in Pu-239 is much more likely to cause it to fission than to be captured and create Pu-240. This is the very reason that substantive breeding is possible, of course. The low Pu-239 capture leaves neutrons available over and above those needed to fission to maintain criticality. The much larger amount of U-238 captures the excess and makes more Pu-239. The effect of breeding additional Pu-239, combined with limited capture producing Pu-240, means the fraction of Pu-240 cannot build up in the plutonium as fast as it does in the moderated-neutron spectrum of a LWR. In the recycled fuel of an IFR this is of no importance whatsoever, as the plutonium of recycled fuel is laden with higher isotopes. But it is important to the plutonium created in the uranium blanket, and the manner in which the uranium-blanket assemblies must be handled.

Plutonium slowly builds up in blanket assemblies, over a period of a decade or so. When removed for processing, the assemblies will be radioactive, and should go to the fuel cycle process to be mixed with fuel assemblies and so maintain the Pu-239 in a mixture of plutonium and higher actinides—always. The characteristics of electrorefining plutonium-bearing uranium assist in making difficult ready separation of plutonium from the uranium mix. As was seen in Chapter 8, to deposit plutonium at all the uranium content of the electrorefiner must be drawn down to a small fraction of the plutonium content. Even then, uranium will be unavoidably present in any plutonium product. This is so even for lengthy blanket fuel electrorefining campaigns where blanket plutonium is allowed to build up in the electrorefiner salt. And this, of course, is exactly the campaign that has been carried out over the years with the EBR-II blanket assemblies described in Chapter 8.

For blanket-only campaigns, very large quantities of uranium must be deposited on solid cathodes while the plutonium content of the salt ever so slowly increases. If plutonium content is to be increased more rapidly, high-plutonium-content fuel must be introduced. This fuel is laden with high isotopes of plutonium and americium, neptunium, and curium, as well as fission products. There is no way around this. If blanket plutonium is wanted, a lengthy campaign of uranium deposition only must take place. Many such runs must be made to build significant amounts into the salt. Weeks and months must pass. There is no way such a campaign could escape the attention of any even rudimentary safeguarding scheme.

Any reasonable processing plan for reactor recycle will include two practical elements: First, when processing spent fuel plutonium will be withdrawn from the electrorefiner regularly in order to minimize criticality constraints. Extensive drawdown of the uranium will be minimized—it is operationally inconvenient for one thing, and as a practical matter, a product high in plutonium will need to be diluted with approximately four times its amount of uranium for use in the reactor in any case. There will be a practical tradeoff, on one hand minimizing the uranium drawdown for plutonium deposition, and on the other, minimizing the stream for cadmium-cathode treatment.

Drawdown is inevitable; there is no other way of collecting plutonium, and by their nature such operations are lengthy, obvious, and verifiable. What goes into the electrorefiner is verifiable in the simplest possible manner, by the counting of discrete individual pieces. Any operation of the cathode processor signals plutonium work. Monitoring of the processor content is carefully and continuously done by instrumentation developed specifically for this purpose. There is no disguising the difference between a process attempting to get at blanket plutonium alone and one optimizing recycle effectiveness. The operations of the electrorefiner, by their nature, are ideally suited to monitoring; they can be changed only slowly, and many signals arise if operations are significantly changed. Finally, after all of this, it is not at all clear just how clean a pure blanket product would be in any case. Certainly there will be a significant quantity of uranium in it, and some small amount of higher actinides, at the very best.

Finally, practicality demands that blanket elements be left in place for as long as possible. Their plutonium contents are always low. Getting at the plutonium requires extensive uranium-only operations, not likely to be deemed the best use of time for electrorefiner equipment and personnel. The longer the blankets stay in the reactor, the less time they need be accounted for, and the more radioactive they will be when they come out (and the higher the proportion of Pu-240). And again, as a practical matter, they will probably be aided by the interspersion of fuel assemblies that bring the plutonium contents in the salt up to practical levels for operation and denature the product with the higher isotopes at the same time.

Thus the (re)processing technology takes over for maintenance of blanket plutonium security from the pure isotopic protection of the fuel itself. It allows recycle of blanket Pu-239 always as a mixture of actinides and fission products. The processing scheme that's ideal for a fast reactor is one that's good enough to do the necessary job of allowing recycle of all the actinides, which, because of the high burnups of fast reactor fuel—20% as we have noted, will not be exceptional—will build up to high levels. Electrorefining does this, and it also processes blankets in an open and obvious manner, suited to easy monitoring and to denaturing of the product as a matter of course. It never gives a really pure product, but it does the job.

12.6.2 Contrast with PUREX Reprocessing

While processing must be good enough to do our job, it's equally true that it's desirable not to have a technology that's too good. A pure plutonium product, low in Pu-240, removes the technological barriers to weapons use—a reactor that can produce quantities of adequately pure Pu-239 and a processing technology that separates plutonium as a very pure product is not desirable for civil nuclear power. It is no coincidence that PUREX produces a pure product—it was developed precisely for the purpose of cleanly separating pure plutonium for the weapons program.

PUREX today is the international standard of reprocessing technology. It is developed. Its development is sunk cost. But its plants have been expensive. The Japanese PUREX plant for reprocessing thermal reactor fuel cost in the neighborhood of twenty billion dollars. New plants will be huge and expensive, it is certain. Pyroprocessing is (relatively) cheap, as will be seen in the chapter on economics. It is inevitable that comparisons be drawn with the established technology, and that technology is aqueous reprocessing; in particular it is PUREX in one form or another.

PUREX is poorly suited to the fast reactor, and we believe that with the developments of the past thirty years or so, it is by no means ideally suited to civil use in general. The high fissile content of fast reactor fuel makes criticality an ever-present concern in the presence of an effective neutron moderator like the hydrogen in water or oil. Hydrogen-bearing liquids are the media of the PUREX process. Aqueous reprocessing plants have suffered unplanned criticalities in the past. [12] The need for dilution and reduction of fissile concentrations, critical for aqueous processing, is much less for the IFR process. A well-designed IFR facility will have severely limited moderating materials and criticality should not be an issue at all. The reactor's concentrated fissile content is matched by a process that maintains those concentrations in making the separations. No hydrogen-bearing material is introduced.

PUREX gives a very pure plutonium product. The IFR spent fuel product is a mixture of plutonium and all the higher actinide elements—neptunium, americium, curium, and so on—as well as uranium and some residual fission products. Uranium is present in amounts ranging from perhaps thirty per cent to amounts in excess of the plutonium amount. This is material that would need reprocessing to purify it, by solvent extraction, by PUREX in fact, to attempt use in weapons. For such purposes it amounts to spent fuel pretty much as it was before processing in the electrorefiner—it contains all the actinides present in the spent fuel, undesirable for weapons, and it remains highly radioactive. Anti-nuclear groups commonly assert otherwise. If they worked on IFR processing, they would do it behind several feet of concrete, looking through feet of leaded glass, with remote handling procedures using manipulators, cranes, and possibly robots. Operations personnel do not use these procedures for enjoyment. Not only is the IFR spent fuel very radioactive, but the product too remains radioactive, and must always be handled remotely. IFR processing adds little to the usability of spent fuel for fabricating weapons.

12.6.3 IFR Role in Plutonium Management

Management of plutonium should be done in complete fuel cycles—where plutonium is recycled and burned for power production and in so doing its amounts are controlled. And that is what the IFR does, of course. An IFR can absorb a large amount of plutonium, and keep it safely inside the reactor core while generating energy or safe in processing in a highly inaccessible fuel cycle facility. The amount of plutonium fuel, the reactor fuel inventory, required to produce a given amount of electrical power is in the range of five tons per 1,000 MWe for a reactor of 1,000 MWe or greater capacity, and in the range of ten tons per 1,000 MWe for a 300 MWe modular reactor. The smaller reactor core has higher neutron leakage and therefore requires a higher fissile enrichment to maintain criticality. In addition to the initial core inventory, two or three annual reloads must be supplied before self-recycle is established. The total startup requirements approach a doubling of the initial core inventory, which is safely retained throughout the reactor life.

The only vulnerable link between the reactor and the processing facility is the short path transferring spent and fresh fuel between the two. This single point is ideally suited to item accountancy and continuous surveillance and monitoring. Where the fuel cycle facility is co-located with the reactor, as was the case for EBR-II and is contemplated for IFRs, transportation system vulnerability is also eliminated.

Recycle of plutonium is much more desirable than stockpiling. Plutonium is produced in all reactors—as a rough guide the amount is about 250 kg of plutonium a year discharged from a 1,000 MWe LWR. This translates to about 90 tonnes a year from existing reactors worldwide. Perpetual safeguarding is required. But with appropriate IFR deployment, plutonium is neither in waste nor in stockpiles. Instead, it is always in the working inventory of nuclear power plants. Instead of growing without end and without energy production, it is maintained safely at the level needed for energy production, no more and no less.

IFRs can be configured to have a conversion ratio (plutonium bred divided by plutonium burned) anywhere in a range from about 0.7 to 1.5. They can therefore consume, can produce excess, or can just maintain. This flexibility can give complete control of plutonium stocks. Configured with a conversion ratio of unity, there will be no net plutonium production—plutonium is burned in the same amount as it is created. In the LWR, plutonium is created and only a portion is burned. The remainder, about 250 kg annually in a 1,000 MWe LWR, as previously noted, is discharged in spent fuel. In an IFR optimized to produce plutonium product (for use in starting up other IFRs), the net annual increase—the difference between plutonium created and plutonium burned in the IFR case—will be about the same or perhaps a little greater than net annual production of plutonium in the LWR fueled with uranium.

12.6.4 Aspects of IFR Self Protection

Remote handling of the fuel product after processing is necessary, making IFR plutonium difficult to handle and far less desirable than clean plutonium or uranium-235 for fabrication of a weapon. Highly enriched centrifuge-produced uranium can be safely handled with gloves only. Plutonium-239 cleanly processed by PUREX can be handled similarly. IFR-processed fuel, principally because of short-lived radioactive decay of some actinides (and trace fission products), remains dangerously radioactive to personnel. The corresponding need for heavy shielding increases the barriers to its unauthorized use, and the absolute necessity of using manipulators to handle IFR-processed plutonium would make the delicate fabrication work of weapons difficult indeed.

12.7 History of the Use of Fissile Material for Weapons

Enough information has been made public now to identify the principal concerns regarding the practicality of "reactor-grade" plutonium in nuclear weapons. The "grade" is a measure of the fraction of the isotopes of plutonium other than plutonium-239 present in the plutonium. Plutonium-240 is the principal isotope at issue. After plutonium-239, it is by far the most abundant plutonium isotope, and it spontaneously emits neutrons that deleteriously affect weapon reliability (e.g. high chance of "pre-ignition" leading to a "fizzle," so-called, when the weapon disassembles before the explosive energy possible can build up), at the very least in the simpler weapon designs.

This entire subject has been obfuscated—mostly for good reason, it can be expected, but sometimes it seems for purposes of making a case that may otherwise lack substance. Anti-nuclear groups point to weapons as their reason for opposing nuclear in general and the IFR in particular. Where truth is shrouded in government and military classification, any assertion stated with conviction seems valid. But there are ascertainable facts here too.

The two U.S. weapons used in August of 1945 used the two different fissile elements—the Hiroshima weapon used uranium-235 and the Nagasaki weapon used plutonium-239. The uranium weapon was considered to be so simple and reliable that its design did not need to be tested, and no test was made before its use in war. (But again, the nation's very best scientists and engineers were responsible.) The plutonium weapon was tested in the New Mexico desert prior to the assembly of a second one that was used on Nagasaki. There is a significant difference in the complexity of design.

Nuclear weapons are detonated by conventional explosives. They compress the fissile material to super-criticality and hold it together long enough for tremendous energy to build up before blowing apart in the huge explosion of a nuclear weapon. Neutrons must be present to initiate the chain reaction. But once neutrons are introduced into the compressed material, the explosive force is very great. It is vital to have the fissile material compressed before neutrons are introduced, intentionally or unintentionally. Once neutrons are present with the fissile material compressed and supercritical, it's all over in an instant. The more supercritical, the greater the explosion. If neutrons come in before the material is adequately compressed, a smaller explosion results, a "fizzle," as has been conjectured for at least one of North Korea's weapons tests, for example.

The sophisticated modern weapons of the weapons states today use plutonium-239 in a fairly pure state. In the earliest form—the Nagasaki weapon, for example—they are constructed as a disassembled plutonium sphere in pieces that are driven together in detonation by conventional explosives. Modern designs simply compact a smaller sub-critical sphere for the same result. Today, naturally fissile uranium, the U-235 isotope, is the likely preferred path to a nuclear weapon for aspiring nuclear weapons states. (Both North Korea and Iran have now assembled uranium enrichment capability using high-speed centrifuges.) This has changed the proliferation picture. It is likely that the easiest and least detectable way to accumulate weapons-usable fissionable material (U-235) is now combined with the easiest way to assemble a weapon (the gun-barrel design, where a separated portion of the uranium-235 "pit" is blown into main part of it in a "gun-barrel" configuration).

The isotopic composition of plutonium and the capabilities of a state or a rogue group need more careful examination than a sweeping statement that "plutonium can be made into weapons." The point at issue here is the place of civilian "reactor-grade" plutonium in weapons, and in particular, the place of it when the weapons designer does not have the extensive scientific, technical, and engineering resources available to a modern national laboratory in a modern state. The situation is not nearly as clear as it might be expected to be. Statements on these issues by authorities in at least some cases have not been particularly helpful. Secrecy is justified, of course, but in some cases it is also convenient to advance a particular political point of view.

In particular, there have been many statements that "reactor-grade plutonium" has been used in nuclear weapons tests, and as evidence of this, the U.S. has released information on its successful test. The fact that two such tests were apparently also attempted by the British in the early fifties was referred to earlier. On examination, the evidence isn't completely convincing that what would be termed "reactor-grade" plutonium today was actually used in any of the three tests.

In 1973, the U.S. released information that in 1962 the U.S. had successfully tested a nuclear weapon using "reactor-grade" plutonium, and additional information was provided in 1993. [13] There has been a continuing controversy over the actual isotopic composition of plutonium—in fact, over whether what later would be understood as reactor-grade plutonium was actually used in the weapon. Evidence in the open literature suggests the likelihood that the plutonium was actually "fuel-grade." If so, it would not be called "reactor-grade" by the later standards continuing to the present day. [13] The difference between the two grades is in the percentage of Pu-240, a very important difference.

Prior to the 1970s, specifically in 1962, there were only two terms in use to define plutonium grades: weapons-grade (no more than 7 percent Pu-240) and reactor-grade (greater than 7 percent Pu-240). At the time of the U.S. test, "reactor-grade" was defined as plutonium with Pu-240 content above the "weapons-grade" 7 percent. In the early 1970s, the term fuel-grade (approximately 7 percent to 19 percent Pu-240) came into use, which shifted the reactor-grade definition to 19 percent or greater Pu-240. [13] Going further, LWR spent fuel has Pu-240 content percentages well up in the twenties, and very substantial fractions of the isotopes Pu-238, Pu-241, and Pu-242 in addition to Pu-239 and Pu-240. It is a very different and much more radioactive isotopic mixture than the composition now termed fuel-grade—which on the lower end of the Pu-240 content at least can be handled without difficulty with gloves and handled routinely in glove boxes.

It is now known that the plutonium used in the test was provided by the United Kingdom under the 1958 United States/United Kingdom Mutual Defense Agreement. [13, 14] It was agreed that a total of about six tonnes of U.K.-produced plutonium would be sent to the United States in return for tritium and highly enriched uranium over the period 1960-1979. Plutonium from the U.K. in 1962 was produced in the U.K.'s graphite-moderated, natural uranium-fueled reactors, no ifs, ands, or buts, for those were the reactors the U.K. had at that time. The commercial versions were called "Magnox reactors" after the magnesium alloy fuel cladding chosen for its low neutron absorption. But still the burnup was sharply limited by the reactivity constraint imposed by its use of natural uranium fuel in the graphite moderator. Buildup of Pu-240 is directly dependent on burnup. The lower the burnup at discharge, the lower Pu-240 content. Once the full fleet of larger Magnox reactors were in operation later, the burnups increased and the plutonium produced had Pu-240 contents approaching 20 percent, but earlier versions would have had lower Pu-240 contents.

At the time of the agreement, the first U.K. commercial Magnox reactors had only just begun construction. The operating reactors in the U.K. were those specifically for weapons-plutonium production at Chapel Cross in Scotland and Windscale in northern England and the closely associated dual-purpose reactors for plutonium production and electricity generation at Calder Hall close by. Argonne National Laboratory was supplied with a few tons of plutonium, which was said to have originated in the U.K., for its fast reactor experiments in the early 1960s. The plutonium had a single uniform isotopic content somewhat above weapons grade, but well less than the Magnox plutonium at the full burnup of the later fleet of commercial reactors. It can be surmised that this too came from the shorter burnup fueling of the Calder Hall or even possibly the Windscale reactors. The point is that the reactor-grade plutonium used in the 1962 test is unlikely to have had plutonium isotopic contents a great deal higher than weapons grade; it was probably on the lower end of fuel grade, and certainly nowhere near those of current LWRs.

Statements by people closely associated with the Carter and Clinton administrations are of some help. In particular, Miller and Von Hippel in Reference [15] state,

> "The information disclosed about this test in 1977 represented a compromise between policy makers in the Carter administration who wished to highlight the proliferation risks of civilian plutonium use and those responsible for protecting classified weapons-design information."

The policy makers referred to in the Carter administration were advancing the policies referred to earlier in this chapter, and would not have been aided in their efforts by highlighting the plutonium isotopic composition now rather disingenuously referred to as "reactor grade." They go on to say,

> "To our knowledge, all U.S. nuclear weapons use weapon-grade plutonium, i.e. plutonium with an isotopic fraction of at least 93.5 percent Pu-239. The same is probably true of the weapons in the arsenals of the other weapon states."

In conclusion, they say,

> "In sum, we are not arguing that a proliferator would not prefer weapons-grade plutonium or highly-enriched uranium to reactor-grade uranium. However, the possible use of reactor-grade plutonium cannot be discounted."

The use of the phrase "possible use of reactor-grade plutonium cannot be discounted" in a context where weapons-grade plutonium or uranium would be preferred would seem to weaken the certainty with which opinions on the use of reactor-grade plutonium have been expressed in the past.

A comment of Luis Alvarez, one of the twentieth century's foremost experimental physicists and an important figure in the development of the bomb, in his autobiography, provides perspective:

> "With modern weapons-grade uranium, the background neutron rate is so low that terrorists, if they had such material, would have a good chance of setting off a high-yield explosion simply by dropping one half of the material onto the other half. Most people seem unaware that if separated U-235 is at hand, it's a trivial job to set off a nuclear explosion, whereas if only plutonium is available, making it explode is the most difficult technical job I know." [16]

The British "reactor-grade plutonium" tests in 1953 very likely had Pu-240 contents only somewhat above weapons grade. Yielding disappointing results, they were probably the tests referred to by the ex-Director of the United Kingdom Atomic Weapons Research Institute at Aldermaston, referred to earlier. ("We tried reactor-grade plutonium a couple of times. We never will again.")

Our purpose here isn't to go into the extensive and for the most part not excessively enlightening literature on the possibility of the use of reactor spent fuel as a weapon, but rather to give the reader a feel for the issues involved. It is obvious, we think, that reactor-grade plutonium from modern reactor spent fuel poses considerable difficulties for all but the most capable of nuclear weapons programs. We also think it obvious that the use of weapons-grade is so much preferable, and for an aspiring proliferating group or nation, uranium-235 so much preferable again, that the link to civilian reactors, properly monitored, is weak to non-existent.

12.8 Monitoring of Processes Always Necessary

In the processes used to provide fuel for reactors, the two vulnerable points are: in the once-through fuel cycle as used in LWRs, at enrichment facilities with their attendant capability to produce highly enriched uranium; and for recycle systems like the fast reactor, the pure plutonium *from reprocessing plants capable of such purification*. Enrichment facilities can produce highly enriched uranium; PUREX reprocessing plants give a very pure plutonium product. The situation with enrichment is straightforward. There is, and there can be, no technical barrier to producing U-235, suitable as produced to use in weapons. Enrichment plants must be carefully safeguarded. The situation with plutonium is more complex. The PUREX plant output is very pure plutonium, but the quality of the plutonium for nuclear weapons depends on the source of the plutonium. From "production reactors" specifically built to produce plutonium for weapons, the plutonium has low contents of isotopes of plutonium above Pu-239, and the PUREX process itself cleans out the non-plutonium higher actinides and fission products. From the two types of civilian thermal reactors, heavy-water-moderated reactors do give lower contents of the higher plutonium isotopes than light-water reactors. The difference has to do both with the higher average energy of the thermal spectrum in light-water reactors, and its much longer burnup, which builds up the higher actinide content. From fast reactors, the plutonium from the fuel has even higher actinide content and radioactivity. Plutonium from IFR fuel is unusable as produced, as documented in a published study by one of the U.S. weapons national laboratories.[17]

The plutonium from the blanket of a fast reactor has much less higher actinide content. Processed in a PUREX plant, it is almost certainly weapons-usable as produced. But in the pyroprocess, the situation as produced is different. The product will contain some higher actinide content—americium particularly, the principal isotope of which is very radioactive, decaying with a half- life of a few hundred years. It will always contain some amount of uranium, and possibly some fission products. The Pu-240 and Pu-241 content will be low. Blanket material will be processed as a practical matter as a blend with fuel material. But normal safeguards and monitoring are necessary.

On-site processing, no off-site transport, no easily concealed diversion streams, no weapons-level purification, and no need for plutonium stocks to build up, represent a pretty sound basis for a start. But there is more.

12.9 Weapons Undesirability: Attributes of IFR Fuel Product—Inherent Self Protection

The IFR pyroprocessed plutonium product is never pure; its chemistry allows plutonium to be extracted always only in a mixture of the minor actinides (neptunium, americium, curium, etc.), uranium and certain of the fission products. The IFR-plutonium mixture has the three properties deleterious to weapons: heat production from the actinides in the mixture, production of spontaneous neutrons, and a high level of gamma radiation.

The characteristics of the IFR fuel form are compared with those of weapons-grade plutonium and reactor grade plutonium in Table 12-1. [17]

Table 12-1. Important Weapons Usability Characteristics

	Weapons-Grade Pu	Reactor-Grade Pu	IFR Actinide Product
Production	Low burnup PUREX	High burnup PUREX	Fast reactor Electrorefining
Composition	Pure Pu 94% Pu-239	Pure Pu 65% Pu-fissile	Pu + MA + U 50% Pu-fissile
Thermal Power watts/kg	2-3	5-10	80-100
Spontaneous neutrons, n/s/g	60	200	300,000
Gamma radiation r/hr at ½ m	0.2	0.2	200

Weapons-grade has the high Pu-239 content of weapons-plutonium production, and the high purity as recovered by PUREX processing. Reactor-grade is a typical LWR spent fuel, with its greater concentrations of the higher Pu isotopes, as recovered by PUREX reprocessing. Essentially, it is the same as weapons-grade except for the isotopic composition. The higher plutonium isotopes give about a factor of three increase in heat production and in spontaneous neutrons, which is extremely inconvenient but which may not be an insurmountable barrier for use as weapons material, depending on the capabilities of the laboratory designing it.

On the other hand, the mixture of IFR actinides has a heat output fifty times higher than weapons-grade plutonium. Such self-heating can cause real problems with the surrounding high explosives, such as melting and perhaps even self-detonation. Spontaneous neutron emission and the gamma radiation level are far above that of weapons or reactor grades—more than a thousand times greater. Neutron multiplication during the assembly will increase the neutron dose even more. Combined with the gamma radiation, the resulting incapacitating dose of radiation would certainly rule out hands-on weapons production. Heat also tends to throw off small tolerances, and stray neutrons interfere with the timing of ignition, key to its effectiveness. Just how undesirable these phenomena are to weapons is obscured by the (justified) classification of information on weapons. But some information is available in published material. A source quoted particularly often is the 1993 article of Carson Mark, ex-Head of the Theoretical Section at Los Alamos.

[18] In it he shows that with a spontaneous neutron source even forty times that of the 1945 Trinity test with weapons plutonium the probability of the expected yield (design yield) would be much less than 1 percent, with a high probability of yields in the neighborhood of a fizzle.

U.S. weapons designers have stated that spent IFR fuel cannot be used to make a nuclear weapon without significant further processing. [17] The IFR actinide mixture could be a feedstock for further processing to recover plutonium, meaning another process must be used on it. But that is precisely the case with the spent fuel itself. The only thing that IFR processing has really added is the ability to put it back in the reactor and burn it as fuel.

12.10 Usability of Pyroprocessing to Acquire Pure Plutonium

Earlier we pointed out that electrorefining process itself is incapable of separating pure plutonium directly usable for weapons production. But could the process be modified to accomplish it? The important fact in assessing this is that the free energies for chloride formation for Pu, Np, Am, and Cm are all in a fairly narrow range, causing them to deposit largely as a group. Element-by-element separation in practical electrorefining isn't possible at any reasonable rate. In theory, even a small difference in free energies can be exploited to separate plutonium, but the rate implied by such a process would be too slow to be of practical value. In practice, it has never been shown that the electrorefining process can somehow be tweaked to produce pure plutonium. (It is difficult enough to achieve the degree of separation between uranium, fission products and the higher actinides that we do. Only starting with pure plutonium throughout the system will give a pure plutonium product, and that is not a useful case.) Assessments by weapons-development experts concluded that the electrorefining process is intrinsically proliferation-resistant and cannot be utilized to produce weapons-usable materials directly. [17, 19-22]

Because of the compactness of electrorefining equipment systems, it has been suggested that the process is more amenable to clandestine operation, even though it produces crude materials requiring further processing. In fact, electrorefining requires high temperature operation, a very high-current power supply, and operations conducted remotely under very pure inert atmosphere conditions. This isn't a process that can be done "in a garage." It's very much more difficult to set up than a crude aqueous chemical separation process that can be done at room temperature in normal atmosphere. The real case is that if conventional aqueous reprocessing is replaced by pyroprocessing in commercial deployment, there certainly is some gain in proliferation resistance.

12.11 Safeguardability

Although proliferation resistant for all the reasons set out above, the fuel cycle facility must have a robust safeguards system. But even in this, the IFR facility has unique characteristics that assist in implementing effective safeguards.

Because all operations in the pyroprocessing facility, including refabrication of fuel, are conducted remotely, inside a hot cell in an inert atmosphere with very limited penetration for access, the containment and surveillance aspect of the safeguards system are straightforward. Authorized access is through airlocks that maintain the purity of the inert-gas atmosphere in the cell. Monitoring the purity of the inert atmosphere itself detects any unauthorized penetration of the cell.

As in any safeguards system, materials control and accountancy is the key to satisfactory safeguards. Comparing IFR processing to conventional aqueous reprocessing, in the latter the spent fuel dissolution tank provides an opportunity to sample a homogeneous solution for the input verification. The absence of such a step in the pyroprocessing facility is an accountancy complication. But while it is certainly true that the aqueous reprocessing plant allows straightforward input verification, it is equally true that after that point accountancy becomes a real challenge: There are literally hundreds of kilometers of piping and thousands of storage vessels of various kinds. Accounting for all of this with accuracy is a considerable task. In existing plants there have been leaks of substantial amounts of material that have gone undetected for a considerable time. [12]

Direct input verification is not feasible by sampling in a pyroprocessing facility. Instead, the input to the electrorefiner is calculated from the known composition of the fuel when it entered the reactor, corrected by calculation of burnup based on detailed physics calculations of the fuel behavior in the reactor. This methodology has been developed over the decades and tested exhaustively against measurement, and assuredly is capable of making such corrections with adequate accuracy. The methodology can be augmented by statistical sampling of the chopped pins, narrowing any uncertainties if need be.

Beyond the input verification step, the majority of the operations in the fuel cycle facility can be relied upon with certainty, as they involve individual discrete items: The cathode products, fabrication ingots, finished fuel pins, etc.—these can simply be counted and weighed. The electrolyte is an exception, but it is amenable to routine sampling and analysis. The nature of the processes and the countable items make a computer-based materials control and accountancy approach both feasible and desirable. In fact, for the EBR-II spent fuel treatment operation in the Fuel Conditioning Facility, such a mass-tracking system has been successfully developed and implemented. [23-24]

Such a system tracks the quantity and location of the materials on an individual isotope basis, including fission products, very nearly in real time. The system also supports the criticality control functions. The individual process steps can be modeled analytically, and with this addition the system can predict the mass flows, and compare and update the calculated results with actual data from the samples given to the analytical laboratory as such data become available.

A mass-tracking system based on these principles could be envisaged for the entire global fast reactor fuel cycle system. With the tremendous advances in super computer speed and almost unlimited data storage capabilities, we can build what will eventually become a massive database system with analytical capabilities to track the fuel materials from the fabricated assembly and pins, through the reactor burnup and changes in composition, through the fuel cycle facility processing steps, through the next reactor loading, or placement in storage. We can start filling in the database with the first prototype or commercial reactor and keep building up the database as more reactors are brought online worldwide. A computer-based system like this can include data on the interface between the containment and surveillance systems and should be able to make the fast reactor and its fuel cycle totally transparent, augmenting and complementing the physical security measures. In theory, such a system could be implemented on existing commercial reactor systems as well, but back tracking the fifty-plus years of historical data for hundreds of reactors would make that impractical. For the fast reactor systems, however, implemented from the new commercial reactor onward, a very useful and workable database system can be made part of the international safeguards.

12.12 The IFR Safeguards and Proliferation Resistant Properties

The characteristics of the IFR system that provide its safeguards and proliferation-resistant properties are inherent in its physics and chemistry. In summarizing, we will distinguish between diversion risk and risk from a nation state.

12.12.1 Risk of diversion Is small

The product is inaccessible due to its high radioactivity. The product of IFR spent fuel processing is a metallic mixture of plutonium, minor actinide elements—neptunium, americium and curium, uranium, and some residual fission products. In the plutonium product uranium is always present—in amounts ranging from perhaps thirty per cent to amounts in excess of the plutonium amount. A pure plutonium product would require reprocessing by some other process unrelated to the IFR process, like solvent extraction—PUREX, in fact, would be the likely choice. In the need to purify it for weapons purposes, the IFR fuel product isn't a great deal different than the spent fuel it came from—it contains all the actinides and it remains highly radioactive.

The product is inaccessible due to its physically inaccessible environment. The process is carried out in a hot cell behind the heavy shielding of the several feet of concrete of the cell walls, looking through feet of leaded glass, with remote handling procedures, tongs, cranes, and at some future time possibly robots. IFR spent fuel is very radioactive, and the product remains very radioactive and must always be handled remotely.

All plutonium in the plant is inaccessible. The large inventory of plutonium safely contained in the reactor core is naturally safeguarded; the inventory in process is safely contained as well; the only vulnerable point is the short path transferring spent and fresh fuel between the two. This single point is ideally suited to item accountancy and continuous surveillance and monitoring. The process facility co-located with the reactor means that any transportation system vulnerability is eliminated as well.

12.12.2 IFR-processed material is highly unlikely to be used by a nation-state for weapons development

IFR spent fuel needs remote handling before processing, of course, but it needs remote handling after processing as well. IFR-processed plutonium is much more difficult to handle than clean plutonium or uranium-235; it is too dangerously radioactive for hands-on fabrication. Highly enriched centrifuge-produced uranium can be safely handled with gloves only. Plutonium-239 cleanly processed by PUREX can be handled similarly. IFR-processed fuel also produces considerable heat. Both radioactivity and heat are to be avoided; more reasonable alternatives are uranium-235 or clean plutonium. Statements by people closely associated with the non-proliferation policies of the Carter and Clinton administrations reinforce the point: "To our knowledge, all U.S. nuclear weapons use weapon-grade plutonium, i.e. plutonium with an isotopic fraction of at least 93.5 percent Pu-239. The same is probably true of the weapons in the arsenals of the other weapon states." [15]

Further to the point, plutonium from the IFR fuel process was stated to be unusable as produced in weapons in a published study by Lawrence Livermore National Laboratory, one of our premier weapons laboratories. [17]

As a weapons material, the IFR spent fuel processing product has three principal undesirable characteristics: high heat production and high levels both of spontaneous neutrons and gamma radiation. The mixture of actinides in pyroprocessed IFR spent fuel has a heat output a factor of fifty or so higher than weapons grade plutonium. Levels of self-heating like this can cause real problems with the surrounding high explosives. Spontaneous neutron emission and the gamma radiation level are three orders of magnitude above that of weapons or reactor grades. Neutron multiplication during the assembly will increase the neutron dose even more. Radiation rules out hands-on fabrication. Again, it is far-fetched to suggest that this product is usable for weapons as is.

It could be used as a feedstock, of course, for further processing by another process; aqueous processing gives high plutonium purity. (Of course, the isotopic distribution of the IFR plutonium is unchanged.) In any case, such a process can equally be used on the spent fuel itself, so what has IFR processing added? For a weapons purpose, only a lot of complication and expense. For its designated purpose, the reason for the process in the first place—to put it in a form to go back into the reactor and burn as fuel.

IFR processing operations on the uranium blanket assemblies differ somewhat from those for spent fuel. Plutonium slowly builds up in the blanket elements, and the plutonium-240 fraction increases very slowly, so the isotopic composition of blanket element plutonium means the onus for protection is placed on the process itself. For electrorefining of uranium blanket assemblies, where plutonium is present only at the few percent level at most, very large quantities of uranium must be deposited on solid cathodes, very slowly increasing the plutonium content of the electrolyte in what must be a very lengthy campaign. If plutonium content is to be more rapidly increased, the high-plutonium-content fuel must be introduced, laden with high isotopes of plutonium and americium, neptunium and curium, as well as fission products. There is no way around this. If blanket plutonium alone is wanted, a lengthy campaign of uranium deposition only must take place. Many runs must be made to build significant amounts of plutonium into the process. Weeks and even months must pass. There is no way such a campaign could escape the attention of even a rudimentary safeguarding scheme.

Operations of the electrorefiner, by their nature, are ideally suited to monitoring. They can be changed only slowly, and a wide variety of signals arise if operations are significantly changed. And in the end, it is not at all clear just how clean a pure blanket product would be. There will still be a significant quantity of uranium in it, and some amount of higher actinides, at best. Electrorefining proceeds in an open and obvious manner, suited to easy monitoring, and to effective denaturing of the product as a matter of course. It never gives a really pure product and it yields excellent IFR fuel. Most process operations will be on the IFR spent fuel, where the assemblies come out in three or four years. Operational efficiency will bring blanket assemblies in along with fuel as they are ready to be processed.

The electrorefining process itself is incapable of separating pure plutonium. For this, introduction of another unrelated process such as PUREX would be necessary. The IFR process itself cannot be modified to do it. Only starting with pure plutonium throughout the system will give a pure plutonium product, and that is not a useful case.

12.12.3 The basic electrochemistry underlying the IFR process illustrates what it can and cannot do

The IFR process takes as input spent reactor fuel and blanket assemblies whose principal element is uranium, with plutonium no more than a third of the uranium (and in blanket elements much less), and with fission products and man-made elements in significant amounts. It uses an electrorefining process somewhat analogous to the processes used in industry for metals like copper and zinc, but in the IFR process with an electrolyte of chloride molten salts. The fission products first are naturally separated from the rest as their electrochemical properties fix them in the electrolyte. The fuel mixture dissolves as chlorides in the electrolyte—uranium chloride, plutonium chloride and so on. A uranium product is gathered first. When the uranium content in the electrolyte has been reduced enough to do so, plutonium and the higher actinides are gathered as a group.

The electrochemical relationships dictate that plutonium will deposit in the cathode used, but only if the uranium chloride content has been reduced very substantially. The uranium chloride must be a fraction of the plutonium chloride, so most of the uranium is collected first on a separate cathode. As long as there is any uranium chloride in the electrolyte, uranium will deposit along with the plutonium and man-made elements. The ratio of plutonium to uranium in the product will vary with the ratio of plutonium chloride to uranium chloride. But the ratio of plutonium to uranium is always less in the product than it was in the electrolyte. A product without uranium requires an electrolyte free of uranium chloride, which after all is the predominant element in the fuel. In practice the electrolyte will always contain some, and a considerable fraction of neptunium, americium, and higher actinides, which also deposit readily with the plutonium.

Thus the electrochemical relationships of the mixture of plutonium and uranium, with higher actinides present always, make separation of a clean plutonium product a practical impossibility. Plutonium simply will not deposit cleanly while uranium chloride is present. There will be significant uranium in the product and the higher actinides deposit with or without uranium present. In the electrochemical cell that is the IFR process, a separated plutonium product is unachievable.

12.13 Summary

In IFR technology the potential for energy production is unlimited, free of greenhouse gas emissions, and its attributes minimize proliferation potential.

How much risk could IFR technology add to the proliferation of nuclear weapons to states not possessing nuclear weapons today? Nuclear weapons have always been developed in secrecy in large laboratories with extensive capabilities. The continuing existence of plutonium is a settled issue; plutonium exists and it will continue to be created in large amounts, with or without IFRs. In all nuclear power, plutonium production is not a matter of choice. All nuclear power plants produce plutonium. Today it generates at least 40 percent of the world's nuclear electricity, and probably more, and stands to generate most of the world's electricity in the future.

Processing of spent fuel is vital to recycle and the massive energy production potential that follows from it. While the processing technique must be good enough for reactor fuel, as electrorefining is, it's equally true that it's desirable *not* to have a technology that's too good. In the IFR process, the high level of radioactivity of the fuel spent fuel ensures remote handling even after processing, making IFR plutonium both difficult to handle and less desirable than clean plutonium or uranium-235 in making a weapon. Highly enriched centrifuge-produced uranium can be handled with gloves only. Plutonium-239 cleanly processed by PUREX can be handled similarly. IFR-processed fuel, principally because of the short-lived radioactive decay of the higher actinides present, remains dangerously radioactive to personnel and produces considerable heat. The corresponding need for heavy shielding increases the height of barriers to its unauthorized use. The absolute necessity of the use of manipulators to handle IFR-processed plutonium rules out delicate fabrication work of the kind mandatory for weapons.

Reactor spent fuel certainly poses considerable difficulties for weapons for all but the most capable of national nuclear-weapons programs. We also think it obvious that the use of weapons-grade fissile material is so much preferable, and for the uninitiated aspiring proliferating group or nation, uranium-235 so much preferable to Pu-239, that the link to civilian reactors that are properly monitored, is weak to non-existent.

The IFR offers on-site processing, no off-site transport, no easily concealed diversion streams, no weapons-level purification, and no need for plutonium stocks to build up. The IFR pyroprocess produces an impure radioactive product, very different from PUREX. Because of the compactness of electrorefining equipment systems, it has been speculated that it is more amenable to clandestine operation even though it produces crude materials requiring further processing. In fact, electrorefining requires high temperature operation and operations conducted remotely with manipulators and tongs under very pure inert-atmosphere conditions.

The nature of the processes and the countable items adapt well to computer-based materials control and accountancy. In fact, for the EBR-II spent fuel-treatment operation at the Idaho National Laboratory's Fuel Conditioning Facility, such a mass tracking system has been successfully developed and implemented. An interesting possibility arises in a mass tracking system based on these principles for the entire global fast reactor fuel cycle system. Supercomputer speeds and almost unlimited data storage capabilities allow a massive database system with analytical capabilities to track the fuel materials from the fabricated assembly and pins, through the reactor burnup and changes in composition, through the fuel cycle facility processing steps, through the next reactor loading or placement in storage. Such a database can begin with the first prototype or commercial reactor and keep on building as more reactors are brought online worldwide.

Finally, returning to the basic protections given by the IFR, the electrochemical relationships of the mixture of plutonium and uranium, with higher actinides present always, make separation of a clean plutonium product a practical impossibility. Plutonium simply will not deposit cleanly while uranium chloride is present. There will be significant uranium in the product and the higher actinides deposit with or without uranium present. In the electrochemical cell that is the IFR process, a clean plutonium product is unachievable.

References

1. Congressional Record, Senate, June 30, 1994.
2. T. Blees, *Prescription for the Planet: The Painless Remedy for our Energy and Environmental Crisis*, 319-343, 2008. http://www.prescriptionfortheplanet.com.
3. T.B. Taylor, "Nuclear Power and Nuclear Weapons," Nuclear Age Peace Foundation, July 1996.
4. S.G. Hecker, quoted in "Visiting North Korea," Physicstoday.org, February 23, 2011.
5. S. M. Younger, *The Bomb, A New History*, Harper Collins Publishers, 2009.
6. IAEA, "International Nuclear Fuel Cycle Evaluation Vol. 1-9," IAEA, Vienna, 1980.
7. Barry Brook, private communication, 2010.
8. U.S. Department of Energy, "Additional Information Concerning Underground Nuclear Weapons Test of Reactor-Grade Plutonium," June 1994.
9. Alright, F. Berkhout, and W. Walker, *World Inventory of Plutonium and Highly Enriched Uranium 1992*, Oxford University Press, 1993.
10. D. Albright, *Separated Civil Plutonium Inventories: Current Status and Future Directions*, Institute for Science and International Security, June 2004.
11. J. Bernstein, *Plutonium: A History of the World's Most Dangerous Element*, Cornell University Press, 2007.

12. T. P. McLaughlin, et al., "A Review of Criticality Accidents, 2000 Revision." LA-13638, May 2000.
13. "Additional Information Concerning Underground Nuclear Weapon Test of Reactor-Grade Plutonium," U.S. Department of Energy, June 1994. http://permanent.access.gpo.gov/websites/osti.gov/www.osti.gov/html/osti/opennet/document/press/pc29.html.
14. "1958 US-UK Mutual Defense Agreement." British American Security Council, http://web.archive.org/web/20061130060326 http://www.basicint.org/nuclear/MDA.htm
15. Marvin Miller and Frank von Hippel, "The Usability of Reactor-Grade Plutonium in Nuclear Weapons: A Reply to Alex deVolpi," *Physics and Society Newsletter*, 26, July 1997.
16. Alvarez, L.W. *Alvarez: Adventures of a Physicist*, Alfred P Sloan Foundation Series Basic Books, 1987.
17. D. L. Goldman, "Some Implications of Using IFR High-Transuranic Plutonium in a Proliferant Nuclear Weapon Program," LLNL Document No. CONDU-94-0199, March 1994.
18. Carson Mark, "Explosive Properties of Reactor-Grade Plutonium," *Science & Global Security*, 4, 111-128, 1993.
19. "Nonproliferation Risks and Benefits of the Integral Fast Reactor," Prepared by International Energy Associates Limited for Argonne National Laboratory, IEAL-R/86-100, 1986.
20. U.S. Department of Energy, "Nonproliferation and Arms Control Assessment of Weapons-Usable Fissile Material Storage and Disposition Alternatives," Draft Report, October 1, 1996.
21. R. Wymer et al., an unpublished report prepared for the Department of State and the Department of Energy, 1992.
22. "Nonproliferation Impacts Assessment for the Treatment and Management of Sodium-Bonded Spent Nuclear Fuel," Office of Arms Control and Nonproliferation, U.S. Department of Energy, July 1999.
23. C. Adams, et al., "The Mass Tracking System: Computerized Support for MC&A and Operations at FCF," *Proc. ANS Topical Meeting on DOE Spent Nuclear Fuel and Fissile Material Management*, Reno, Nevada, June 16-20, 1996.
24. R. W. Benedict, et al., "Material Accountancy in an Electrometallurgical Fuel Conditioning Facility," *Proc. ANS Topical Meeting on DOE Spent Nuclear Fuel and Fissile Material Management*, Reno, Nevada, June 16-20, 1996.

CHAPTER 13

ECONOMICS

13.1 Fast Reactor Capital Cost

13.1.1 What can be learned from fast reactor construction experience to date?

Some notion of likely cost competitiveness can be gained from past fast reactor construction experience, but the information available is limited. It can be said that the capital costs per MWe of the early fast reactors built around the world were much higher than those of LWRs. But the comparisons are not by any means direct and unambiguous. In comparison to the LWR, every difference between the two adds a cost increment to the fast reactor. With one significant exception, they were much smaller in size and electrical capacity than the LWRs built for commercial electricity generation. There were only a few of them. They were built as demonstration plants, by governments underwriting fast reactor development. There was basically one demonstration per country, with no follow-on to take advantage of the experience and lessons learned. Nor were they scaled up and replicated. The LWR had long since passed the stage where first-of-a-kind costs were involved, and had the advantage of economies of scale as well. Further, their purpose was commercial, with the attendant incentive to keep costs down. None of this has applied to fast reactors built to the present time.

Experience with thermal reactor types, as well as other large-scale construction, has shown that capital cost reduction follows naturally through a series of demonstration plants of increasing size once feasibility is proven. This has been true in every country, with exceptions only in the periods when construction undergoes lengthy delays due to organized anti-nuclear legal challenges. But this phased approach of multiple demonstration plants is no longer likely to be affordable, and in any case, with the experience worldwide now, it is probably unnecessary for a fast reactor plant today. Estimating the “settled down” capital cost potential is not an easy task without such experience. Nevertheless, as the economic competitiveness of the fast reactor is taken to be a prerequisite to commercial deployment, we do need to understand the capital cost potential of the fast reactor and what factors influence it.

The earliest fast reactors designed and built in the 1950s give essentially no usable cost information. They were small, and there were just a few of them: EBR-II at 20 MWe; Fermi-1 at 61 MWe; and DFR in the U.K., at about 15 MWe. The principal experience is with the oxide-fueled demonstration plants that came on in the 1970s. France built the 250 MWe Phenix reactor, operational in 1974 and only taken out of service in 2009. Germany built SNR-300, completed, but never operated due to anti-nuclear sentiment in that country. The U.K. built and operated the 270 MWe Prototype Breeder Reactor, PFR. It was troubled by problems in the non-nuclear portion of the system and has been closed down for many years. Japan built and operated Monju, a 300 MWe reactor, shutdown after a relatively minor sodium leak in 1995, and returned to operation in May of 2011. The U.S. built and operated a test reactor, FFTF, with a thermal power of 400 MWth, which operated for a decade without any problems. Because it had no electrical generation capability, it was shut down when need for fuel irradiation experience in a fast neutron environment lessened and there were no further U.S. plans for fast reactor introduction of any kind. The U.S. demonstration plant, CRBR, at 375 MWe, was ready for construction, but was cancelled by the Carter administration, and was eventually terminated in 1983.

In the West, France was the exception. The French followed their demonstration plant with a plant of commercial size, and of the nations that had active fast reactor development programs they went further toward the goal of commercialization of fast reactors than any of the others. Following a 40 MWth experimental fast reactor Rapsodie in 1967 and the 250 MWe Phenix in 1974, a 1,240 MWe full-scale demonstration reactor, SuperPhenix, began operation in 1985. It did not benefit significantly from Phenix experience, as its design had begun almost concurrently with Phenix. It was constructed by Novatome, a largely French European consortium, under a turnkey contract. The fixed price for the nuclear steam supply system was about $1 billion. [1] EdF, the French national utility, was responsible for the balance of the plant, which was about 40% of the total cost. The capital cost of SuperPhenix was reported as 2.1 times that of French PWRs of the time. [2]

In parallel to the SuperPhenix construction, Novatome was also developing a design for a 1,500 MWe SuperPhenix-II. They estimated a 20% cost reduction due to elimination of first-of-a-kind factors and a 17% reduction due to scaling to the larger size. These two factors alone would have reduced the capital cost per kWe basis to about 1.4 times the French PWR costs. [1-2] The conceptual design of SuperPhenix-II also identified substantial reductions of construction commodities relative to SuperPhenix on a per kWe basis. They are shown in Table 13-1. [3]

But SuperPhenix did show that a near-commercial-size fast reactor was more expensive than an LWR. Somewhat blurring the issue even for SuperPhenix, however, was the fact that it was not a strict follow-on from the totally French designed Phenix reactor. The latter was a simple design; the SuperPhenix design was considerably more complex. Nevertheless, estimates from SuperPhenix experience form the most direct comparison we have.

Table 13-1. Structural mass of SuperPhenix-II relative to SuperPhenix based on per kWe comparison

Reactor block	0.52
Fuel handling and storage system	0.21
Intermediate circuits and auxiliaries	0.55
Primary sodium pumps	0.33
IHX and steam generators	0.80
Decay heat removal system excluding stacks	0.57

The scaleup from 250 MWe to 1,240 MWe was a gigantic step and assuring that the reactor would actually operate satisfactorily was the primary goal. But with the cost reduction the principal goal for the follow-on SuperPhenix-II, significant reductions in construction commodities were possible, as illustrated in Table 13-1. This estimate was made in 1985, when the SuperPhenix-II design was still in early development stage. After an extraordinarily troubled history of protests, court cases, violence, and low-level sabotage including a rocket attack by Greens, as well as some technical problems, SuperPhenix operation was finally terminated once and for all in 1997. A new government had come in with Green participation, and that fact, combined with SuperPhenix's erratic operating record, was sufficient to end its operating life; decommissioning began the following year. The prospects for SuperPhenix-II died with it. Nevertheless, the work that is summarized in Table 13-1 does illustrate where the potential is for cost reduction in fast reactor designs in a mature economy of fast reactors. It suggests that the potential for economic competitiveness for the first fast reactors built is approachable, and is likely in a larger mature economy of fast reactors.

Apart from the French fast reactor program, the next largest fast reactor plant ever built is Russian BN-600, which started operation in 1980. (An earlier demonstration plant, BN-350, had a satisfactory history, producing power as well as filling a desalination mission.) The BN-600 capital cost in $/kWe was reported to be 1.5 times that of the Russian LWRs. As follow-on, three BN-800 plants were planned to be constructed but were abandoned in the chaos of the breakup of the Soviet Union. In 2006, construction of one BN-800 was re-initiated at Beloyarsk. The updated cost estimate for BN-800 per kWe basis is 0.9 of BN-600. [4] In parallel, a larger BN-1800 design is being developed, and a preliminary estimate indicates that the capital cost per kWe would be about 0.48 of BN-600. This does indicate the potential for competitiveness with LWRs. [4]

India, too, has successfully operated a small Fast Breeder Test Reactor (42.5 MWth/12 MWe); it has been operating since 1985, and has included demonstration of a full carbide-fueled core. India probably will be the first nation to commercialize fast reactors. In 2004, India started construction of a 500 MWe Prototype Fast Breeder Reactor (PFBR). The capital cost for this project is estimated at $622 million. [5] The construction completion is targeted for 2014. Following this project, four more similar 500 MWe units are planned in two twin units at two different sites. The capital cost for these units are estimated at $544 million each. [5] With different materials and labor rates and financing structure, it is difficult to judge how this would translate to a plant built in the U.S. India's current commercial reactors are small 200 MWe heavy water reactors, and the 500 MWe fast reactors should compete favorably in India in $/kWe. However, the fully-developed LWRs to be introduced later may have significant capital cost advantages there as well.

13.1.2 Generic comparison with LWR

Comparison of fast reactor capital cost with the capital cost of commercial LWRs is not straightforward either. First, the part that should be straightforward, that of identifying the capital cost of commercial reactors, isn't straightforward at all. U.S. LWRs were built twenty or more years ago, under wildly varying construction environments, some prior to the anti-nuclear campaigns of cost increases, some during the height of them, and a few after. Comparisons between PWR, BWR, heavy water reactors, and gas-cooled reactors are not straightforward either, even though, with the water reactor types, we are dealing with actual experience. Comparison with yet-to-be-designed fast reactors involves more uncertainty. However, the details of the makeup of capital costs do provide useful insight.

The Department of Energy's Energy Economics Data Base (EEDB) defines a code of accounts for estimating and categorizing such cost components. [6] For illustrative purposes, a reference PWR capital cost breakdown developed for the EEDB is presented in Table 13-2. [7] Since the database was generated in the 1980s, the absolute dollar amounts have little relevance to today, so the cost breakdown is expressed in terms of percentage of the total direct costs.

The normalization doesn't allow direct comparison of the total cost of the two reactor types. It shows how the costs of a nuclear plant apportion between the various elements of the plant. A brief explanation of the nature of these costs is given below.

Direct costs include those construction and installation costs associated with the operating plant structures, systems, and components. Account 21 is the site improvement and all the reactor buildings and the balance-of-plant buildings and structures. The rest of the direct cost accounts are associated with the equipment systems, the equipment itself, and such things as the costs of transportation and insurance, provision for shipping fixtures and skids, startup and acceptance testing equipment, on-site unloading and receiving inspections, and installation.

Table 13-2. Typical Capital Cost Components Normalized to Total Direct Costs

EEDB Account	Account Description	EEDB LWR	Typical Fast Reactor
21	Structure and Improvements	26	13
22	Reactor Plant Equipment	30	56
23	Turbine Plant Equipment	24	14
24	Electrical Plant Equipment	11	11
25	Misc. Plant Equipment	4	3
26	Heat Rejection System	5	3
	Total Direct Costs	100	100
91	Construction Services	21	19
92	Home Office Engineering and Services	33	44
93	Field Office Engineering and Services	16	9
94	Owner's Costs	-	-
	Total Indirect Costs	70	72
	Total Direct and Indirect Costs	170	172

Indirect costs begin with account 91, for construction services, which includes costs for temporary construction facilities at the reactor site, construction tools and equipment, payroll insurance and taxes, and payments to federal, state, and local governments for taxes, fees, and permits. Account 92, for home office engineering services, includes costs of engineering and home office services that are specific to the site. These costs include engineering and design, procurement and expediting activities, estimating and cost control, engineering planning and scheduling, and the services of home office QA engineers and staff personnel engaged in work on the project. Account 93, for field office engineering and services, includes the construction management activities associated with on-site management of construction, site QA/QC, plant startup and test procedures, and the supporting costs for these functions. Account 94 include the owner's staff for project management, integration, licensing, QA/QC, etc., the initial staffing and training of operations, maintenance, supervisory, and administrative personnel, the initial stock of spare parts, consumables, and supplies, and the initial inventory of sodium and other capital equipment.

Controlling indirect costs is of the utmost importance, and those have varied widely from project to project. In past U.S. experience, the indirect costs dominated the direct costs in some plants, with the indirect costs running up to ridiculous levels, actually higher than the direct costs. Interest on the borrowed capital kept adding on as plant construction was held up by one legal challenge after another. A key to capital cost competitiveness for the fast reactor, as for any other, is keeping the indirect costs down to a reasonable level; in particular, construction hold-ups cannot be tolerated. In maturity, there is no reason why the indirect costs for fast reactors would be any higher than those of LWRs.

The most effective measure to reduce the indirect costs is to standardize the plant design so that site-specific engineering and design costs are avoided. The seismic isolation system can help to keep the standard design applicable for a variety of potential sites regardless of the site-specific seismic design spectra. The seismic isolation system itself can be fine-tuned to cope with any site-specific seismic design criteria, leaving the plant structural design as a standardized plant. [8-9]

As for the direct cost components, the reactor equipment cost may be higher for the fast reactor because of higher-temperature structural materials and additional equipment associated with the intermediate heat transport system. On the other hand, the containment building and structures can be reduced in size and in the commodity amounts, because high pressure containment is not required for fast reactors. The balance of the equipment systems, such as the turbine, could be slightly less for fast reactor because of a higher thermal efficiency and hence reduced thermal output for any given electrical output. On balance, the capital cost for fast reactors should be in the same range of variations that exist for LWRs.

In a recently completed commercial feasibility study done by the Japan Atomic Energy Agency, the capital cost for their 1,500 MWe JAEA Sodium Cooled Fast Reactor (JSFR) was estimated to be less than that of an equivalent size PWR. [10-11] A significant reduction in the construction commodities and building sizes was achieved by a number of design changes, such as combining the IHX and pump into one unit, shortening the piping length with advanced alloys, and reducing the number of loops with large components.

Overall, though, because of first-of-a-kind costs, capital cost competitiveness is unlikely in the first few fast reactor plants. Too much focus should not be placed on the capital cost reduction for the early reactors. Risk in terms of large sodium component reliability and system engineering is more important than the economies of scale that push toward larger reactor sizes. Initial fast reactors should be in the range of 600 MWe sizes before scaleups begin. Economies of scale will naturally push to a larger size in a mature economy. Even for the mature LWR industry, the reactor size has been in the 1,000–1,300 MWe range and the scaleup to 1,500–1,800 MWe size is only now being planned for the next evolutionary plants, after thousands of reactor-years experience with the current generation.

Most importantly, fast reactor design should exploit inherent properties of sodium and inherent safety characteristics so that the system as a whole is highly reliable, easy to operate, and has assured longevity. Favorable economics will follow in a mature fast reactor economy. If the first commercial plant isn't cost-competitive, initial deployment will have to be driven by considerations of national policy, prudent planning, resource scarcity, perceptions of future costs, and other such factors. Capital costs for the initial few reactors will be dominated by indirect costs, and it is crucially important to establish a project infrastructure that contains them. The EBR-II project model had a small cadre of experts (less than one hundred) who were fully responsible, from the initial concept through engineering, detailed design, procurement, installation, and final acceptance testing. [12] A modern fast reactor, whatever its size, will require a somewhat larger cadre, but expertise is paramount, and in size it should be on the order of few hundred engineers only. A sound project organization is vital, not only for the success of the initial demonstration project, but also as the model for construction infrastructure in maturity.

13.2 LWR Fuel Cycle Cost

13.2.1 The Components

Individual LWR fuel cost components vary slightly, depending on the fuel type, burnup level, and other factors; however, the total LWR fuel cycle cost is rather insensitive to the specific fuel data. Here we will take the reference case assumed in the 2003 MIT study.[13] The fresh fuel is enriched to 4.5% U-235 to achieve 50,000 MWD/T burnup, and the spent fuel composition is 93.4% uranium (of which 1.1% is U-235), 5.15% fission products, 1.33% plutonium and 0.12% minor actinides. The cost assumptions, also taken from the 2003 MIT study (with the exception of uranium price), are listed in Table 13-3. The MIT study assumed a uranium price of \$30/kgU (=\$11.54/lbU_3O_8) which was the prevailing price at the time. In the future \$30/lbU_3O_8 is more realistic and we have adopted it as the reference value in our analysis. (The conversion factor is 1.18 for U to U_3O_8 and 2.2 for kg to lb, and hence is 2.6 for \$/lbU_3O_8 to \$/kgU.)

The fuel cycle cost methodology practiced in regulated utility accounting is rather complex. Cash flows occurring at various times are present-worthed, capitalized at fuel loading, and depreciated as allowed by the tax code. In our analysis, we will treat all cost components as expenses (not capitalized) and will ignore the time differences of cash flows, except for a major difference of ten years or more which occurs at the backend of the fuel cycle, as will be discussed later. Without much loss in accuracy, this simpler methodology makes it easier to quantify the impact of any individual component and to compare different options for the backend of the fuel cycle, as well as to understand the fast reactor implications.

Table 13-3. Cost Assumptions

Uranium Ore, $/lb$U_3O_8$	30
Conversion to UF_6, $/kgU	8
Enrichment, $/SWU	100
Fabrication, $/kgU	275
Disposal Fee, mills/kWhr	1
Reprocessing, $/kgHM	1,000
MOX Fabrication, $/kgHM	1,500

In the above table UF_6 is uranium fluoride, a gas that enables the enrichment process to be carried out. MOX is mixed uranium/plutonium oxide. The SWU—"separative work unit"—is a measure of the amount of uranium processed and its degree of enrichment. Mills are defined as a tenth of a cent. HM, heavy metal, is uranium and all isotopes heavier than uranium.

The resulting fuel cycle cost is presented in Table 13-4 in terms of $/kgHM and mills/kWh. The 5.5 mills/kWh in Table 13-4 is consistent with current average U.S. fuel cycle costs of 5 mills/kWh. Conversion to UF_6, enrichment, and fabrication services are well proven and highly competitive and their price is expected to be stable (if anything, prices may drop, especially for new enrichment technologies). However, although the uranium spot market price had been in the neighborhood of $10/lb for 15 years through 2003, since then the spot price rose above $130/lb briefly and has now settled down in the $70/lb range, as illustrated in Figure 13-1. [14]

Table 13-4. LWR Fuel Cycle Cost

	$/kgHM	mills/kWh
Uranium	660	1.66
Conversion	70	0.17
Enrichment	770	1.94
Fabrication	275	0.69
Disposal Fee	400	1.00
Total	2175	5.46

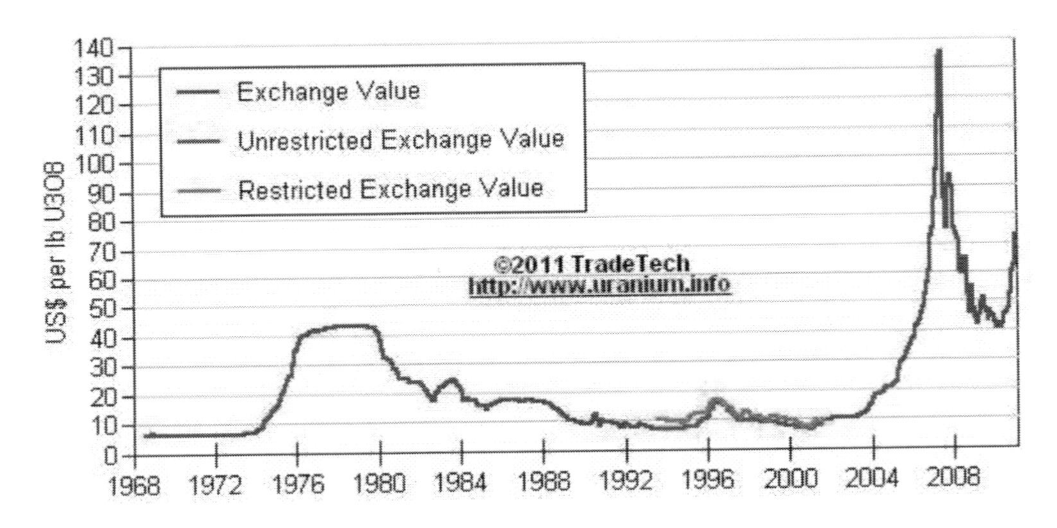

Figure 13-1. Uranium spot market price trend (Source: TradeTech [14])

13.2.2 Effect of Uranium Price

Uranium prices for long-term contracts have been less volatile. The uranium price increased sharply in the mid-1970s, from about $8/lb to above $40/lb, when the large expansion of nuclear power was in progress. When nuclear power plant orders were delayed or canceled in the late 1970s and early 1980s, the uranium price started a steady decline, which lasted about twenty-five years. The uranium prices for long-term contracts, which supply about 85% of the demand, have been rather opaque because there is no terminal market quotation and full details of contract clauses are seldom released. Figure 13-2 illustrates the differences between the long-term contract prices and the spot-market prices as estimated by Euratom for the European market. [15] The Euratom spot-market prices track the U.S. spot market very closely. As can be seen, the long-term contract price decline has been less drastic. The spot-market prices started to increase during 1996–97, closing the gap with long-term contract prices. However, when uranium from weapons stockpiles became available in the late 90s, it depressed the spot-market prices. The effect lasted a few years, but since 2004 the uranium price has increased sharply. The recent uranium price trend purchased by the U.S. utilities is summarized in Figure 13-3. [16] There is no question that the sharp increases in spot-market price and the continued expansion of nuclear energy around the world will exert pressure on the long-term contract prices as well. This is evidenced in the U.S. market, as shown in Figure 13-3; however, it is impossible to predict the future price levels.

For the analysis presented here we assumed $30/lb as the reference case. The impact of varying uranium price on the fuel cycle cost is illustrated in Figure 13-4. The sensitivity of fuel cycle cost to uranium price can be seen in the following: If the uranium price were to escalate fivefold above this assumption, to $150/lb, then the fuel cycle cost component will more than double, to about 12 mills/kWh and have a significant impact on the LWR fuel cycle cost.

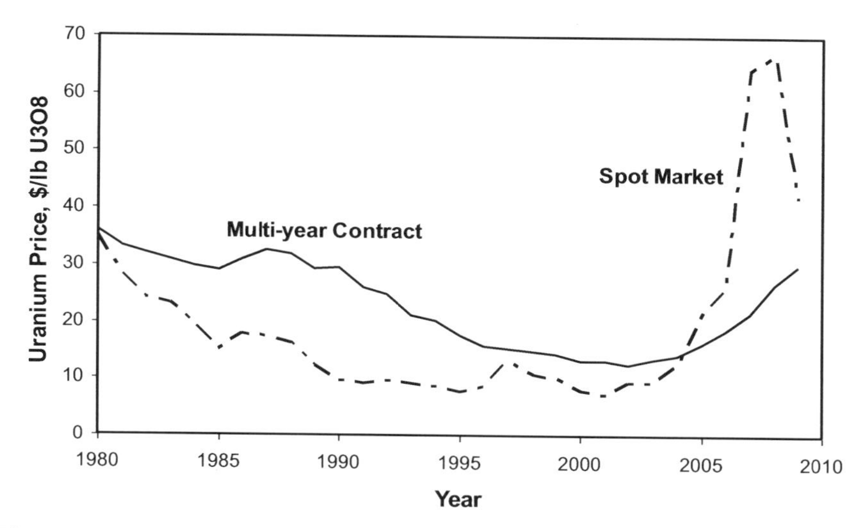

Figure 13-2. Euratom annual average uranium prices: spot market vs. long-term contracts (Source: Euratom Annual Report [15])

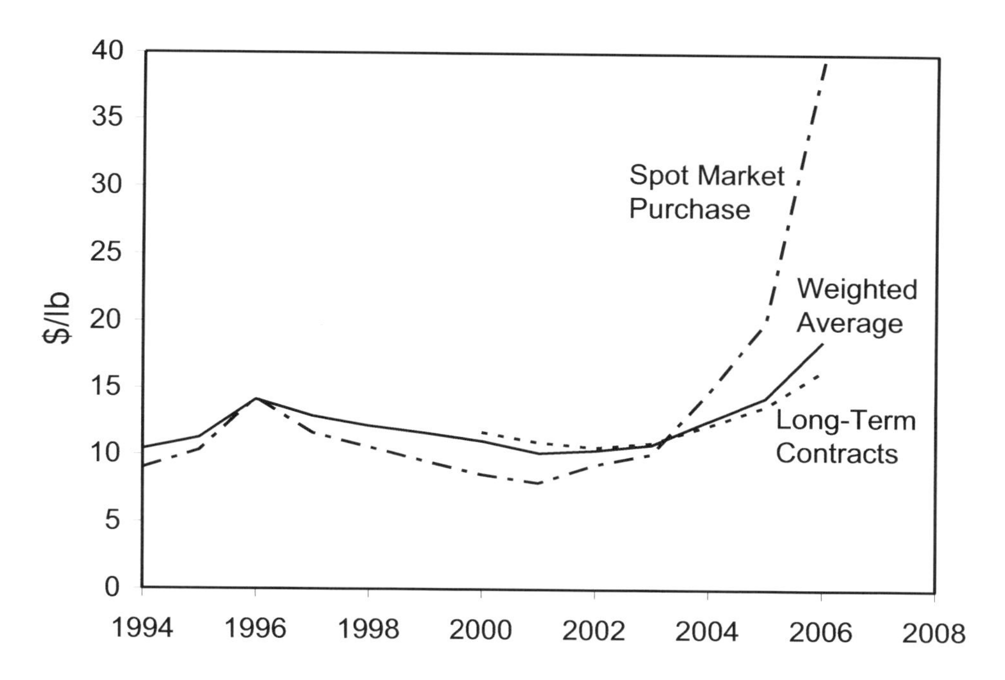

Figure 13-3. Weighted-average price of uranium purchased by the U.S. utilities

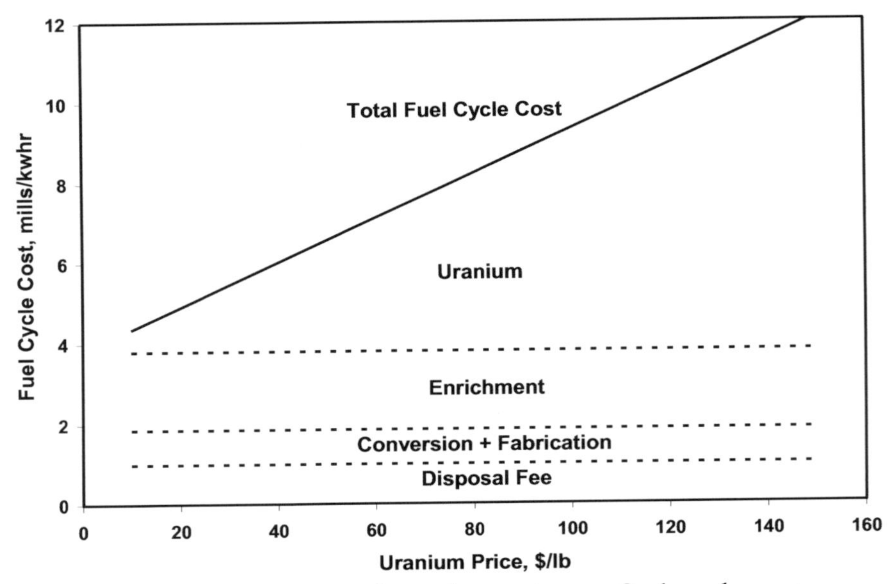

Figure 13-4. Impact of uranium price on fuel cycle cost

13.2.3 Effect on Recycle in LWR

Uranium price also has a significant impact on economic incentives to recycle. In the economics of limited recycle as proposed and currently practiced in the LWR, it is important to recognize the quite different perspectives on the closed fuel cycle in Europe and Japan. Single-pass reprocessing to recover Pu and create fresh MOX fuel leads to a cost model different from the U.S. situation. We will look at the European and Japanese cost models first and then come back to the question of the recycle economics for the U.S. situation.

The first situation we will look at is recycle of the uranium made available after reprocessing. In the recycle system, the reprocessing and subsequent waste disposal costs are levied against the fuel batch as it generates electricity. Since reprocessing is carried out after a significant cooling in storage, this future cost is present-worthed using a set discount rate. If we assume a delay time of ten years and a discount rate of 5%/yr, then the present worth of reprocessing cost would be around 60% of the future cash outlay. High-level waste volume is compacted by reprocessing, and the disposal cost may be reduced from that of direct disposal of spent fuel. We arbitrarily assume that the disposal cost could be reduced to one half of the direct disposal fee and this cost is also present-worthed since the cost occurs in the future. For the U.S. situation, the waste disposal fee of 1 mill/kWh is levied at fuel discharge; hence it is the present cost and cannot be discounted.

The fuel cycle cost for the limited closed fuel cycle of the LWR is presented in Table 13-5, and compared with the once-through fuel cycle cost. Because of the added reprocessing cost, the closed fuel cycle cost is about 15% higher than that of the once-through fuel cycle. However, this increased cost can be partially offset by the recycle credits.

Table 13-5. Comparison of closed fuel cycle vs. once-through costs, \$/kgHM

	Closed Fuel Cycle	Once-Through
Uranium	660	660
Conversion	70	70
Enrichment	770	770
Fabrication	275	275
Reprocessing	610	-
Disposal Fee	120	400
Total	2505	2175

Fresh low-enriched uranium fuel cannot be burned down completely in fissile U-235 in a single pass. At discharge, it contains approximately 20% of the initial natural uranium equivalent value and about 5% of its separative work unit (SWU) value. However, recycling of the reprocessed uranium is not straightforward. If reprocessed uranium is used as feedstock in the reenrichment process, the non-fissile U-236 will have built up sufficiently that the U-235 enrichment has to be raised about 15% to counteract the reactivity penalty caused by the absorption cross-section of U-236. This results in a separative work unit penalty, offsetting the savings in natural uranium. U-236 is a nuisance from a reactor physics point of view, but harmless otherwise. More serious is the problem of U-232 from alpha decay of Pu-236 in reprocessed high-burnup uranium. The U-232 buildup is only a trace amount —0.5 to 5 parts per billion depending on the burnup level. But U-232 undergoes a series of alpha decays with very energetic gammas, harmful to personnel. Contamination concerns in the enrichment plants and the fabrication lines itself will prevent recycling of reprocessed uranium. However, it can be used in mixed oxide fuel fabrication plants where plutonium amounts can be selected to neutralize the U-236, or in fast reactor fuels where none of this matters. Therefore, the reprocessed uranium is typically left in interim storage until a higher uranium price can justify its use.

13.2.4 Recycle of Plutonium as Mixed Oxide Fuel

Although reprocessed uranium is not recycled, some plutonium is recycled in selected reactors as mixed oxide (MOX) fuel. However, the reactivity worth of plutonium in the thermal spectrum of an LWR is only about half that of U-235. Full plutonium recycling saves a natural uranium equivalent of 10-15% only. Since the MOX fabrication penalty is substantial (about five times more expensive than uranium fuel fabrication), the economic incentive to recycle MOX fuel is weak. The MOX fuel cycle cost is compared with uranium fuel in a closed fuel cycle in Table 13-6.

Table 13-6. Comparison of MOX with UOX in Closed Fuel Cycles, $/kgHM

	UOX	MOX
Uranium	660	78
Conversion	70	8
Enrichment	770	0
Fabrication	275	1500
Reprocessing	610	610
Disposal Fee	120	120
Total	2505	2316

If reprocessing and MOX fabrication and associated plutonium reprocessing facilities have been constructed and their capital costs have been amortized, then a closed fuel cycle is affordable, although there is still some economic penalty. On the other hand, if such infrastructure is not available, then there is no economic incentive to reprocess and recycle in LWRs. That is the present situation in the U.S.

Another way of looking at the economic incentives for reprocessing and recycle is to consider current spent fuel disposition. If spent fuel is disposed directly to a repository, the 1 mill/kWh or $400/kg (at 50,000 MWD/T burnup) will be sufficient. If the uranium price escalates, what level would it have to reach in order to provide a serious incentive to recycle? As illustrated in Figure 13-5, assuming a reprocessing cost of $1,000/kg, the uranium price has to escalate to about $120/lb for recycle to be economically viable. Alternatively, if the reprocessing cost is reduced to $500/kg, recycling can be economic even at a uranium price of $40/lb.

From an economic point of view only, LWR fuel cycle closure cannot be justified under today's economic parameters, nor is it expected to be in the foreseeable future. Nevertheless, in France and Japan, somewhat limited plutonium recycling is being carried out, suggested as an interim step toward a longer-term full fuel cycle closure with fast reactors.

Another potential justification for LWR recycle could be for waste management purposes. But for any significant impact on waste management, the long-lived actinides must be removed from the waste stream and burned in the reactor. This isn't what's currently being done, and more importantly, the thermal spectrum of LWRs will not burn the actinides efficiently, as we have discussed previously.

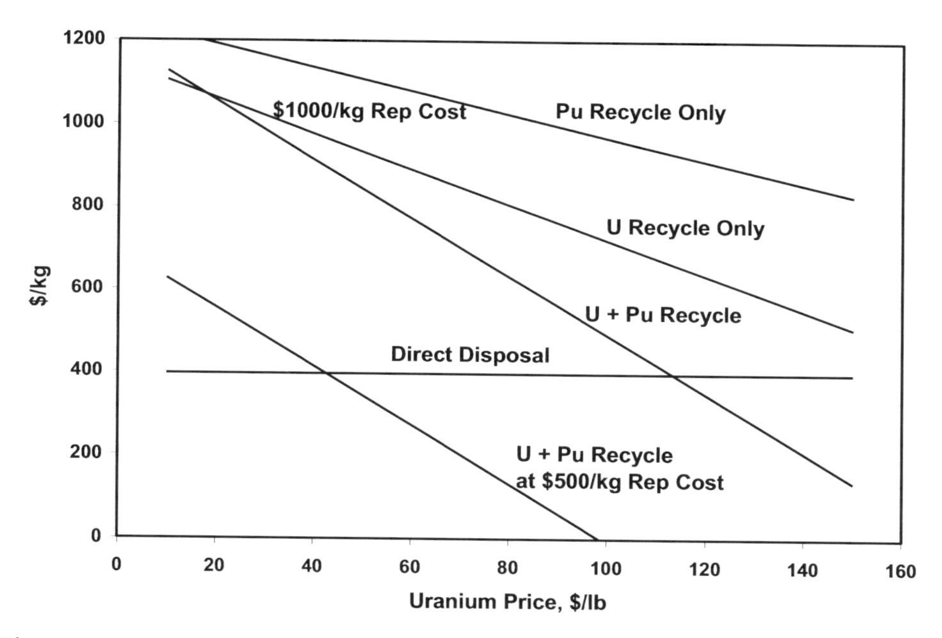

Figure 13-5. Spent fuel disposition cost: direct disposal vs. recycle in LWR

13.3 Fast Reactor Fuel Cycle Closure

In contrast to the LWR, fast reactor fuel cycle closure is mandatory for a number of reasons, most notably fissile recovery, resource extension, and waste management. The question then is its economics. Besides eliminating the short- to medium-term requirements for fresh uranium fuel, there are also indirect economic benefits from the resource extension and waste management advantages as well, which will help to offset the direct cost of the fuel cycle closure using fast reactors. But the reprocessing cost itself is the most important factor. An economically viable fuel cycle will help to reduce the economic barrier to early deployment of fast reactors. The choice of reprocessing technology is at the center of this.

The IFR fuel cycle based on metal fuel and pyroprocessing should be superior in economics to conventional oxide fuel and aqueous reprocessing. Aqueous reprocessing has been commercialized for LWR spent fuel processing in Europe and Japan. In principle, the same aqueous reprocessing can be applied to handle the fast reactor oxide spent fuel. However, fast reactor fuel does raise some additional technical challenges. Fast reactor spent fuel has a much higher content of actinides—on the order of 20% compared to the 1% typical of LWR spent fuel. High-actinide oxide fuel does not dissolve as readily in nitric acid as the uranium-based fuel, and importantly, criticality constraints limit the process equipment sizes. Fast reactor spent fuel processing must deal with much higher-burnup spent fuel, and hence higher specific radioactivity, stainless steel cladding instead of zircaloy, and residual sodium, all of which result in additional engineering challenges. Laboratory-scale demonstrations have been conducted at Marcoule in France, Dounreay in the U.K., and Tokai in Japan. A pilot scale project, TOR (5-10 T/yr), has been successfully demonstrated at Marcoule. It is clear that the same degree of economies of scale as contemplated for the LWR reprocessing plants cannot be achieved for the fast reactor processing plants with oxide fuel. Both the U.K. and France planned a 60 T/yr throughput rate as a target for the next set of pilot-scale plants.

Further, remote fabrication of actinide-containing oxide fuel is extremely difficult. Uranium oxide fabrication is done currently with hands-on operation and requires precision milling of the pellets. Mixed oxide fabrication is done remotely, but maintenance also requires hands-on operation. Remotizing both fabrication and maintenance operations for actinide-containing fuels is a major challenge and would certainly sharply impact economics.

On the hand, IFR metal fuel easily fabricates remotely using injection casting. It is compatible with pyroprocessing—both are compact processes. The two could make improvements in kind in the economics of fast reactor fuel cycle closure. Although many experts agree on the potential for improvements, exact quantification is a difficult task because of the difference in technology maturity. Aqueous reprocessing is much more fully developed, especially for LWR reprocessing. On the other hand, remote fabrication is better developed for metal than for oxide fuel. In addition to a large difference in the technology maturity level, the two processes are too radically different in all aspects to allow a direct comparison of one to the other.

Conceptual designs for both, under the same ground rules, using a bottom-up approach, must first be developed. Credible estimates of capital costs can then be developed. Fortunately, comprehensive conceptual design efforts were carried out in the mid-1980s to allow such comparison. Argonne National Laboratory developed a detailed pre-conceptual design of commercial-scale fuel cycle facility based on pyroprocessing and metal fuel fabrication to serve a 1400 MWe fast reactor. [17] Oak Ridge National Laboratory and Hanford Engineering Development Laboratory jointly developed an equivalent fuel cycle facility based on aqueous reprocessing and mixed oxide fabrication to serve the same size reactor. [18]

The pyroprocessing-based fuel cycle facility involves only a few processing steps, and all processing equipment systems are extraordinarily compact. There are dramatic simplifications and cost reductions in all three areas of reprocessing, refabrication, and waste treatment. The comparisons of these two facilities are summarized in Table 13-7.

Table 13-7. Summary Comparison of Fuel Cycle Facility to Serve 1,400 MWe Fast Reactor

	Pyroprocessing	Aqueous Processing
Size and Commodities		
Building Volume, ft^3	852,500	5,314,000
Volume of Process Cells, ft^3	41,260	424,300
High Density Concrete, cy	133	3,000
Normal Density Concrete, cy	7,970	35-40,000
Capital Cost, $million (2011$)		
Facility and Construction	65.2	186.0
Equipment Systems	31.0	311.0
Contingencies	24.0	124.2
Total	120.2	621.2

The capital cost of the pyroprocessing-based fuel cycle facility is a factor of five less than that of an equivalent fuel cycle facility based on aqueous reprocessing. This large difference is easily understandable in the comparison of the size and material requirements of the equipment systems, which are reduced by a factor of ten, and construction commodity amounts reduced by a factor of five to ten. This comparison was based on significant engineering efforts, on the order of two hundred man-months. This was the most comprehensive attempt to date comparing the economic potential of pyroprocessing to aqueous reprocessing. Since then, several cursory evaluations have been made, but none comes close to this one in technical detail.

The comparison can be discounted on the basis that the aqueous processing technology is established whereas the pyroprocessing technology has only partly been demonstrated at the scale required. However, a distinction should be drawn between uncertainties in technology details and uncertainties in cost estimates. The electrorefining step requires further engineering development, and detailed process parameters, such as current density and applied voltage, may evolve through optimization. However, these refinements won't affect the electrorefining approach, the equipment size, or the facility layout. Cost estimates will not change by any appreciable amount. The differences in technology are different in kind, and they result in large differences in the kind of process equipment, the number of pieces, and the sizes of it. This point is illustrated in Table 13-8.

Experience with the post-1994 EBR-II spent fuel treatment project lends support to the likely economic advantage of pyroprocessing over aqueous processing for fast reactor application. The high concentration of plutonium or actinides, on the order of 20-30 % of heavy metal, dictates a small-batch or small-vessel operation for criticality control, a natural fit to the batch operation mode of pyroprocessing. Further development work will be focused on optimizing the process chemistry details, but the overall equipment size or complexity should not be impacted. The simplest way of saying all this is that if the process works as expected it will be far more economic than the aqueous process. But more work is needed to firmly establish the process.

Table 13-8. Comparison of Reprocessing Equipment

Pyroprocessing	Aqueous Reprocessing
1 Disassembly/Chop Station	104 Tanks
1 Anode Loading Station	17 Centrifuges and Contactors
1 Halide Slagging Furnace	11 Strippers and Separators
1 Ingot Cleaning Station	2 Steam Generators
1 Sample and Ingot Weight Station	8 Filters
2 Electrorefiners	7 Vaporizers and Evaporators
2 Cathode Processors	9 Traps
3 Storage Stations	15 Retention and Recovery Beds
1 Analytical Sampling Station	6 Washers
1 Analytical Lab Equipment	1 Absorption Column
1 In-Cell Support Equipment	1 Compressor
	1 Purification Still
	1 Acid Fractionator
	2 Digesters
	1 Rotary Dissolver
	1 Fuel Cleaning Chamber
	1 Disassembly Station
	1 Shear
	1 Reduction Kiln
	1 Denitration Kiln
	1 Co-Denitration Kiln
	Large amount of Piping and Connectors

Remote fabrication based on injection casting was demonstrated as far back as the 1960s. The new Fuel Manufacturing Facility (FMF) at Argonne-West, which became operational in March 1987, was constructed to the upgraded safeguards requirements, but it also allowed scale up of the casting process itself. The EBR-II injecting casting furnace had been operated on a 5-kg batch, but two 10-kg batches were provided for this facility. As it turned out, this facility alone had sufficient throughput capability to supply the fuel for both EBR-II and FFTF. The total cost for the facility construction and the equipment systems was only $4 million. (This can be compared with the costs of the remote fabrication line (the SAF line) for oxide fuel for FFTF, higher by an amount that may have approached or exceeded a hundred million dollars. It's just so much more difficult to do.) Although the FMF was designed to fabricate uranium fuels only, it was estimated that an additional $1 million would have been sufficient to qualify the facility for plutonium fuel fabrication. This, along with a low-cost refurbishment of Fuel Conditioning Facility, strongly suggests favorable economics for metal fuel cycle closure.

13.4 IFR Fuel Cycle Cost

The fuel cycle facility pre-conceptual design from 1985, discussed above, was sized to serve just one 1,400 MWe fast reactor. In a mature industry, one larger regional fuel cycle facility should serve several fast reactors, possibly on the same site, in which case some further economies of scale can be realized. The capital cost estimate of $120.2 million presented above may be optimistic, but with economies of scale achievable in maturity as well, we will adopt a $100 million capital cost figure for a 1,000 MWe plant as the basis for the fuel cycle cost analyses presented below.

The traditional breakdown of the fuel cycle cost for LWRs isn't directly applicable to the IFR fuel cycle. For the IFR fuel cycle facility, the reprocessing and refabrication are done in the same hot cell and costs can't be segregated. The ownership of the facility itself is an open question. It may be operated by the utility itself, or it very likely could be subcontracted to an independent specialty contractor.

For the IFR fuel cycle, four components contribute to the cost: fuel cycle facility capital fixed charges, fuel cycle facility operating and maintenance costs, driver and blanket hardware supplies, and waste disposal fee.

The fuel cycle facility is capitalized the same way as the reactor plant. The initial capitalized investment includes the construction costs, interest during construction, and owner's cost for startup. Traditionally, a higher fixed charge rate is used for fuel cycle facilities, reflecting higher market risks and different capital structure. However, the IFR fuel cycle facility is an integral part of the reactor plant and it is appropriate to use the same fixed charge rate as the reactor plant. Here we assume a capital fixed charge rate of 15 % per year, which includes amortization of the capital investment as well as applicable taxes, and so on. For the assumed capital cost of $100 million, the annual fixed charge cost will be $15 million.

The operating and maintenance cost of the fuel cycle facility includes costs for process personnel and support personnel, estimated to be $6 million, and process consumables, spare parts, and utilities, estimated to be $4 million, for a total of $10 million per year.

The cost for driver and blanket assembly hardware components, such as cladding, end plugs, wires, duct hardware, etc., is separated from the process consumables because these items are independent of the process operations. This cost component can be treated as expense. The costs for control rods and shield assemblies are not included here as part of the fuel cycle cost, because they are included in the reactor plant's operating and maintenance cost. With the initial setup costs put aside, the hardware expenses are estimated to be $6 million per year.

The IFR fuel cycle cost is then summarized in Table 13-9. We assumed the disposal fee would be half of the LWR direct disposal fee of 1 mill/kWh because of the major reduction in the radiological lifetime.

Table 13-9. IFR Fuel Cycle Cost Components

	$million/GWe-yr	mill/kWh
Capital fixed charges	15	1.90
Operating and maintenance	10	1.27
Process consumables, etc.	6	0.76
Disposal fee	4	0.50
Total	35	4.43

Once an IFR starts up, with an initial fissile inventory, the reactor will be self-sustaining in fissile fuel, and hence will not be subject to any cost escalation due to scarcity of fuel over its lifetime (there is sufficient fertile material available for many centuries of operations). In contrast, the LWR fuel cycle cost is subject to escalation depending on the uranium resource or enrichment costs, as illustrated in Figure 13-6. As such, the uncertainty band for the IFR fuel cycle cost is assumed to be 50% of the reference case.

Just as imported LNG plays a backstop role on the natural gas price, IFRs can backstop uranium price escalation. If we take $50/lb for U_3O_8 and $100/SWU, the fissile value in the low-enriched uranium used in LWRs is equivalent to about $40/gm-U235. In comparison, the recovery cost of bred plutonium in the fast reactor blanket is on the order of $15/gm-fissile, assuming a marginal cost for blanket processing of $300/kg and a 2% plutonium buildup in the blanket, both numbers reasonable today. With fast reactors deployed on an economically competitive basis for electricity generation, fissile material production is a byproduct. In essence, the use of fast reactors is equivalent to adding at least a hundredfold uranium resource base at the current uranium price. The fissile value of enriched uranium as a function of the uranium ore price is plotted in Figure 13-7 and compared with the cost of plutonium recovery as a byproduct in fast reactors over a wide range in incremental blanket processing costs. (The one-for-one substitution of plutonium for U-235 will be less valuable in the LWR and more valuable in the fast reactor because of their physics characteristics, but the general point of backstopping uranium cost is accurate.)

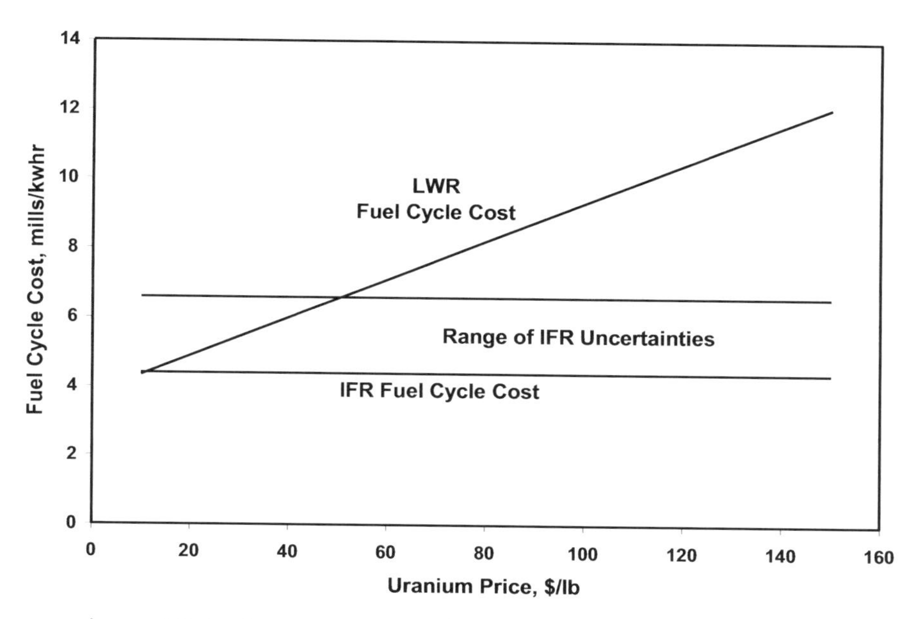

Figure 13-6. Comparison of the IFR fuel cycle cost with LWR as a function of uranium price

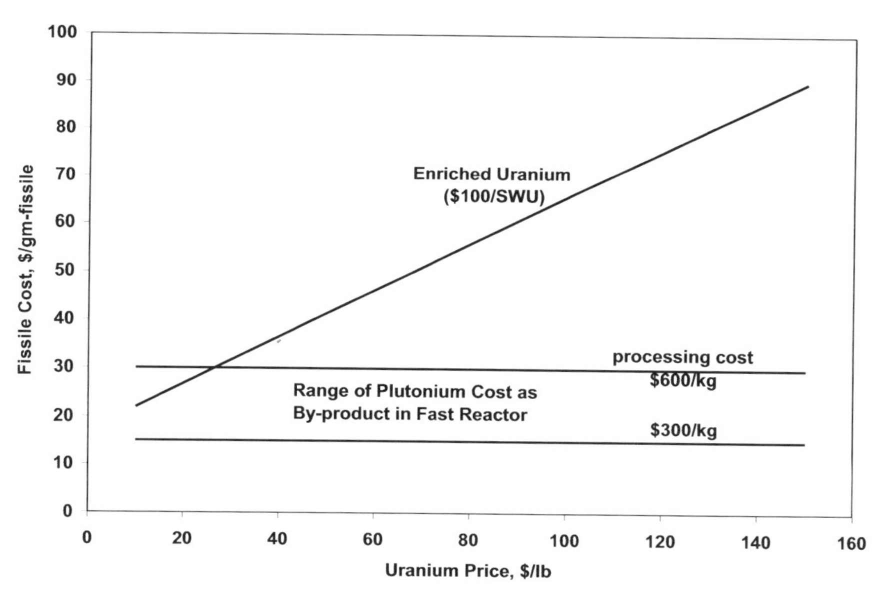

Figure 13-7. Comparison of equivalent fissile cost between enriched uranium and plutonium recovered as by-product in fast reactors

13.5 Application of Pyroprocessing to LWR Spent Fuel

The potential for major economic improvements for the fast reactor fuel cycle closure leads to the obvious question of the applicability of pyroprocesses to LWR reprocessing economics as well. The main difference between the LWR and the fast reactor spent fuel is in their actinide contents. The LWR spent fuel has a low actinide content, 1–2%, depending on discharge burnup. Pyroprocessing is particularly attractive at low throughput rates. Aqueous reprocessing, on the other hand, can take advantage of the economies of scale for LWR spent fuel processing. Therefore, the first question is whether pyroprocessing can achieve the necessary economies of scale for economic processing of LWR spent fuel.

Pyroprocessing for the fast reactor is compatible with remote refabrication in the same hot cell. For LWR aqueous reprocessing, this is not practical and the reprocessing plant is separate from the fabrication facility. Pyroprocessing then needs to be economically competitive with the stand-alone large-throughput aqueous reprocessing plants. The currently operating La Hague reprocessing complex in France consists of the UP2 plant, which was upgraded from 400 to 800 T/yr throughput and a new UP3 plant with an additional 800 T/yr throughput capacity. The two combined were reported to cost $8.2 billion in 1987 [19], or $16 billion in 2011 dollars. The Rokkasho plant in Japan, with an 800 T/yr throughput, is estimated to have cost around $20 billion.

Pyroprocessing is fundamentally more amenable to batch processing than to continuous processing. This does not necessarily imply that scaling up involves only multiple units. As discussed in Chapter 10, a single electrorefiner incorporating planar electrode arrangement can be designed for a 500-kg or even 1,000-kg batch size, which leads to an annual throughput rate of a hundred metric tons. And in fact a pre-conceptual design for a pilot-scale pyroprocessing facility at the100 ton/yr throughput for LWR spent fuel was developed at Argonne, as was described in Chapter 10.

Based on this work and the extrapolation of the previous refurbishment work for the EBR-II Fuel Conditioning Facility, it is estimated that such a facility could be constructed for about $500 million. A rough breakdown of this estimate is as follows:

Engineering................	150
Construction...............	130
Equipment systems	120
Contingencies	100
Total	$500 million

Any further scaleup can be achieved by duplicating the process equipment systems and some economies of scale where they come in. It is plausible that a pyroprocessing facility with an 800 ton/yr throughput could be constructed at around $2.5 billion, far below the capital cost experience of large aqueous reprocessing plants discussed above.

13.6 System Aspects

The previous sections discussed the economics on the basis of a single reactor. In this we ignore the significant impacts of introducing IFRs to the total system economics, costs such as those associated with uranium resources and repositories.

In a growing nuclear economy over the longer term, the impact on overall uranium resource requirements of introducing IFRs would be substantial. The total uranium requirements can be kept at a reasonable level and price. On the other hand, with LWRs only, meeting the same electricity demand will require very large uranium resources, going on into higher-cost uranium resources (i.e., lower-grade ores and more expensive associated mining and processing operations). Such a tradeoff can be quantified in a linear programming (LP) optimization of the entire system, where an optimum reactor mix is sought to minimize the total system cost. Such a system modeling was utilized during the International Nuclear Fuel Cycle Evaluation (INFCE) studies [20] and is currently being re-established to analyze the optimum introduction of fast reactors to minimize the total system cost. [21]

This new system modeling has been utilized to illustrate the system aspects that aren't covered in economics comparisons of individual reactors. For this purpose, we will consider the worldwide nuclear capacity scenario, which will be presented in Section 14.7.

The following assumptions were made for the economic data used in the LP optimization:

- The LWR capital cost is $3,000/kWe, O&M cost is $120 million per GWe-yr, and fuel cycle cost is $30 million per GWe-yr excluding the uranium cost.
- The IFR O&M cost is $120 million per GWe-yr, fuel cycle cost is $45 million per year including recycling cost, and the capital cost is varied relative to LWR.
- The uranium price is $30/lb$U_3O_8$ up to 6.3 million MTU (metric ton of uranium), $100/lb up to 16.7 million MTU, and unlimited amounts at $200/lb. The first cutoff represents the identified resources and the second undiscovered resources.
- The analysis is done on a constant dollar basis.
- The capital cost is amortized at an average (equity and debt) rate of 10% per year over a 60-year lifetime.
- The future benefits/costs are discounted at 3% per year.

These assumptions represent a case where the electricity generation cost of an IFR is slightly higher than that of an LWR at the uranium price of $30/lb. Therefore, on a single reactor comparison basis, the IFR should not be introduced for economic optimization. However, Figure 13-8 illustrates that even if the IFR's capital cost is 30% more expensive than that of the LWR, the total discounted system cost would be less because the introduction of IFRs provides substantial savings in uranium costs over the long term.

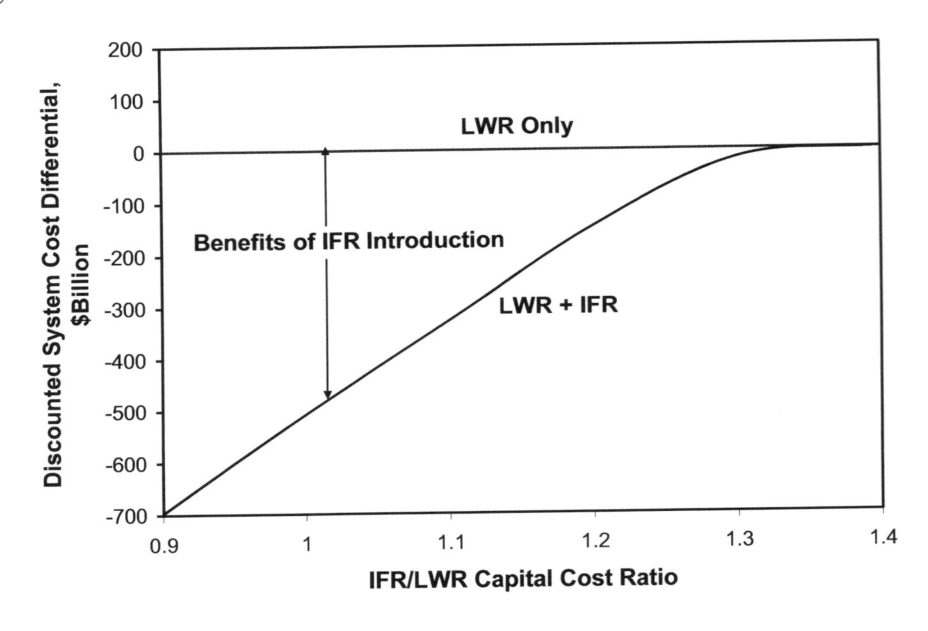

Figure 13-8. Benefits of IFR on discounted system cost for various IFR/LWR capital cost ratio

If the IFR's capital cost is at parity with the LWR's, then the total discounted system cost benefit is on the order of $500 billion for the example shown in Figure 13-8. As the capital cost ratio increases, the benefit is reduced because of a higher capital cost, but the system cost is still minimized by maximum introduction of IFRs. The IFR introduction is constrained in the same manner as described in Section 14.8; namely, the introduction rate is limited through 2050. If this constraint were relaxed, then more IFRs would be introduced earlier. The maximum IFR introduction is maintained up to the capital cost ratio of 1.2. Beyond that, the optimum IFR introduction date is delayed, and at the capital cost ratio of 1.4 or greater, no IFR introduction is optimum.

Although the specific numbers will change with a different set of assumptions, the LP optimization provides a powerful message. Even if IFRs are more expensive to build today, it is far better to start building them now in order to reduce the large uranium requirements that we would face otherwise. Such decisions cannot be made by individual utilities. It would have to result from national policies benefitting future generations.

Possibly even more important, particularly in the nearer term, is the system-wide impact on the high-level waste repositories. In addition to reducing the lifetime from hundreds of thousands of years to a few hundred years, the pyroprocessed waste streams have no long-term decay heat to limit how the waste packages can be placed in a given space. The removal of actinides improves the repository space utilization by a factor of five or so. If such benefits are quantified into the LP system model, then even earlier IFR would be justified even at higher capital costs. More generally, without satisfactory resolution of the waste issue, substantial expansion of LWR capacity will be difficult, if not completely impossible.

13.7 Summary

The capital costs of the fast reactor demonstration plants built in several countries in the late 1960s and early 1970s provide little usable information on the likely cost of fast reactor construction today—especially of an IFR-type design—and the very early fast reactors of the 1950s provide none. All were first of a kind, financed as a part of ongoing fast reactor development. The best information available is from the construction of the 1,240 MWe SuperPhenix in France through the 1980s, shut down in the 1990s by concentrated anti-nuclear campaigning. This very large plant, very much first of a kind, cost about twice as much as a comparable LWR in France at that time (which were mature costs, France having built dozens of PWR units in the previous ten years). Estimates of the costs of a followon fast reactor plant put the costs about 40 percent higher than the French LWR, and further design changes dropped the estimated differences in cost still further. All in all, the limited experience suggests costs considerably higher for the first one, and in keeping with other construction experience, lessening rapidly with repetition.

Cost breakdowns of both fast reactor and LWR construction underline the importance of indirect costs to the total cost of either reactor. The U.S. experience in the 1970s of huge costs resulting from delays from anti-nuclear legal maneuvering, made the cost a plant almost totally dependent on these "indirect" costs, and had little to do with labor and materials. Any cost difference between fast reactors and LWRs will be completely obscured by such actions if they are allowed in any new reactor construction today. Fine distinctions in cost between reactor types are of interest academically, but are meaningless in the real world, where indirect costs are determined by the speed of construction, the quality of engineering management, and discernment in design choice in such first-of-a-kind construction. Prudence in risk in early versions should be preferred over economies of scale.

In the fast reactor, where recycle is mandatory in any case, the IFR fuel cycle should be economic. Cost comparisons between aqueous and IFR processing heavily favor the IFR cycle. The actinide content of high-burnup fast reactor fuel makes remote processes a necessity. The IFR processes, with metal fuel and pyroprocessing, make remote operations relatively easy. Remote maintenance in an aqueous system and remote fabrication of actinide-laden oxide fuel is both difficult and expensive.

The fuel cycle cost for an IFR plant, based on a rather detailed conceptual design of an IFR fuel cycle facility, has been estimated to be about 4.5 mills/kWh. This low cost is the direct result of the small quantities of materials for construction and the very few pieces of equipment used in the IFR process. The situation can be summarized by saying that the capital costs are not really subject to doubt; if the process is developed to work as planned, the costs should be pretty much as estimated.

Fuel costs for the LWR are currently about 5 mills per kWh, but escalate as uranium prices increase. The IFR fuel cycle costs serve to put a limit on future costs of fissile material, as its costs are not dependent on uranium prices. In the LWR, recycle is questionable from both cost and operational standpoints for uranium recycle, as U-232 and U-236 build up in the uranium, complicating uranium reuse, and there is the limited economic incentive as well for mixed plutonium-uranium-oxide fuel.

There is incentive to use the IFR process for LWR spent fuel as well. Even though an additional step is necessary in reducing the oxide to metal prior to electrorefining, the costs of a hundred ton per year facility are estimated very roughly to be five hundred million dollars, a number that promises good economics for this process as well. Development remains to be done. But again, if the process can be proven, the economics will be there.

Finally, there is real benefit in easing LWR waste management and a related benefit in the cost of nuclear power as a power-generating system. Lessening pressure on the uranium resource and lessening the demands on repositories leads to economic benefits that, while not included in plant cost estimates, are very real. The economic promise of IFR technology is very real.

References

1. M. Rosenholc, NOVATOM, unpublished presentations, November 1981.
2. M. Rapin, "Fast Breeder Fuel Cycle: World and French Prospects," *Proc. BNES Conf. on Fast Reactor Fuel Cycles*, 1981.
3. M. Barberger, "The French Nuclear Power Program," unpublished paper, 1985.
4. A. Zrodnikov, "The Closing of Nuclear Fuel Cycle and Role of Fast Reactor in the Innovative Development of Large-Scale Nuclear Power in Russia," presentation at International Workshop on Future Nuclear Systems and Fuel Cycles, Karlsruhe, Germany, September 1-2, 2005.

5. S. C. Chetal, "India's Fast Reactor Programme," presentation at International Workshop on Future Nuclear Systems and Fuel Cycles, Jeju, Korea, September 7-8, 2006.
6. "Guide for the Economic Evaluation of Nuclear Reactor Plant Designs," NUS-531, 1969.
7. "Nuclear Energy Cost Data Base—A Reference Data Base for Nuclear and Coal-Fired Power Plant Generation Cost Analysis," DOE/NE-0095, 1988.
8. "Large LMFBR Pool Plant," unpublished report, Rockwell International Corporation and Argonne National Laboratory, 1983.
9. J. S. McDonald, et al., "Cost-Competitive, Inherently Safe LMFBR Pool Plant," *Proc. American Power Conference*, 46, 696, 1984.
10. *Phase II Final Report of Feasibility Study on Commercialized Fast Reactor Cycle Systems*, Japan Atomic Energy Agency and Japan Atomic Power Company, March 2006.
11. M. Ichimiya, T. Mizuno and S. Kotake, "A Next Generation Sodium-cooled Fast reactor Concept ans its R&D Program," *Nuclear Engineering and Technology*, 39, 171-186, 2007.
12. L. J. Koch, *Experimental Breeder Reactor-II: An Integrated Experimental Fast Reactor Nuclear Power Station*, Argonne National Laboratory.
13. *The Future of Nuclear Power*, MIT, 2003.
14. TradeTech website: www.uranium.info.
15. Euratom Supply Agency, *Annual Report 2009.* http://ec.europa.eu/euratom/ar/ar2009.pdf
16. Energy Information Administration, "Uranium Purchased by Owners and Operators of U.S. Civilian Nuclear Power," data released on May 16, 2007.
17. M. J. Lineberry, R. D. Phipps, and J. P. Burelbach, "Commercial-size IFR Fuel Cycle Facility: Conceptual Design and Cost Estimate," unpublished report, Argonne National Laboratory, 1985.
18. W. D. Burch, H. R. York and R. E. Lerch, "A Study of Options for the LMR Fuel Cycle," unpublished report, Oak Ridge National Laboratory, 1986.
19. F. Chenevier and C. Bernard, "COGEMA Expands LaHague," *Nuclear Engineering International*, 41, August 1987.
20. C. E. Till and Y. I. Chang, "Application of an LP Model to Breeder Strategy Studies," *Proc. ANS Topical meeting on Computational Methods in Nuclear Engineering*, Williamsburg, VA, April 23-25, 1979.
21. J. Lee, Y. H. Jeong, Y. I. Chang and S. H. Chang, "Linear programming Optimization of Nuclear Energy Strategy with Sodium-cooled Fast reactors," *Nuclear Engineering and Technology*, 43, 383-390, 2011.

CHAPTER 14

IFR DESIGN OPTIONS, OPTIMUM DEPLOYMENT AND THE NEXT STEP FORWARD

The IFR technology allows great flexibility in design of its nuclear characteristics to configure its actinide fuel to breed excess fuel, to self sustain, or to burn the actinides down. Perfectly practical deployments can answer almost any future electrical energy need. The design of the IFR core can emphasize one or another important characteristic, to a degree at will. In particular, the important characteristics of breeding performance and the amount of fissile inventory can be selected by the designer to give any breeding performance desired, over a wide range. Before getting in to design considerations, we begin by describing the appearance of an IFR plant. Then because choices for the coolant other than sodium have been brought forward recently we review alternative coolants, helium gas or lead and lead-bismuth alloys, and compare their characteristics to sodium's, concluding that sodium remains much the best choice. We then go on to look at the physics principles underlying breeding, showing the possible breeding characteristics of the world's principal reactor types. Concentrating then on IFR design itself, we show that fuel pin diameter is the important variable in determining breeding in the core, and with this in mind, the way the thermal, hydraulic, and mechanical constraints must be accommodated. Finally, we go on to discuss tradeoffs that enter in further balancing the requirements for an optimum design.

We then turn to the experience with fast reactor development in the past, examine the various problems and difficulties that arose, and contrast that experience with the thirty years of faultless operation of EBR-II. For all these years, EBR-II has acted as a pilot demonstration of today's IFR technology, and in the ease of its operation and its flexibility, as shown by the multitude of experiments carried out on it, lies proof that IFR technology will provide a very high level of reliability, maintainability, operability, and longevity.

Coming to the important subject of the effect of IFR deployment—the reason for its development in fact—we show the importance of IFR deployment on world uranium needs, based on the World Nuclear Association's estimates of nuclear capacity over this century, in a system study that illustrates the need for early large scale deployment of IFR technology.

Finally, construction of a pilot pyroprocessing plant is first and foremost. It is the most practical and most necessary first step in implementing IFR technology. We discuss the path to be taken. Powerful nations—China, Russia, India—increasingly are moving ahead with the fast reactor. The IFR technology can be expected to be superior, but demonstrating it, if it is to be done, should be done in the U.S. and it should be done by ourselves—possibly in collaboration with international partners, but on our own initiative.

14.1 What Will an IFR Look Like?

An IFR will look much like any other nuclear power plant: a site of several acres, a reactor building enclosing the reactor in containment, most of the site occupied by service facilities of various kinds, and the usual cooling towers and electrical switchyard array. One of the buildings, however—probably adjacent to and connected to the reactor building—will be a fuel cycle facility to process spent fuel and re-fabricate new fuel remotely. The EBR-II fuel cycle facility was a hot cell with several-feet-thick concrete walls and leaded glass windows at the work stations for the manipulators. The future fuel cycle facility will be a hot cell with similar shielding, fully automated for remote operation and maintenance with overhead cranes and robotics. The hot cell has an inert gas atmosphere, its interior is highly radioactive, and it is never entered. A containment dome over the reactor portion is likely, but the reactor or reactors may be also placed below ground level, depending on the design (as has been suggested in a small modular reactor approach).

The reactor core, its coolant pumps and associated piping to heat exchangers, and the heat exchangers themselves are submerged in a large reactor vessel filled with liquid sodium, its surface blanketed with inert argon gas, in this pool design. Outside this primary reactor vessel is a guard vessel, supported independently of the primary reactor vessel. The gap between the two vessels, typically 6–9 inches, also contains inert argon gas. The gap allows inspection by robotics of the continued integrity of the reactor vessel and provides space for the instrumentation for sodium leak detection. The vessel volumes are selected to assure that even if sodium leaks to the guard vessel, the core remains submerged and the heat exchangers and pumps still operate normally in removing heat.

The purpose of the in-vessel intermediate heat exchangers (IHX) is to transfer the heat to non-radioactive sodium, which will then carry it from the vessel to the steam system. In this way all radioactivity is kept within the primary vessel. Should there be a failure in the sodium-to-water steam generator, no radioactivity will be involved.

The entire assemblage is referred to as the "primary system," and the whole of it, the entire radioactive portion of the plant, is surrounded by radiation shielding of concrete several feet thick. The necessary penetrations for the secondary non-radioactive sodium flowing from the in-vessel heat exchangers, and for electrical connections, are made through the top of the vessel. The vessel top may be substantial enough, as it is in EBR-II, to provide all the shielding necessary to make it part of the working floor and allow complete access by operating personnel and others during full power operation. There are no penetrations in the walls of the vessel.

Much of safety comes from assuring that the core is always submerged in the sodium pool. This guarantees that it is continuously cooled, shutdown or not. Further backup can be provided, if desired, by selecting the diameter of the thick shielding around the primary system to buttress this feature.

Although the IFR plant would look like today's commercial reactors, a very important difference is that the IFR is cooled with sodium rather than water (paramount in giving its breeding capabilities). We will now discuss the coolant choice and related physics principles and get into IFR design principles and options.

14.2 Coolant Choice Revisited

The first requirement for a fast reactor coolant is that its neutron energy moderating power should be low enough to maintain a fast neutron spectrum. This rules out hydrogenous coolants. High heat transfer coefficients, a high heat capacity, high boiling temperature, and low melting temperature, in addition to compatibility with the structural materials and both chemical stability and stability under both irradiation and high temperatures, are among the most important characteristics of the coolant.

There is no ideal coolant superior in all criteria. Sodium has been selected as a good compromise and has been the choice for essentially all fast reactors operated to date. The properties of sodium are listed in Table 14-1 compared with some alternative fast reactor coolant options. [1]

Also included in Table 14-1 is a sodium-potassium eutectic alloy, NaK, whose principal advantage is its lower melting point than sodium itself. The Na(56%)K(44%) eutectic melts at 19°C and the Na(22%)K(78%) eutectic at –11°C. Fluid at room temperature therefore, NaK was the primary coolant in EBR-I and the Dounreay Fast Reactor, the first fast reactors (in the West). But the other important thermal properties of NaK—the boiling point, specific heat, and thermal conductivity—are all inferior to those of sodium, and sodium has been used as the primary coolant in all subsequent fast reactors. However, NaK has been and likely will be used in the shutdown heat removal systems required for emergency situations where the normal heat sink is unavailable to transport the residual heat in the reactor system to a heat dump (typically an air dump heat exchanger). The low melting temperature of NaK is important in assuring that the system is always available, even without an active heating system.

Table 14-1. Properties of selected fast reactor coolant types

	Sodium	Na(56%) K(44%)	Na(22%) K(78%)	Lead	Pb(44.5%) Bi(55.5%)	Helium
Density, g/cc	0.82	0.76	0.74	10.41	10.02	0.00018
Viscosity, centipoise	0.23	0.19	0.18	1.69	1.19	0.018
Boiling point, °C	881	826	784	1737	1670	-
Melting point, °C	98	19	-11	327	125	-
Specific heat, J/gK	1.26	1.04	0.875	0.145	0.146	5.20
Heat capacity, J/ccK	1.03	0.79	0.65	1.51	1.46	0.0009
Thermal cond, W/mK	64	28	26	15	14	0.15

In recent years, there have been suggestions for utilizing lead alloys or helium gas as alternative fast reactor coolants [2] so we will discuss the tradeoffs involved with these alternative fast reactor coolant options.

14.2.1 Lead Alloys

Both lead and lead-bismuth eutectic have been proposed as coolants. As shown in Table 14-1, the melting point of lead is 327°C as compared to sodium's 98°C. The lead coolant has to be kept at much higher temperatures, requiring high-temperature structure materials. The alternative lead-bismuth eutectic with a melting point of 125°C has the drawback that the neutron capture in bismuth results in polonium-210, which is highly radioactive and can cause radiation doses to operators and maintenance crews. Except for these differences, the remaining discussions apply to both lead and lead-bismuth eutectic.

Although the specific heat of the lead alloys is about an order of magnitude less than that of sodium, the density is more than an order of magnitude greater. As a result, the volumetric heat capacity is about the same, and the lattice design parameters might be thought to be similar to sodium cooling. However, the coolant pressure drop across the fuel assembly is proportional to the coolant density, and if a similar lattice design is adopted for lead coolant, a very high pressure drop is inevitable, and with it unacceptably high pumping power requirements, as the latter are proportional to the pressure drop. For sodium-cooled fast reactors, the pumping power requirement is on the order of 2–3% of the reactor output, and a requirement an order of magnitude greater would be unacceptable.

The pressure drop is proportional to the square of the coolant velocity. If the coolant velocity reduced to one third, say, the pressure drop would be reduced by a factor of nine. However, the coolant volume fraction would have to increase by a factor of three. This means a very loose lattice design. Grid-type spacers would be necessary, as conventional wire-wrap spacers would not maintain the pin bundle geometry. The high coolant volume fraction combined with the higher density means that the total lead coolant mass could be on the order of fifteen times that of sodium for an equivalent reactor size. Typical 1,000 MWe sodium-cooled reactors would have a sodium inventory on the order of several hundreds of tons, and hence an equivalent lead-cooled reactor could end up with a total lead inventory in excess of ten thousand tons.

The heavy coolant mass introduces two design implications. First, the reactor vessel and structural components would need increased thickness to support the heavier static weight. More importantly, the heavy liquid mass would be a major design challenge for seismic events. These factors are likely to impose a constraint on the size of the reactor vessel and hence the reactor power.

However, it is the compatibility of liquid lead with the reactor's structural materials that is of the greatest concern. Lead has substantial corrosive effects on the structural materials due to its capacity for dissolving metals. For example, the solubility of nickel in lead is about 37,000 ppm at 600°C, and metal solubility in bismuth is worse than in lead. The Russian Alfa-class submarines employed lead-bismuth-cooled fast reactors, and they found that a protective oxide layer on structural components (particularly cladding) was the most effective approach to deal with the corrosion problem. But the oxide layer is not static. Above 550°C, it can become thick and unstable. The high-velocity coolant and high shear stress due to viscosity can erode it. Oxygen levels have to be controlled continuously by adding PbO balls in the coolant flow path, at the same time removing oxygen by bubbling hydrogen gas in a helium carrier to form water. [3] A non-homogeneous oxygen distribution results in non-uniform coatings. Crevice corrosion can occur.

Liquid metal embrittlement is also a concern, caused by a decrease in the cohesive strength of structural materials due to lead adsorption. Martensitic steel, such as 9Cr-1Mo alloy, is particularly susceptible to liquid metal embrittlement if intimate contact between liquid lead and steel is made below 400°C. Maintaining the oxide layer to prevent direct contact between the steel and liquid lead eliminates this problem as well. As an alternative to the oxide film barrier on steels, refractory metals have been proposed that are known to show resistance to the action of lead, but their mechanical properties are not as suitable in reactor applications.

It is suspected that the lead corrosion problem was the main culprit in shortening the life of the Alfa-class submarines. A prototype Alfa-class was launched in 1972, but a coolant freezing accident destroyed the reactor and it was dismantled in 1974. Six additional of the Alfa-class were launched between 1977 and 1983. Eventually, four of seven Alfa-class submarines experienced reactor failures. [4-5] In 1982, for example, a steam generator failure leaked about two tons of lead-bismuth into the reactor compartment, damaging the reactor irreparably.

Lead alloys will be considered as an option if a very high outlet temperature is required for specific advanced applications, always providing the corrosion problem is satisfactorily solved. However, for conventional power reactor applications, its disadvantages make it unlikely to replace sodium as the coolant of choice.

14.2.2 Helium Gas

Gas-cooled fast reactors were considered in early days but abandoned in the late '70s. However, the interest in helium gas cooling has resurfaced in recent years. [2] Thermophysical properties of helium are also listed in Table 14-1. The specific heat of helium is about four times that of sodium, but the density is very low—five thousand times less. The volumetric heat capacity, therefore, is less by about three orders of magnitude. Even if helium is pressurized to 100 atmospheres, the coolant velocity would have to be increased by a factor of ten for the typical lattice design parameters used in sodium cooling. The helium coolant velocity required is in the 100 m/s range, introducing the risk of flow-induced vibration, and attention to this is necessary in the reactor's structural design. The thermal conductivity is very low—four hundred times less than that of sodium—so the heat transfer is poor even at high coolant velocity and the cladding surface needs to be roughened to increase the effective surface for heat transfer.

Certainly, from a thermal-hydraulics point of view, helium is not a good coolant for the high specific power of fast reactors, and justification for its use would have to come from elsewhere. One rationale for considering a helium-cooled fast reactor stems from the fact that it could be a natural extension of thermal-spectrum high-temperature gas-cooled reactors. For thermal-spectrum reactors, graphite is employed as moderator and the large volume of graphite in all such reactor designs provides a passive heat sink lasting for a period of days in the event of loss of coolant, for example. For fast reactors, such a graphite heat sink is not possible; coolant must be available, and active safety systems have to be relied upon to shut down the reactor and remove the decay heat. For a high-pressure, high-velocity helium coolant system, loss of coolant flow and even loss of pressurization are credible possibilities. Even for protected (with scram) transients, in the fast reactor an active emergency cooling system might be required to prevent core melting. Anticipated transients without scram will result in core disruptive accidents.

In order to provide some buffer for the anticipated transients, new fuel types are being developed, that could provide some partial heat capacity, including cermet-type fuel and alternate fuel configurations. [6] However, their irradiation performance characteristics have not yet been fully explored.

14.2.3 Sodium

In addition to advantages in thermophysical properties, another unique characteristic of sodium as coolant is its compatibility with the metals used for the reactor structures and components. Radioactive corrosion products are not formed in any significant amounts, which, circulating and depositing around the system, would otherwise make access for maintenance difficult. Access for maintenance in the sodium-cooled system is easy and radiation exposures to plant personnel are expected to be very low. The noncorrosive coolant also implies reliable sodium component performance and improved plant availability. Experience in EBR-II has shown that components submerged in a sodium pool tend to exhibit a better reliability than similar components exposed to an argon or air atmosphere.

Sodium has been used as coolant in all sixteen fast reactors that have been operated around the world, other than the NaK-cooled EBR-I and DFR, because of its excellent properties as a fast reactor coolant. Experience has been excellent, as will be discussed more in detail in Section 14.7. The only drawback of sodium is its chemical reactivity in air and water, an issue quite manageable without causing safety consequences, and this aspect has been discussed in detail in Section 7.12. In summary, sodium is likely to remain as the coolant of choice for future fast reactors.

14.3 Physics Principle of Breeding

The principle underlying adequate breeding is the neutron economy available only in fast spectrum reactors. The fundamental parameter that gives the fast reactor superior breeding is the high value of the average number of neutrons emitted for each neutron absorbed by a fissile isotope, commonly designated as η or eta value, in a fast neutron spectrum. The breeding ratio (BR) is conventionally defined as fissile production divided by fissile destruction over the fuel life, but to show its components it can be expressed alternatively in terms of the neutron balance written below. All the components are normalized to the neutron absorption in the fissile isotopes.

$BR = \eta + \varepsilon - 1 - A - L - D$, where

η = number of neutrons emitted by fission in fissile isotopes,
ε = number of neutrons consumed by fission in fertile isotopes,

A = number of neutrons absorbed in non-fuel materials,
L = number of neutrons lost by leakage, and
D accounts for loss of fissile isotopes by decay.

Of the ($\eta + \varepsilon$) neutrons available, one is needed to maintain the chain reaction, and after absorption in structural materials and leakage is accounted for, the balance are available for capture in fertile isotopes. The values of the parameter η for various fissile isotopes are presented in Figure 14-1 as a function of the energy of neutrons causing fissions. The neutron yield in a thermal spectrum (energies <0.1 eV) is just over two and fairly constant, whereas in fast spectrum (>100 keV) it continues to increase as neutron energy increases. This is the key to high breeding potential.

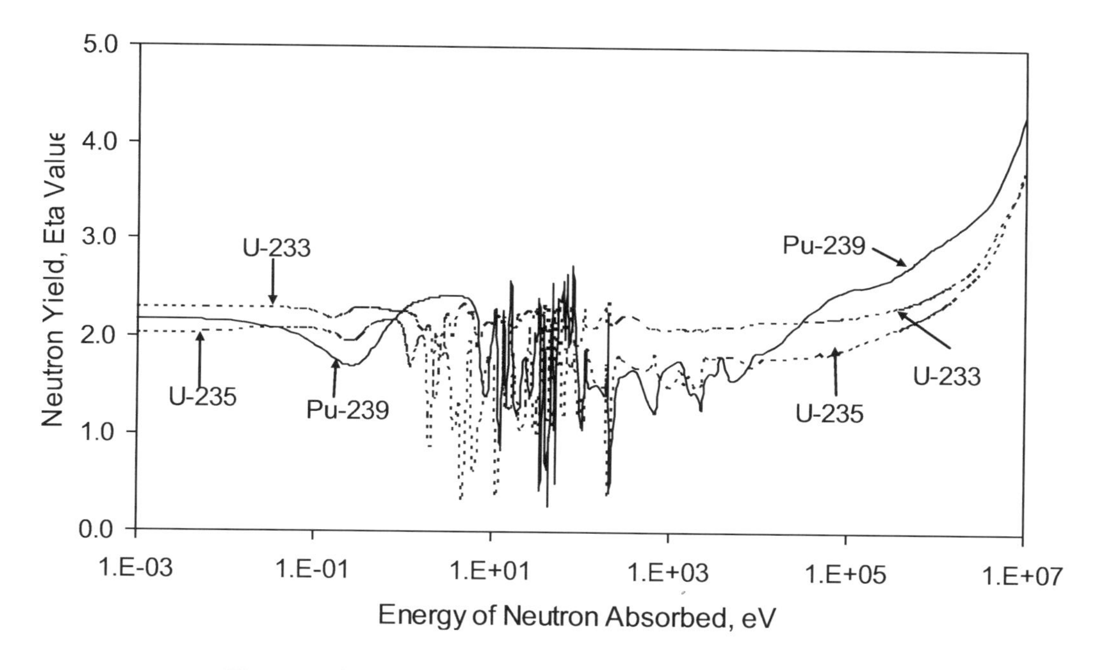

Figure 14-1. Neutron yield vs. neutron spectrum

The neutron energy spectra are presented in Figure 14-2 for a typical IFR and light water reactor. The LWR spectrum has dual peaks—one around 1 MeV and the other around 0.1 eV. Although the thermal flux is smaller in magnitude, the thermal neutron cross sections (probability of reactions with matter) are two to three orders of magnitude greater than for fast neutrons, and hence the neutronic characteristics are dictated by thermal neutrons in the LWR spectrum. For the IFR, most of the neutrons are in the few hundred keV range and higher, and fast neutrons dictate the neutron economy.

In thermal spectrum reactors, while the Pu-239 η appears to have a higher value than that of U-235, in fact the average value is less than that of U-235 because of low values for Pu-239 over a considerable range of energy above the thermal region. U-233 has the highest value in a thermal spectrum at around 2.3. Therefore, to the extent a thermal breeder is feasible, it is based on U-233 fueling (precisely what is proposed for thorium-232-based reactors). Indeed, the naval reactors program in decades past investigated the breeding ratio in a U-233-fueled LWR. It was shown to be just incrementally over unity, between 1.00 and 1.001. For this, parasitic capture from fission products and structural materials and leakage have to be kept very low.

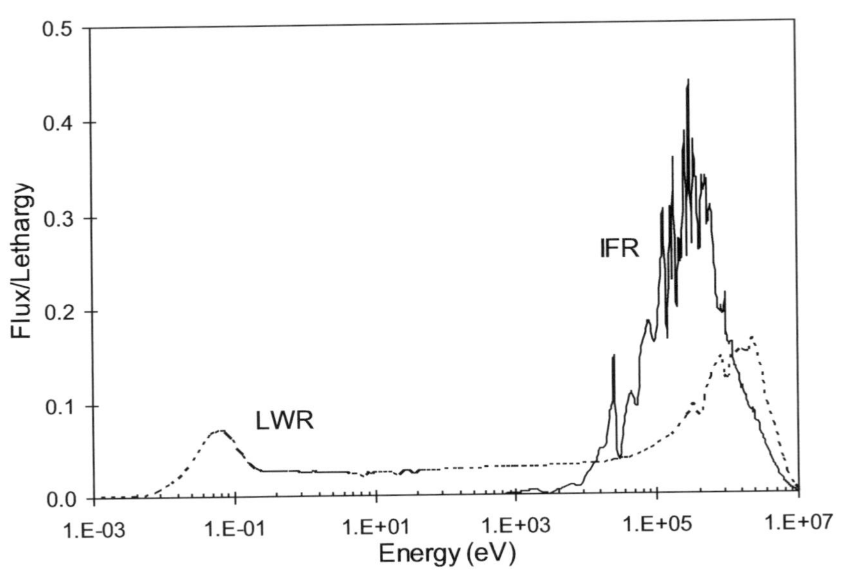

Figure 14-2. Comparison of neutron energy spectra between LWR and IFR

Some typical neutron balances are compared in Table 14-2 for the world's significant reactor types. [7] The conversion ratio (corresponding to the breeding ratio in fast reactors) for the LWR is 0.59. The heavy water reactor (HWR) achieves a higher conversion ratio at 0.74, so its uranium resource utilization is better as well. The sodium-cooled fast reactor (SFR) in the comparison is oxide-fueled, and as shown even the oxide-fueled fast reactor can achieve a reasonably high breeding ratio, due to its higher η- and ε-values.

14.4 Core Design Principles and Approaches

14.4.1 Effects of Fuel Volume Fraction

The steady increase in the η-value as the neutron spectrum hardens (more high-energy neutrons than low-energy neutrons) is shown in Figure 14-1. Lightweight materials, such as the coolant and other structure materials that tend to slow neutrons, should be minimized in core designs for high breeding. Put another way, to maintain a hard spectrum, the fuel volume fraction needs to be maximized in the core lattice design.

Table 14-2. Comparison of neutron economy for various reactor types

	HWR	LWR	SFR
η	2.03	1.92	2.28
ε	0.02	0.09	0.36
$\eta + \varepsilon - 1$	1.05	1.01	1.64
Losses: Structure	0.09	0.03	0.16
Coolant	0.03	0.08	0.01
Fis. Prod.	0.11	0.16	0.06
Leakage	0.08	0.15	0.05
Decay	-	-	0.03
Subtotal	0.31	0.42	0.31
Excess Neutrons (CR or BR)	0.74	0.59	1.33

A high fuel volume fraction is more easily achieved in a hexagonal lattice arrangement than in the rectangular lattice adopted in all light water reactors. For a given lattice pitch to pin diameter ratio, the hexagonal lattice gives about a 15% higher fuel pin volume fraction. Fuel pins thus arranged in a hexagonal array inside a hexagonal assembly duct are shown in Figure 14-3. The hexagonal fuel assemblies are again arranged in a hexagonal core layout, as shown in Figure 14-4. The edge assemblies are absent so that the core can fit into the smallest possible cylindrical reactor vessel.

For a given hexagonal lattice, a high fuel volume fraction is achieved by increasing the fuel pin diameter and reducing the lattice pitch (which reduces the coolant area). The structural volume fraction is minimized by reducing the cladding thickness and the hexagonal duct wall thickness. These design parameters are subject to thermal-hydraulic and mechanical design constraints which will be discussed below. But we will look at the effects of pin diameter on volume fractions first, as presented in Table 14-3. Only the pin diameter is varied, and the other design parameters are constrained by thermal-hydraulic and mechanical design constraints presented in the next section.

As shown in Table 14-3, a small change in pin diameter has a much greater impact on the fuel volume fraction, which in turn has a significant impact on the breeding ratio and other core performance characteristics. The volume fractions do not add up to unity, since the bond sodium is ignored as it tends to get pushed out of the core to the upper plenum region as the fuel swells out to the cladding.

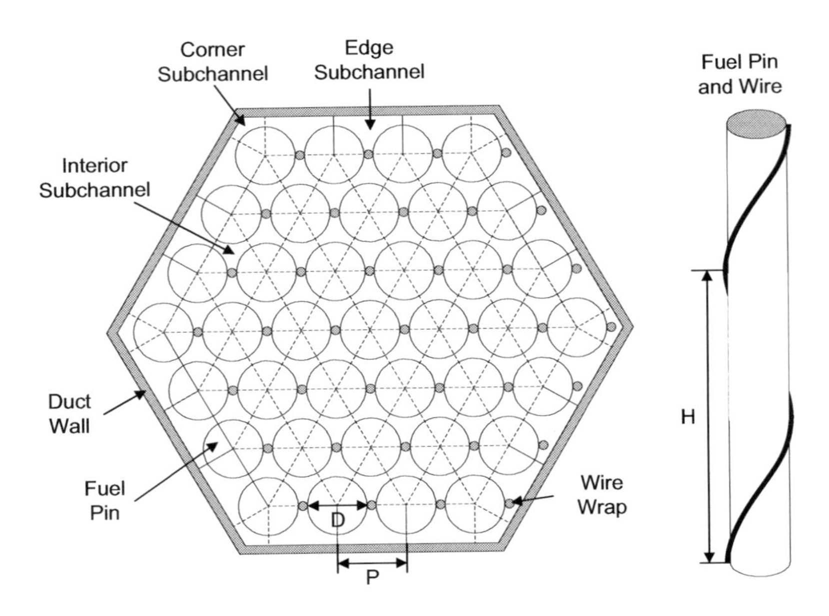

Figure 14-3. Fuel pin arrangement in a hexagonal duct

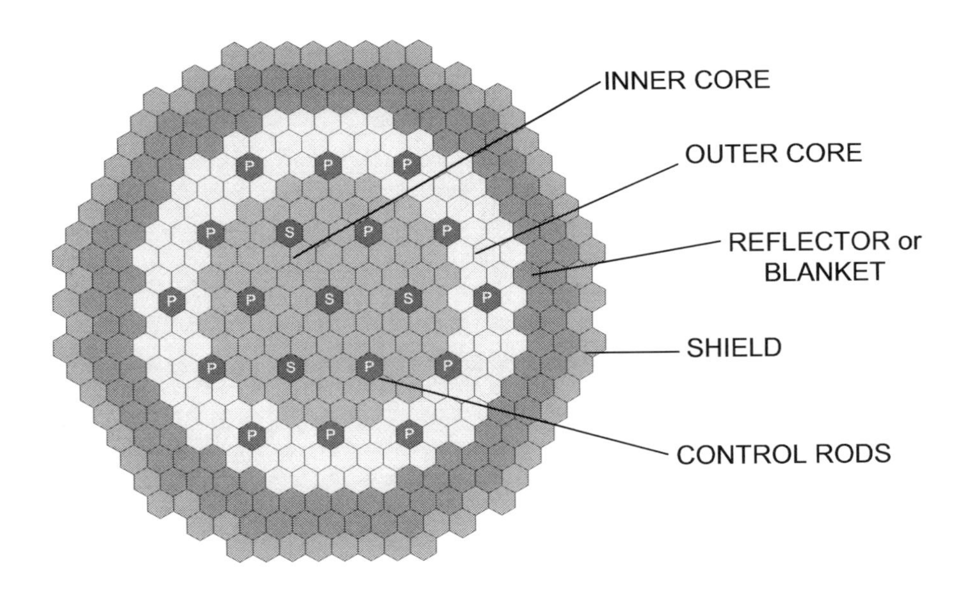

Figure 14-4. Typical hexagonal core layout

Table 14-3. Effects of fuel pin diameter on volume fractions

Pin OD cm	p/d ratio	Lattice Pitch, cm	Fuel V.F.	Coolant V.F.	Structure V.F.
0.9	1.269	16.2	0.279	0.377	0.203
1.0	1.217	17.2	0.415	0.345	0.189
1.1	1.177	18.2	0.347	0.316	0.178
1.2	1.147	19.2	0.376	0.290	0.168

14.4.2 Thermal-Hydraulic and Mechanical Design Constraints

The basic principle of thermal-hydraulic constraint is that the power generated in the fuel pins or the core has to be removed by the coolant and transported to the power conversion system. The heat generation rate balance equation is given by:

$P = \rho A v C_p \Delta T$, where

P = power,
ρ = coolant density,
A = coolant area,
v = coolant velocity,
C_p = coolant heat capacity, and
ΔT = coolant temperature rise through the core.

In the equation above, ρAv is the mass flow rate. The power (heat generation rate) in the fuel, P, equals the heat removal rate by the coolant—the coolant mass flow rate multiplied by heat capacity and temperature rise. The above balance applies to the total reactor power or to the average fuel pin. For our discussion, we will consider a unit lattice cell of a fuel pin surrounded by coolant. The total reactor power divided by the number of fuel pins and by the height gives an average linear power or heat rate, a convenient design parameter because it determines the cladding and fuel temperatures for a given coolant temperature. The radial temperature rises across the cladding and the fuel are proportional to the linear heat rate and inversely proportional to the thermal conductivities of cladding and fuel, respectively. The peak fuel and cladding temperatures directly impact the fuel performance margins, and there is a generally an accepted peak linear heat rate for any given fuel type. For metal fuel, a peak linear heat rate of 45 kW/m or an average linear heat rate of 30 kW/m are reasonable values. This linear heat rating is not a limit, but is representative of the values arrived at by balancing various tradeoffs.

The selection of a reference linear heat rate is a starting point for core design iterations. The normally accepted value for the coolant temperature rise through the core is around 150°C. For higher temperature rises, the coolant mass flow rate can be reduced, but this results in a higher thermal stress on cladding and a greater difference in outlet temperatures between adjacent assemblies. This can lead to thermal striping (a random joining of hot and cold flows, leading to cyclical stresses that contribute to fatigue cracking) on upper internal structures.

14.4.3 Design Tradeoffs

As discussed above, the selection of linear heat rate and temperature rise involve tradeoffs. Tradeoffs continue as the design progresses. For a given linear heat rate and temperature rise, the mass flow rate is determined by the heat balance equation. The coolant area of the unit lattice cell is then given by the mass flow rate divided by the coolant velocity. Higher coolant velocities reduce the coolant area and increase the fuel volume fraction. However, the pressure drop increases (proportional to the square of the velocity), which in turn increases pumping power requirements and reduces natural circulation tendency. Therefore, the coolant velocity is restricted by the need to keep the pressure drop below a given target value. For the design parameters presented in Table 14-3, a linear rate of 30.8 kW/m, a temperature rise of 155°C, and a pressure drop of 15 psi were taken. For a given pin diameter, the coolant area can be calculated and the unit lattice configuration is fixed to arrive at the volume fractions of fuel, coolant and structures.

A cladding thickness of 0.05 cm, hexagonal duct wall thickness of 0.3 cm, and wirewrap spacers as illustrated in Figure 14-3 are reasonable choices, as is the choice of 169 pins per fuel assembly. The number of pins for n-hexagonal arrays is given by $3n(n\text{-}1)+1$, and illustrated in Table 14-4. If the number of pins per assembly is increased to 217, the structural fraction is further reduced and the fuel volume fraction is increased, as shown in Figure 14-5. The assembly weight is also increased, and this places additional burden on the in-vessel fuel handling machine. Also the number of the assemblies for the entire core is reduced, which results in less flexibility in optimizing the number and location of the control rods. In general, a smaller reactor has a smaller number of pins per assembly, and as the reactor size increases the number of pins per assembly increases.

In general, the higher fuel volume fraction design gives better neutron economy, resulting in lower fissile enrichment, a higher internal conversion ratio, and a reduced reactivity swing during burnup. By minimizing the initial excess reactivity requirements, the control rod requirement is reduced and accidental reactivity insertion events are more easily handled. In general, the excess reactivity requirements for the IFR are much less than those of thermal reactors, as illustrated in Figure 14-6. In a thermal spectrum, the reactivity change between refueling intervals is rather large, and in addition, the buildup of fission products such as Xe and Sm consume significant amount of reactivity. The thermal reactor core must have a large excess reactivity at the beginning of a cycle, controlled by the control rods and burnable poisons. The excess reactivity requirement in IFR cores can be made to be very small. Among other advantages, this eliminates the potential for serious reactivity insertion accidents.

Table 14-4 Number of pins per assembly in hexagonal arrays

Number of rows	Total number of pins
1	1
2	7
3	19
4	37
5	61
6	91
7	127
8	169
9	217
10	271

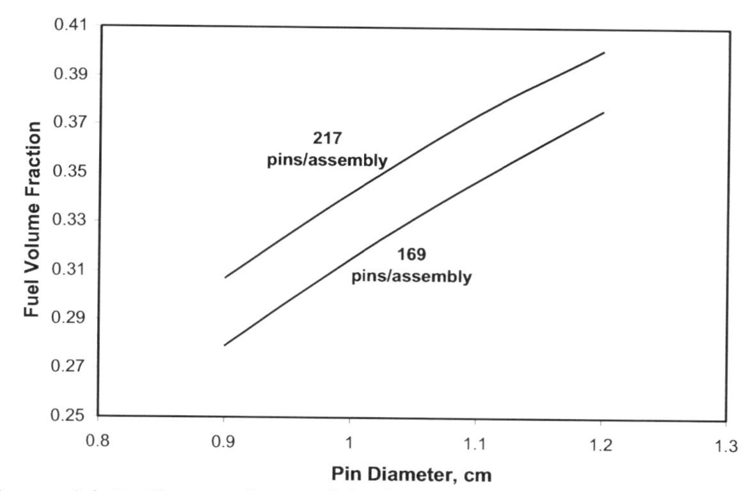

Figure 14-5. Comparison of fuel volume fraction as a function of pin diameter between 169 vs. 217 pins per assembly

A higher fuel volume fraction design, on the other hand, has penalties in terms of an increased fissile inventory and an increased heavy metal inventory and larger core size overall. Also, the lower specific power means the residence time of the fuel is increased to reach a given target burnup level. The high energy (>100 keV) neutron dose, also called fluence (integrated neutron flux over time) is increased with lengthened residence time, and it is the controlling parameter for the cladding lifetime rather than the burnup itself.

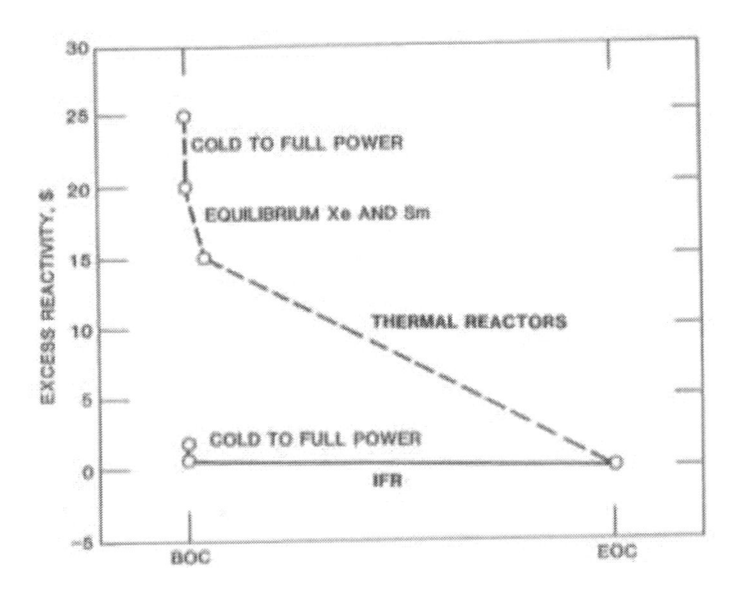

Figure 14-6 Comparison of excess reactivity requirements between IFR and thermal reactors

The thermal-hydraulic, mechanical, and neutronic design constraints do not fix the core design parameters since there are more degrees of freedom in the IFR core design than in the thermal reactors. The core design process, therefore, is one of continued iteration to balance the tradeoffs to meet the overall core performance goals.

A typical core, as illustrated in Figure 14-4, will be made up of fuel pins about 1 cm in diameter, enclosed in hexagonal ducts about 20 cm across, running the full height of the core and blankets. There are two hundred or so pins in each duct; the ducts channel the coolant flow and allow flow to be controlled by each individual duct. The coolant exit temperatures are controlled in this way to maximize thermal efficiency. For this, the core exit temperatures need to be as high as possible, and uniformly so from every duct, but kept within the limits placed on fuel pin cladding temperature (550 to 600°C).

The reactor core region fuel pins are about 20 percent enriched in fissionable materials. The blanket regions surrounding the core are made up of assemblies of pins containing uranium only, generally somewhat larger in diameter than in the core, as satisfactory pin temperatures can be still be maintained at the much lower powers they generate. Their principal purpose is breeding, but they can also give some latitude in design of the safety-related properties of the core if, as is sometimes done, they are made part of the core itself. Mostly, however, the blanket assemblies will "blanket" the core, surrounding it. Their purpose is to catch the neutrons leaking from the core which otherwise would be wasted. The thicker the blanket, the more neutrons captured, and the higher the breeding. But as the radial neutron leakages from the core are a lot less than from the top and bottom (the diameter of the core as a rule is at least double the height), the optimum radial blanket thickness even for maximum breeding is only two or three layers. And for self-sustaining reactivity only, perhaps one layer of external blanket is necessary; a net burner of actinides will have no uranium blankets at all, only steel shielding. The core layout presented in Figure 14-4 is for a 350 MWe reactor core. Size is determined by the number of fuel assemblies; the assembly design itself probably remains the same.

14.4.4 Reactor Size Effects

The unit lattice cell design principles described in the previous section are applicable independent of the reactor power. Larger reactors require more unit cells or more fuel pins and assemblies. The core size can grow in both radial and axial directions. But when the core size grows to about three feet in height, any further expansion is in the radial direction. Even with a three-foot active fuel column length, the overall assembly length could be more than twelve or thirteen feet when the gas plenum, upper and lower shields (or blankets), and inlet and outlet nozzles are added. Since the spent fuel has to be handled under sodium when transferred across the core, the sodium pool depth and the vessel height have to be increased as the active core height increases. But the three-foot optimum is not set in concrete. Both the height and the diameter can differ with the same design for electrical output if different properties of the core are emphasized. But a typical core diameter would be about ten feet for a 350 MWe plant and some fifteen feet or so for a 1,000 MWe, both with core heights of about three feet.

A key question in the effects of reactor size is whether the inherent safety features demonstrated in the small EBR-II can be achieved in larger reactor sizes, or if indeed there is a size limit for such behavior. Some inherent safety characteristics sought in other reactor concepts, such as the radiative heat removal, do depend on the reactor size. However, the IFR's inherent safety is more or less independent of the reactor size. As explained in detail in Section 7.8, this independence is due to the reactivity feedback mechanisms themselves being quite independent of the reactor size. The net result is that the margins to coolant boiling during unprotected loss-of-flow and unprotected loss-of-heat-sink events are very similar independent of reactor size, as presented in Table 7-1.

One important effect of reactor size is its effect on the specific fissile inventory requirement. Since the unit cell design remains the same, the total amount of heavy metal increases directly proportional to the reactor power. However, the fissile mass does not follow this proportionality, since the fissile enrichment (ratio of fissile mass to heavy metal mass) is reduced due to the reduced neutron leakage of a larger core. More neutrons are then available inside the core for *in situ* breeding, which further reduces the enrichment level. The amount of fuel needed for a given power goes up sharply toward the low end of the reactor power range. Figure 14-7 shows that the actinide inventory is almost ten tons per GWe for 300 MWe versus about five tons for a 1 GWe reactor. These numbers are for the amounts loaded in the core only. The amounts actually involved must include the inventory out of the core as well—that is, the spent fuel being processed and fabricated into new fuel. For a fuel loading that is self-generated from a reactor and simply maintains its reactivity, two annual reloads, which is half to two thirds of the amount in core, must be added for recycling to be established. In Figure 14-7 the acronyms denote different designs: two versions of the PRISM reactor design of General Electric [8-9], and three national design studies of the Department of Energy [10-12].

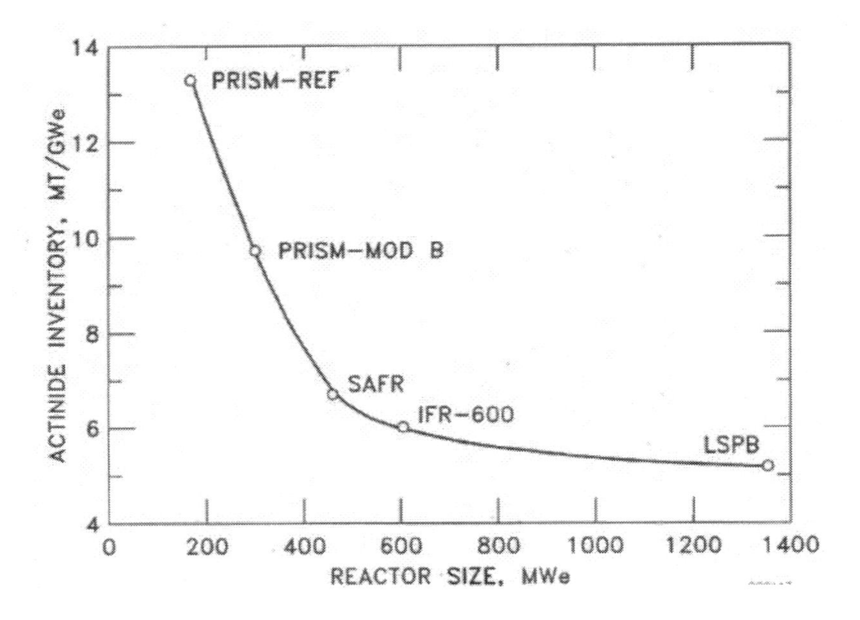

Figure 14-7. Initial core inventory as a function of reactor size

As for the IFR fuel management scheme, a portion of the fuel is replaced annually. A three year fuel residence time is typical, so a third of the core is refueled each year. For a 1,000 MWe reactor, ten to fifteen tons of heavy metal (actinides, but unburned uranium principally) contained in the core fuel and blanket assemblies are recycled each year. IFR fuel pins have proven burnups of 150 megawatt-days per kilogram of fuel, or about 15 % of the initial heavy metal, and no failures have been found to date. Over the three-year period, about 15% of the initial uranium and plutonium in the fuel pins will have fissioned, so full use of an initial loading takes about twenty years.

The lifetime of a fuel pin is limited not only by the evolving composition of the fuel itself (principally the increase in bulk of the fission products) but also by the irradiation damage to the fuel cladding. A "hard" neutron spectrum (a preponderance of high energy neutrons) like that of the IFR, desirable for properties like breeding, does increase such damage. But three- or four-year residence times for fuel are perfectly satisfactory, and in fact are typical for all power reactors.

14.5 Considerations for Burner vs. Breeder

There have been suggestions that higher actinides created in other reactors should be destroyed by irradiation in reactors with a fast neutron spectrum. Certainly at present, when plutonium stocks from LWRs are building up worldwide, and some weapons plutonium stocks having been declared surplus, a case can be made for reactor configurations that burn more of these actinide elements than they create. The IFR core can be designed as a burner or breeder of fissile isotopes, or indeed of all actinides. All are fissionable in the IFR neutron spectrum. If the core shown in Figure 14-4 is surrounded by reflector assemblies (steel components that tend to reflect neutrons back into the core, thus conserving neutrons), the core itself will not breed enough to break even in fissile conversion, and the core becomes a net burner of fissile isotopes and total actinides. If the reflector assemblies are replaced by blanket assemblies of depleted uranium, leakage neutrons are captured by the uranium and produce plutonium that can later be harvested by pyroprocessing and used to manufacture new fuel pins. If adequate blankets are provided (including axial as well as radial), the breeding ratio (the ratio of fissile production to fissile destruction) can substantially exceed unity. Self-sufficiency only, a breeding ratio of unity, indeed any burner or breeder configuration desired, is easily achieved by such blanket adjustments during planned refueling outages at any time during the reactor's lifetime.

As a rule of thumb, an LWR operating for thirty years generates enough actinides to start up an IFR. The LWRs currently operating worldwide, a capacity of about 375 GWe, operating for a sixty-year lifetime, will have generated sufficient actinides to start up about 750 GWe IFRs. Even if only a fraction of the LWR spent fuel is reprocessed, a large number of IFRs can be built without the need yet to breed. The penalty of such breeding is only that blanket assemblies will have to be processed, which can mean more heavy metal throughput than for the core fuel itself. If reflectors are installed, this cost can be avoided until such time as breeding is required.

There are two figures of merit that have been put forward for designers concentrating their attention on actinide burning instead of breeding. [13] Clearly there is incentive to burn actinides in IFRs. Where burning actinides instead of replacing them by normal breeding has been emphasized, the net actinide destruction rate in terms of kg/GWe-yr is used as a figure of merit. This emphasizes non-fertile fueling (little or no uranium in the fuel), and even non-reactor options such as accelerator-driven subcritical systems emerge as possible optimum strategies.

An idea of the sensitivity of design to burning actinides is given by noting what happens when neutron leakage from the core is increased by decreasing the core height. Halving the core height from three feet, say, to a foot and a half, approximately halves breeding. Instead of fuel being completely replaced by breeding, only half of that burned is replaced. At 1,000 MWe, three to four hundred kilograms of actinides are destroyed per year. The reactivity decreases steadily over the year of operation because of the steep drop in fissile replacement, where a three-foot core height design would maintain its reactivity. The effect of this is to force higher enrichments to maintain criticality. This means that reactivity must be held down initially by increasing the worth of control rods. Inherent safety properties protecting against inadvertent control rod withdrawal may be degraded by this, depending on the magnitude of the increase in rod reactivity worth. The much-increased leakage does decrease the reactivity worth of the coolant, but the inherent shutdown properties of the IFR mean that the reduced coolant worth doesn't play a part in accidents in any case. Neither can it hurt. However, such design does compromise other core-performance characteristics. Maximizing net destruction in this way is synonymous with minimizing breeding ratio or conversion ratio, and is counterproductive to the long-term sustainability of nuclear energy. This aspect will be illustrated further in the next section as we look at future growth scenarios.

The other figure of merit, also not necessarily useful, is the ratio of actinide-generating reactors (LWRs) to actinide-burning reactors (IFRs). The desirability of minimizing the fraction of actinide-burning reactors in a deployment of both implicitly assumes that actinide recycling is the important goal and that the actinide-burning reactors (IFRs) are not economic. The goal of actinide recycling is to minimize the amount of actinides destined to go to permanent waste repositories. But it's better to design the IFR to actually utilize actinides as fuel in a manner that minimally penalizes core performance, fuel performance, safety performance, and economics. Except as a temporary convenience, actinides cannot be simply accumulated and stored; they must either be destroyed or go to a repository. If actinides are utilized in the IFR reactor and its associated fuel cycle system, not disposed of as waste, they are a valuable resource. In that case net production should be maximized, and that is the figure of merit. As a practical matter, if there exists a large inventory of actinides from LWRs, then the initial IFR need not breed, in which case the IFR core will operate without external blankets. Such cores can be called burners in a literal sense, but the burning of actinides *per se* is not the goal. In the long term, breeding is required not only for nuclear capacity expansion but also for supplying low-cost fissile materials for thermal reactor if the uranium price continues to escalate. This would have the additional advantage of obviating the need for further enrichment facilities. (See discussions associated with Figure 13-7 in Section 13.4.)

Another way of saying all this is that the most important characteristic of the IFR is its ability to breed new fuel. The IFR's metallic fuel gives the designer a substantial advantage to start with. We have discussed the importance of a high fuel-volume fraction for breeding. For a given fuel-volume fraction, a higher fuel density is preferred for an efficient neutron economy. The theoretical densities of various fuel types are summarized in Table 14-5.

Table 14-5. Theoretical density of various fuel types, g/cc (Pu/U ratio of 20/80)

	Fuel form density	Heavy metal density
Oxide	11.1	9.7
Carbide	13.6	12.8
Nitride	14.3	13.4
Metal	19.2	19.2
U-Pu-10%Zr	15.7	14.1

As shown in Table 14-5, metal fuel has the highest heavy-metal density among all fuel types, and as a result it has much superior neutron economy to other fuel types, as shown in Table 14-6. A wide range of breeding ratios achievable by various fuel types are illustrated in Figure 14-8. [7] The breeding ratio using any fuel material can be made greater or lesser by design. All these designs are optimized for breeding. Altering the fuel pin diameter is a simple way of altering the fraction of the core content that is fuel (versus coolant and structure). The higher the fuel fraction, the higher the breeding ratio, and the manner in which breeding is affected is shown in the figure below. Metal fuel is always superior to oxide; carbide is in between, but all vary over a range. In the past, breeding ratio was always defined for fissile isotopes only (fissile production/fissile destruction), but with the inclusion of the non-fissile actinides as eminently fissionable fuel for the IFR it has become customary to speak of total actinides produced divided by total actinides destroyed instead. The distinction, however, is rather small, even for the IFR.

Higher breeding ratios generally imply higher fissile inventories, so there will be an optimum between the two when the truly practical measure of breeding is used; that is, how long does it take to double the starting inventory? The "doubling time" optimum for the metal fuel of the IFR is given in the Figure 14-9. [7] The doubling time is calculated for a large IFR with the minimum ex-core inventory, which gives the maximum growth potential. Metal fuel can achieve a doubling time well below ten years if required by a very rapid construction program.

Table 14-6. Comparison of neutron economy for various fuel types in fast reactor

	Oxide	Carbide	Metal
η	2.283	2.353	2.450
ε	0.356	0.429	0.509
$\eta + \varepsilon - 1$	1.639	1.782	1.959
Losses: Structure	0.158	0.131	0.127
Coolant	0.010	0.009	0.008
Fis. Prod.	0.055	0.058	0.058
O, C, Zr	0.008	0.009	0.025
Leakage	0.046	0.051	0.082
Decay	0.031	0.029	0.032
Subtotal of losses	0.308	0.279	0.332
Excess Neutrons (BR)	1.331	1.503	1.627

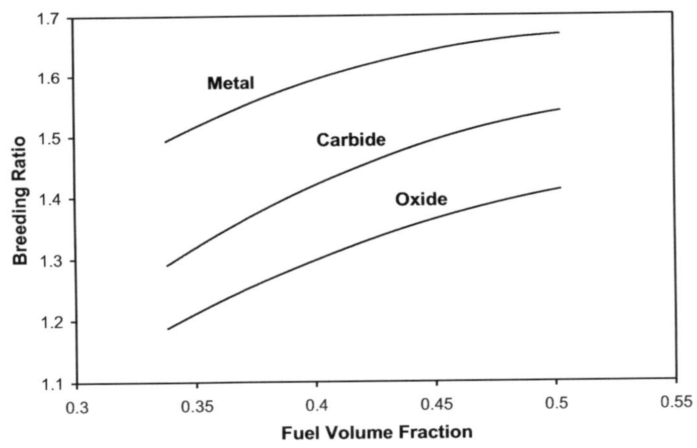

Figure 14-8. Range of breeding ratios for various fuel types

14.6 Design Principles of Long-Life Core

In recent years there have been suggestions for designs of very long-life core concepts. To achieve a long-life core, the neutron economy has to be pushed to the limit in order to breed enough fissile materials *in situ* to maintain the reactivity for a lengthy period without refueling. Therefore, long-life cores are possible only in fast spectrum, and preferably with metal fuel and high fuel volume fractions. In theory, the reactivity can then be maintained for a long time if the neutron economy is maximized by appropriate designs.

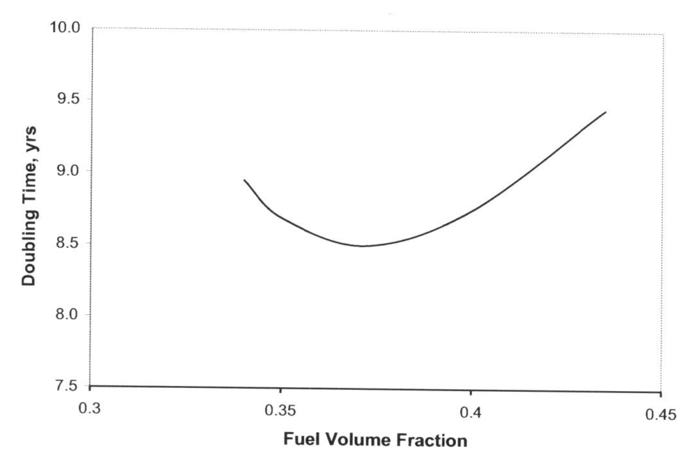

Figure 14-9. Range of compound system doubling time for metal fueled IFR

However, there are three constraints that have to be met in the design as well. The first is the fuel burnup limit. With the cladding materials developed and tested to date, a peak burnup limit of 200,000 megawatt-days per tonne, MWD/T, is about the most that can be envisioned. About 200 MeV energy is released per fission, or about 1 MWD when 1 gram is fissioned. The 200,000 MWD/T burnup then means that 200 kg has been fissioned per tonne of heavy metal, 20% of the initial heavy metal fuel, and converted into fission products. The gaseous fission products are collected in the upper plenum. The gas pressure stresses the cladding, and combined with irradiation-induced creep, it results in radial strain on the cladding, which limits its lifetime. The solid fission product accumulation can also have detrimental effects at higher burnups, and therefore 20% is the accepted burnup limit, depending on the specifics of the fuel pin design.

For conventional designs, such a burnup limit is reached in about five years of irradiation. To achieve a long-life core for a given burnup limit, the specific power has to be derated. Since the burnup in MWD/T is given by the specific power (MW/kg) multiplied by the full power days of operation, the long life can be achieved only by lowering the specific power. For example, if a thirty-year core life is desired, compared to a conventional five-year core (in practice one fifth of the core is refueled each year), the specific power must be derated to one sixth, and the actinide inventory provided at the beginning of the long-life reactor's operation would be six times the fuel required for a typical IFR. Put another way, six reactor cores are installed initially and burned slowly at one sixth burnup rate to achieve the thirty-year core life. Because of this large core requirement, most of the long-life cores proposed are for a power in the 5–50 MWe range. Although the heavy metal loading is directly proportional to the core life, the fissile loading does not increase in the same proportion, since the neutron leakage is reduced in larger cores and the fissile enrichment is reduced accordingly. The fissile enrichment can be reduced substantially in a similar fashion, as illustrated in Figure 14-7, depending on the reactor power.

Parenthetically, the uranium resource utilization cannot be noticeably improved by once-through long-life cores. As a reference point, a high-burnup LWR fuel requires 4.5% enrichment to achieve 5% burnup. About 88% of the original uranium ends up in the tailings during the enrichment process, then called depleted uranium. Of the 12 % loaded into the reactor, only 5% is fissioned, resulting in a 0.6% utilization of the original uranium. A large fast reactor would typically require 15% enrichment, instead of the 4.5% which would mean that 96.5% is discarded as depleted uranium. Even a 20% burnup of the remaining 3.5% results in an overall utilization of 0.7%—not much different than the 0.6% of the LWR's once-through cycle. This example case is already on a flat side of Figure 14-7, and therefore the best one can expect by extreme designs of long-life cores would be a little over 1% uranium utilization.

The second design constraint for long-life cores is the fast neutron (>0.1 MeV) fluence limit. The fast neutron fluence (the neutron flux multiplied by irradiation time) causes damage to the cladding material. The fluence limit is in the same range as the burnup limit for conventional design conditions, and hence normally no particular attention is given to the fluence. However, for long-life cores, the fluence continues to increase with time. The cladding strain due to irradiation-induced creep will continue beyond the conventional design limit and will become the limiting constraint for the long-life cores. Most long-life cores reach fast neutron fluences on the order of 3–4 times those of conventional designs.

The third constraint involves the thermal-hydraulics. The cladding lifetime is very sensitive to temperature, which sets a peak temperature limit. Within this limit, the coolant outlet temperature needs to be as uniform as possible among assemblies in order to maintain a high average temperature. In conventional core designs, this is achieved by providing a few orificing zones of assembly inlet nozzles to match the coolant flow to power. Because the assembly power shifts during irradiation, it is impossible to achieve a constant power-to-flow ratio for all assemblies. For larger long-life cores, the power shift will be more prominent and widespread throughout the core, and it will be much more challenging to develop a workable orificing scheme that will remain effective throughout the long core life.

In summary, the neutronic and burnup constraints can be met readily by judicious design choices, but the fluence constraint cannot be met unless a "magic" cladding material is developed. The thermal-hydraulic orificing constraint is also a tough challenge, especially for large cores.

14.7 Worldwide Fast Reactor Experience and Current Status

In earlier chapters the fast reactor experience in the U.S.—EBR-I, EBR-II, Fermi-1, FFTF—has been discussed. After the very early success of fast reactor development in the U.S., fast reactor construction was undertaken elsewhere as well. Including the U.S. reactors, almost twenty fast reactors have been built and operated around the world. The experience is summarized in Table 14-7. [14-17]

Table 14-7. Summary of Fast Reactor Operations

Country	Reactor	MWth/MWe	Operation Period
U.S.	EBR-I	1/0.2	1951-63
	EBR-II	62.5/20	1964-94
	Fermi-1	200/61	1965-72
	FFTF	400/0	1980-92
Russia	BR-5/10	8/0	1958-02
	BOR-60	60/12	1969-
	BN-350	1000/150	1973-99
	BN-600	1470/600	1980-
France	Rapsodie	40/0	1967-83
	Phenix	590/250	1974-99
	SuperPhenix	3000/1240	1985-97
Japan	Joyo	140/0	1978-
	Monju	714/300	1994-
UK	DFR	72/15	1963-77
	PFR	600/270	1976-94
Germany	KNK-II	58/21	1972-91
India	FBTR	42.5/12	1985-
China	CEFR	65/20	2011-

A few early fast reactors that were somewhat outside of the main line of development toward commercial power reactors are not listed in Table 14-7. The sodium-cooled fast reactor was used to power the submarine Seawolf from 1957 to '58. Atomics International operated the 6.5 MWe Sodium Reactor Experiment (SRE) from 1957 to '64. The Los Alamos Molten Plutonium Reactor Experiment (LAMPRE) was a homogeneous reactor with molten plutonium fuel and sodium coolant, which operated at 1 MWth from 1961 to '63. A 20 MWth Southwest Experimental Fast Oxide Reactor (SEFOR) in Arkansas was operated by General Electric from 1969 to '72.

Then there were sodium-cooled fast reactors that were constructed but never operated: Italy cancelled the 120 MWth PEC at an advanced stage of construction in 1987; Germany completed the construction of SNR-300 (762 MWth/327 MWe) in 1985 but never operated it; and the U.S. cancelled the Clinch River Breeder Reactor (1,000 MWth/380 MWe) while its NRC licensing process was nearing a satisfactory completion.

Fast reactors listed in Table 14-7 have had a mixed record of operation. Largely these were first-of-kind demonstration plants built in each country. Design mistakes were made occasionally; there were component failures, particularly in the non-nuclear portion of plants, a few sodium leaks and fires, and, in EBR-I and Fermi-1, the partial core meltdowns we have described earlier. On the other hand, the decades-long success of EBR-II, operating on a shoestring budget, and the lessons taken from the mistakes made elsewhere, lend confidence that fast reactors, when properly designed and operated, and with the experience accumulated by a more mature industry, should be safe, reliable, and easy to operate and maintain.

Sodium difficulties included various leaks (mainly in the secondary systems and in the steam generator systems), contamination of the coolant itself, and the effects of aerosol deposits. Sodium leaks in the secondary system piping are easily detected and the resulting smokes or fires are extinguished without undue difficulty or consequence. As discussed in Chapter 7, the largest sodium fire was in the Japanese demonstration reactor, Monju. There was a long delay in restart. It was not technical in nature. An unrelated fire in a low-level waste plant at Tokai, an unrelated criticality accident in a subcontractor's operation, a high court ruling nullifying the original licensing base, opposition of local communities, and so on, compounded over the years to delay restart for a prolonged interval. Monju finally restarted its operation in May 2010.

A few reactors were plagued with leaks in steam generators during the early years of their operation. Most leaks were at the tube-to-tube sheet welds, a significant fraction due to manufacturing (welding) defects. Some were due to material defects and lack of post-weld heat treatment. Manufacturing techniques are as important as material selection to reliable components. Some of the component failures were due to the selection of advanced materials that had little testing in sodium at elevated temperature. Early steam generators of stainless steel experienced stress corrosion cracking. To date, 2-1/4Cr-1Mo has been the dominant choice for steam generators and has had a very satisfactory operating experience. The 9Cr-1Mo is thought to be an improvement over 2-1/4Cr-1Mo for future applications and has received the ASME code qualification. For the primary system, both stainless steel types 316 and 304 have been successfully used. For fuel assemblies, ferritic HT-9 has proven to be the most reliable, low-swelling material for cladding and hexagonal ducts.

EBR-II stands out in all of this as a well-designed reactor which operated very successfully for a long time, and could have gone on operating with no identifiable limit to its life. The ingenuity of EBR-II design tends to get overshadowed by the metal fuel development success, the landmark inherent safety tests, and the closed fuel cycle demonstration that were discussed at length in earlier chapters. But the engineering of EBR-II needs to be given much credit. EBR-II was built under the assumption that it would operate for a few years before it was overtaken by the next step in scaleup as an EBR-III. Instead, it operated for decades; its long life attests to the quality of its design, and its operational successes clarified features that are important and desirable in future plants.

The compatibility of sodium with steel is an important trait. When the reactor-grade sodium was loaded initially in EBR-II, extra drums of sodium were provided for makeup if needed. The primary sodium was used for over thirty years without a drop of makeup. Impurities were removed by cold traps in routine operation. When the reactor vessel was drained after shutdown, the chalk marks made on the inside vessel wall during the initial fit-up were clearly visible on the remote camera. There was no sign of corrosion of any kind. The experience illustrates vividly the complete compatibility of sodium with structural materials, even to the extent that components submerged in sodium maintain their pristine condition for the full thirty-year lifetime.

Non-corrosive coolant contributes to reliable component performance. Steam-generator performance is particularly important to fast reactor operation because it is this one component that contains both sodium and water and the two must not come in contact. The steam generator tube failures in LWRs are caused by water chemistry and the accumulating corrosion products in the shell-side crevices. In sodium-cooled fast reactors, non-corrosive sodium flows through the shell side and corrosion product accumulation in crevices is minimal. The steam is contained in the tubes, where the simple geometry prevents corrosion product accumulation. The tubes in the EBR-II steam generators were double-walled and straight. Over thirty years of continuous service accumulated without a single tube leak.

Minimal corrosion-product accumulation also implies easy access for maintenance and a situation where radiation exposures to plant personnel can be kept to a minimum. Corrosion products are radioactive. No exposures are expected from maintenance and inspection of the steam generator, turbine generator, steam and feedwater pumps and equipment, and other parts. It is no coincidence that the occupational exposures at EBR-II and other sodium-cooled fast reactors have been about an order of magnitude less than those of LWRs. Sodium component reliability also means improved plant availability. Even with the frequent refueling of EBR-II due to fuel testing—an average of five times a year—and shutdowns to accommodate various irradiation tests and experiments, a high capacity factor, above 80%, was achieved in the later years of the EBR-II operation.

EBR-II experience demonstrates that if the next sodium-cooled fast reactors are designed thoughtfully, and as designs evolve further, sodium-cooled fast reactors can have a very high level of reliability, maintainability, operability, and longevity. The basic design approach must be a simple, forgiving design, avoiding complexity and avoiding layers of safety systems, and made possible by its inherent safety features.

Turning to the future prospects of fast reactors, some interest in fast reactors has been seen in recent years around the world, following the hiatus of the past two or three decades. In Russia, where BN-600 has been operating successfully over thirty years, and the construction of BN-800, originally planned in the 1980s, but caught up in the political maelstrom of the early '90s, was resumed a few years ago. It is now scheduled to be on-line around 2014. The China Institute of Atomic Energy (CIAE) has constructed the China Experimental Fast Reactor (CEFR), rated at 65 MWth, 20 MWe (the capacity of EBR-II), which achieved criticality in 2010 and first power operation in 2011. China plans to construct two additional 800 MWe fast reactors by 2020 in cooperation with Russia. In India the Fast Breeder Test Reactor (FBTR) reached initial criticality in 1985 and has been operating successfully, including demonstration of a full carbide-fueled core. India is now building a 500 MWe Prototype Fast Breeder Reactor (PFBR), with commissioning targeted for 2014. Four more units of the same size are planned for completion by 2020. Two sites have been designated for twin units each. India may well be the first nation in the world to establish a commercial fast reactor economy. Both France and Japan envision commercial fast reactors starting in 2045–2050 and to that end they plan prototype demonstration projects in the 2020–2025 time frame. South Korea is also developing a prototype sodium-cooled fast reactor toward the construction by 2028. [18-19]

14.8 Typical Deployment Scenarios

It is important to understand the effects of IFR deployments. To give an idea of its effect, we have considered deployment in the plausible nuclear growth scenario depicted in Figure 14-10. The World Nuclear Association surveyed member countries and compiled low and high estimates for nuclear capacity. [20] In 2030 the low is 552 GWe, the high case 1,203 GWe. In 2060, the low and high cases are 1,136 GWe and 3,488 GWe, and in 2100 they are 2,050 GWe and 11,000 GWe. In Figure 14-10, we took an average of the low case and the high case. This is not a forecast; our objective is to illustrate the limitations and impacts of a high IFR deployment rate on uranium resources.

In Figure 14-10, nuclear expansion in the near term is assumed to be provided by the license extension (to sixty years) of the existing reactors plus the 1,575 GWe of new LWRs needed to meet the capacity assumed through 2050, taking into account the replacement of retiring LWRs and the amount of fast reactor introduction possible by that time. The IFR capacity is arbitrarily constrained to 10 GWe in 2030, 30 GWe in 2040, and 80 GWe in 2050 to reflect a realistic introduction rate. Afterward, the IFR introduction is constrained only by the availability of actinides for the startup inventory. This constraint limits the total IFR capacity that could be started up with the LWR actinides to about 3,300 GWe. The balance of the capacity demand must be met by breeding in the operating IFRs. New IFR capacity beyond 2070 can easily be met by breeding in the IFRs that were started with the LWR actinides.

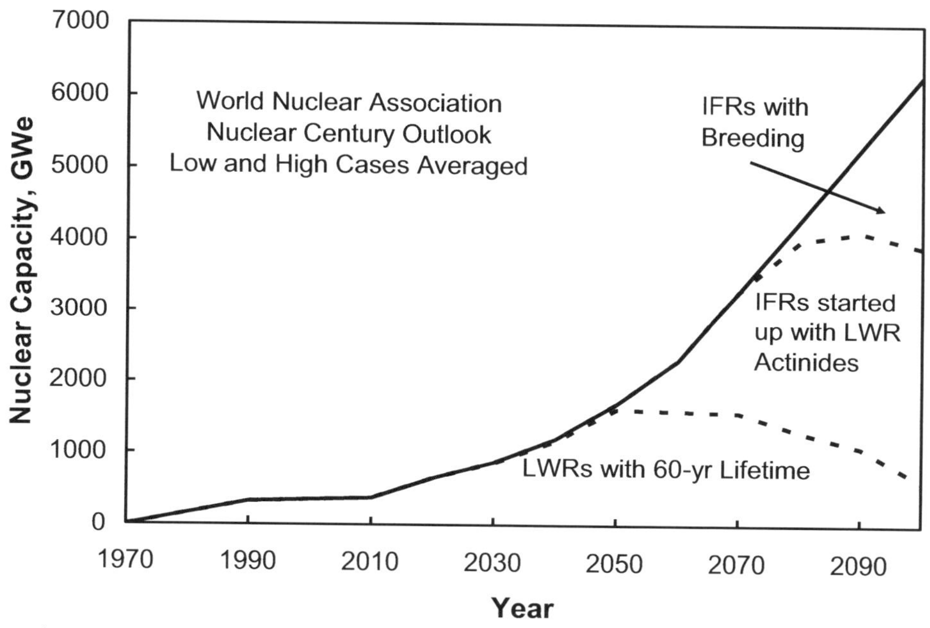

Figure 14-10. Example scenario for worldwide nuclear energy growth

The recently published book by Tom Blees, "Prescription for the Planet" [21] as a principal theme, focused attention on the necessity of IFR technology to substitute for carbon-based fuels in combating climate change. A much faster growth rate was assumed by the author, but even that rate can be justified by optimizing IFR design for a minimum doubling time in the range of 8–9 years, possible with the IFR metallic fuel.

The impact of IFR deployment on the uranium resources is illustrated in Figure 14-11, for the nuclear energy growth scenario assumed in Figure 14-10. IFR deployment can cap the cumulative uranium requirements just above the "Identified Resources" and "Undiscovered Resources" combined. [22] If the nuclear capacity in Figure 14-10 was to be met solely by LWRs, the uranium requirements would rise rapidly and go beyond the Undiscovered Resources category by the year 2070, and would of course continuously increase by very large annual amounts beyond that. Undiscovered Resources refers to uranium that is thought to exist on the basis of indirect evidence and geological extrapolation. The existence, size, and recovery cost of such resources are speculative. In fact, it is reasonable to suggest that amounts only up to the limits of Identified Resources category should be taken as the limit of uranium resources at any given time, because commitments to build nuclear capacity must be made on the basis of confidence in the availability of uranium resources over their entire lifetimes.

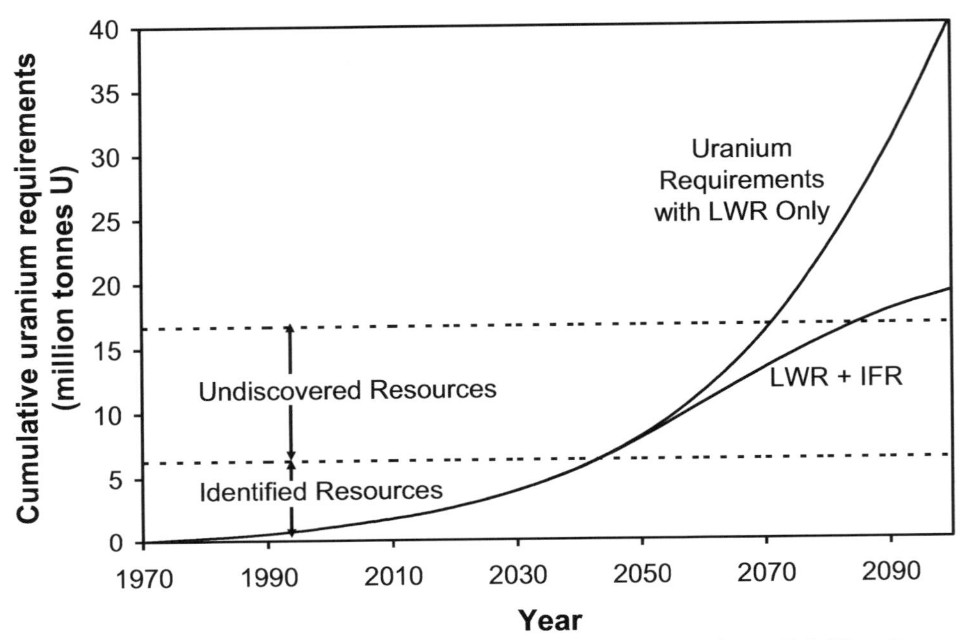

Figure 14-11. Uranium resources requirements and availability for nuclear growth scenarios with and without IFRs

A widely accepted perception is that there is a lot of cheap uranium. (This is so in circles where breeder development is unacceptable; however, in anti-nuclear campaigning generally the opposite view is held—there uranium is said to be scarce.) In practice, most utilities have long-term uranium supply contracts. When there are gaps in these long-term contracts, purchases of small quantities are made in the spot market. The five hundred metric tons of highly enriched uranium from the Russian weapons excess material is being blended down and has flooded the uranium market with a temporary glut of cheap uranium. However, the five hundred metric tons is equivalent to ninety thousand metric tons of natural uranium, and represents less than the two-year requirements of uranium for currently operating reactors.

What this simple system study indicates is that if large-scale deployment of IFRs starts as late as 2050, needs may not be met if the nuclear demand is as high as our example and if uranium resources remain as we understand them today.

14.9 How to Deploy Pyroprocessing Plants?

In discussing the IFR and pyroprocessing in earlier chapters, we have assumed that the fuel cycle facility will be co-located with the reactor plant, as was done for the EBR-II and its fuel cycle facility. This is the logical choice for the initial IFRs deployed. Because of the criticality constraint, pyroprocessing and refabrication equipment systems are sized naturally to serve a single reactor or two. The co-located fuel cycle facility also eliminates the transportation of spent and new fuels. It is the obvious choice. But as more IFR plants are built, there may be an advantage in a regional fuel cycle center that will serve several IFR plants, especially within the same utility grid and in geographical proximity. Some economies of scale can be achieved and further improvements made in terms of operational flexibility and capacity expansion, at the expense of possible objections to transport.

In pyroprocessing LWR spent fuel, the obvious question is whether the processing of LWR spent fuel can be or should be done in the same facility as for the IFR. The LWR spent fuel contains on the order of 1% fissile actinides vs. ~20% in the IFR. The criticality constraint that limits the equipment size in IFR processing can be relaxed for LWR spent fuel processing. To a first order of approximation, the LWR spent processing batch size can be increased by some factor approaching twenty or so, given by the difference in fissile fractions, and will need some such increase to be economically viable. LWR processing also requires the new front-end step for oxide to metal conversion described in Chapter 10. Some process steps could share equipment systems, product consolidation, and waste processing, for example; however, even those processes would require multiple equipment systems. All in all, the requirements and the optimization possible are quite different, and it seems best to take the equipment systems for LWR spent fuel processing and IFR fuel processing as two independent systems. Combining the two complicates materials accountability as well, and seems unlikely to show any significant countering advantage. If economies of scale are possible, the IFR fuel cycle facility should be designed for multiple IFRs, and LWR spent fuel processing plants should be free to handle as much throughput as necessary.

The related question is whether LWR spent fuel processing should then be centrally located or regionally dispersed. Currently, LWR spent fuel in the U.S. is accumulated at about two thousand tons a year. The LWR reprocessing capacity needs to be in the range of three thousand tons/year to handle the annual discharge as well as a start on reduction of the existing inventory. Currently operating aqueous reprocessing plants typically have an eight hundred tons/year throughput capacity. But a large central plant with two to three thousand tons/year is also possible, and would have some economies of scale advantages. The Barnwell reprocessing plant being built in the '70s was designed for fifteen hundred tons/year.

In pyroprocessing plants, the benefits of economies of scale will saturate at a much lower throughput rate. There is little incentive to go to a super-large central facility. The current commercial reprocessing throughput rate of eight hundred tons/year is probably a reasonable target. Once pyroprocessing has been demonstrated at pilot scale in the range of one hundred tons/year, scaleup to eight hundred tons will involve manufacturing multiple equipment systems and capitalizing on economies of scale achievable in support facilities. Three or four regional pyroprocessing centers can be envisioned, located to minimize the transportation. Sequential construction of these facilities would be natural to allow lessons learned from the first to be incorporated into the subsequent facilities. A modularized construction approach would be advantageous.

Fabrication of the initial startup metal fuel for IFRs can best be handled in the individual fuel cycle facilities servicing the IFRs. The actinide product ingots and uranium ingots can be shipped from the LWR spent fuel pyroprocessing plants to the IFR fuel cycle facilities and fabricated into metal fuel pins using systems in place to handle the recycle. The LWR pyroprocessing plant has no need to duplicate the fabrication system, and fuel design may well vary from IFR plant to plant.

14.10 Path Forward on Deployment

The bottom line, of course, is not what might be best, but how to start and proceed from where we stand today. We need to introduce a large IFR capacity sufficient to take care of future energy demands, and at the same time we need to solve the nuclear waste problem. The natural question can be posed this way: If IFR/pyroprocessing is so advantageous, why have other countries who have had strong fast reactor programs, and have constructed fast reactors, not adopted it? France and Japan have maintained strong fast reactor development programs, supported by long-term national policy and commensurate R&D funding. Both countries believe commercial fast reactors will be needed by the 2045–50 time frame, and a prototype fast reactor demonstration project of the best technology needs to be constructed around 2025. However, since they have invested so much in the conventional technology of oxide fuel and aqueous reprocessing and have a multi-billion-dollar facilities infrastructure, they cannot simply abandon what they have and turn to a path that, in processing at least, is certainly not fully proven. The burden of proof for IFR technology remains on the U.S. If the superiority of the IFR technology is proven here, the U.S. energy economy will be the winner. Other countries will certainly adopt it. The U.S. will have the lead to influence the manner in which it is deployed.

Other countries, in fact, are actually constructing fast reactors today. India and China, as we have mentioned, have both developed fast reactor technologies more or less on their own and their efforts are now focused on completing the construction projects. They may see the merits of metal fuel and pyroprocessing, but they don't have the technology base we have developed here. They would welcome opportunities for technical cooperation on the IFR technology, but we have no program now to share development, as we have had periodically in the past. Demonstrating IFR technology, if it is to be done, will be done in the U.S., by ourselves or in collaboration with international partners, but on our initiative.

Domestically, it is said that large amounts of funding have been spent on advanced reactor concepts; so if the IFR is so promising, why don't we let the utilities choose to build one? Very importantly, of course, any new advanced reactor concept will have to bear the first-of-a-kind costs and risks. Compared to commercialized standard designs, any new reactor demonstration project will require large efforts in design and engineering, until a new standard design emerges with construction of a few plants. A new manufacturing infrastructure may also be necessary, and its setup costs have to be absorbed by the demonstration project. There are much greater licensing uncertainties, too, to consider. The Nuclear Regulatory Commission is planning to promulgate a new regulation applicable to advanced reactors, but this rulemaking process will take time.

The utility industry cannot order any new advanced reactors until the economic and licensing uncertainties have been resolved. The government will have the principal responsibility for demonstration of economic viability and licensibility of new reactor concepts, but can accept it only if sufficient justification and incentive is plain. The IFR concept has potential for such huge benefits in the long term. But today, the demonstration projects are well beyond the capabilities of private-sector financing and beyond their planning horizon.

We need to demonstrate the IFR reactor plant with its own recycling facility, and demonstrate the application of pyroprocessing to LWR spent fuel. The near-term demonstration projects must be approached in stages. In the first stage, a pilot-scale (100 tons/yr) demonstration of the oxide-to-metal pyroprocessing facility for LWR spent fuel should be given priority. It is crucial for decisions on alternate spent fuel management strategies. Leaving spent fuel in interim storage for an indefinite period is unsatisfactory, both politically and technically, and aqueous reprocessing is very expensive and does nothing to reduce the long-term radiotoxicity of the spent fuel. Both bring proliferation concerns. Pilot-scale demonstration of the LWR fuel pyroprocess would be a step forward in a nuclear renewal by assuring that LWR spent fuel can be dealt with sensibly, and at the same time, the feasibility of the IFR processes can be convincingly demonstrated.

An idea of the procedures necessary for a pilot-scale (100 tons/yr) pyroprocessing plant estimated at roughly five hundred million dollars is as follows:

A design is to be done in two phases. The first is a two-year preliminary design phase producing an initial construction cost estimate allowing an informed decision on proceeding further, along with a Preliminary Safety Information Document (PSID) to allow an early informal review by the NRC. Their safety evaluation report will be utilized in the next phase to refine the design and complete the safety case. This second phase will produce a detailed design and the cost estimate necessary for the decision on whether to proceed with construction. A Preliminary Safety Analysis Report (PSAR) will be prepared at this time for NRC review for the necessary construction permit. The final design incorporates the as-built drawings, and Final Safety Analysis report (FSAR) is prepared for NRC review for an operating license. NRC licensing will be based on 10CFR Part 70, Domestic License of Special Nuclear Material (Fuel Cycle Facilities).

The project should take nine years. The first two are required for the preliminary design and cost estimate; the next two for the detailed design, PSAR, and detailed cost estimate. Licensing review is carried on throughout, from the end of the second year onward; however, it is assumed that the construction permit is issued after a two-year formal review following the PSAR submission. The final design carries on through the eighth year, construction having begun after the sixth year. The construction period is three years.

The IFR reactor demonstration can then follow. Successful demonstration of pyroprocessing provides the basis for moving ahead with a complete IFR system demonstration. This should provide impetus for actually dealing with the long-lived actinides from the LWR spent fuel and for the beginnings of U.S. reactor construction for the long term, as other large and increasingly powerful nations have begun.

14.11 Summary

The flexibility of its nuclear characteristics allows the IFR to breed excess fuel, to self-sustain, or to burn actinides down. Perfectly practical deployments can answer almost any future electrical energy need. The design of the IFR core can emphasize one or another important characteristic, to a degree at will. In particular, the important characteristics of breeding performance and the amount of fissile inventory can be selected by the designer to give any breeding performance desired, over a wide range. The design of the core, its inventory of fuel, and its mass flows differ between cores designed for breeding or self-sustaining and those designed to intentionally destroy actinides. However, conversion between the breeder and the burner modes can be done at any time during the reactor's lifetime simply by replacing assemblies during planned refueling outages.

To briefly review the important choice of coolant, the alternative coolants to sodium—helium gas or lead and lead-bismuth alloys—in their different ways are inferior to sodium in their thermal/hydraulic properties. Sodium's compatibility with the metals of reactor structures and components is important too, and is a characteristic not shared by lead and lead alloys. Radioactive corrosion products are not formed in any significant amount, radiation exposures to plant personnel are very low, and access for maintenance is easy. Sodium reaction with air or water, its principal disadvantage, can be easily handled by proper design. Sodium leaks, principally in the non-radioactive secondary systems but occasionally in the steam generator systems, in the first-generation demonstration plants were handled as a practical matter without much difficulty. Easily detected, the resulting smoke or flame can be extinguished without significant consequences. Major problems resulted in only one case—the Japanese demonstration plant, MONJU, where anti-nuclear campaigning and related political matters extended the shutdown interminably after the relatively minor cleanup.

We then went on to look at the physics principles underlying breeding, showing the principles involved in practical design for high breeding. The possible breeding characteristics of the world's principal reactor types were shown. Then, concentrating on IFR design itself, we showed that fuel pin diameter is the important variable in determining breeding, in the core, and with this, the way the thermal, hydraulic and mechanical constraints must be accommodated. Finally, we discussed the tradeoffs that enter in further balancing the requirements for an optimum design.

We then turned to the experience with fast reactor development in the past and examine the various problems and difficulties. Operational experience with fast reactors has generally been with first-of-a-kind demonstrations in each country as each took its own path. There have been a variety of difficulties. As with any other reactor type, or for that matter with any engineered system, there were design mistakes, component failures, sodium leaks and fires, partial core meltdowns, and so on.

We contrasted that experience with the thirty years of operation of EBR-II, which after a few-year initial shakedown period, were faultless. Remarkably easy operation and its flexibility in carrying out the multitude of experiments it did over the years attest to its insightful design. For all those years EBR-II acted as a pilot demonstration of today's IFR technology. It could have gone operating indefinitely, and it provides concrete evidence of the very high level of reliability, maintainability, operability, and longevity inherent in IFR technology. A sodium-cooled fast reactor designed along the lines of EBR-II—pool, metallic fuel—can be a simple, forgiving system, avoiding complexity, and avoiding layers of safety systems. The operational success of EBR-II clarified the design features that will be important and desirable in future plants.

A system study based on the World Nuclear Association's estimates of nuclear capacity over the twenty-first century illustrated the need for early large-scale deployment of IFR technology. If demand is met by LWRs, the uranium requirements will go beyond the "Identified Resources" before the year 2050 and beyond the Undiscovered Resource amounts by 2070. If IFRs are introduced in number by the year 2050, uranium resource usage will level out, but for the reasonable combination of LWR and IFR capacity taken in the study it will still exceed the "Undiscovered Resources" amounts by 2090. Commitments to build nuclear plants will be made only if there is confidence in the availability of uranium fuel over their lifetimes, not just the amounts required to start up. Assuming a sixty-year reactor lifetime, these needs will be felt by 2030. IFR introduction in numbers should take place well before 2050.

In looking at how to proceed in beginning to introduce IFRs, we concluded that the construction of a pilot pyroprocessing plant should come first. This is a practical and most necessary first step in implementing IFR technology. Not only will it demonstrate the recycle technology for the IFR, but it will also begin the development of a viable process for treatment of LWR spent fuel. It should pick up support on this basis as well. The cost of a hundred ton a year pilot plant should be in the range of five hundred million dollars over a nine-year period, with construction over the last three years of the project.

Other major nations, such as China, Russia, and India, increasingly are moving ahead with the fast reactor. Each has its own infrastructure and its own alliances. IFR technology can be expected to be superior, but demonstrating it, if it is to be done at all, should be done in the U.S. and by the U.S., possibly in collaboration with international partners, but on U.S. initiative. The IFR reactor demonstration can then follow. An IFR demonstration plant opens the way to actinide transmutation, and at the same time starts us on the road to plentiful energy for future generations.

References

1. J. G. Yevick, ed., *Fast Reactor Technology: Plant Design*, The MIT Press, 1966.
2. GIF-002-00, "A Technology Roadmap for Generation IV Nuclear Energy Systems," U.S. DOE Nuclear Energy Research Advisory Committee and the Generation IV International Forum, December 2002. http://nuclear.energy.gov/genIV/documents/gen_iv_roadmap.pdf
3. N. Li, "Lead-Alloy Coolant Technology and Materials—Technology Readiness Level Evaluation," Proc. Second International Symposium on Innovative Nuclear Energy Systems, Yokohama, Japan, November 26-30, 2006.
4. N. Polmar, *Cold War Submarines: The Design and Construction of U.S. and Soviet Submarines, 1945-2001*, Potomac Books Inc.

5. Federation of American Scientists, “Project 705 Lyra: Alfa class Attack Submarines,” http://www.fas.org/man/dod-101/sys/ship/row/rus/705.htm.
6. M. K. Meyer et al., “Fuel Development for Gas-Cooled Fast Reactors,” *J. Nuclear Materials*, 371, 281-287, 2007.
7. C. E. Till et al., “Fast Breeder Reactor Studies,” ANL-80-40, Argonne National Laboratory, 1980.
8. L. N. Salerno et al., “PRISM Concept, Modular LMR Reactors,” *Nuclear Engineering and Design*, 109, 79-86, 1988.
9. C. E. Boardman et al., “A Description of the S-PRISM Plant,” *Proc. 8th International Conference on Nuclear Engineering (ICONE-8)*, Baltimore, MD, April 26, 2000.
10. J. E. Brunings et al., “Sodium Advanced Fast Reactor (SAFR) for Economic Power,” *Proc. American Power Conference*, 48, 683-687, 1986.
11. Unpublished design study for a 600 MWe IFR at Argonne National Laboratory, 2002.
12. The Large-Scale Prototype Breeder (LSPB) was a 1320 MWe plant design project sponsored by DOE and managed by EPRI’s Consolidated Management Office during 1982-85.
13. National Research Council, *Nuclear Wastes: Technologies for Separations and Transmutation*, National Academy Press, 1996.
14. E. Waltar and A. B. Reynolds, *Fast Breeder Reactors*, Pergamon Press, 1981.
15. IAEA, “Fast Reactor Database: 2006 Update,” IAEA-TECDOC-1531, IAEA, Vienna, 2006.
16. American Nuclear Society, *Controlled Chain Reaction: The First 50 Years*, 1992.
17. American Nuclear Society, *Proceedings of LMR: A Decade of LMR Progress and Promise*, 1991.
18. Dohee Hahn, et al., “Conceptual Design of the Sodium-Cooled Fast Reactor KALIMER-600,” *Nuclear Engineering and Technology*, 39, 193, 2007.
19. Dohee Hahn, et al., “Advanced SFR Design Concepts and R&D Activities,” *Nuclear Engineering and Technology*, 41, 1, 2009.
20. World Nuclear Association, “The WNA Nuclear Century Outlook - Averting the Danger of Catastrophic Climate Change: Is the Nuclear Renaissance Essential?” 2010. http://www.world-nuclear.org.
21. Tom Blees, *Prescription for the Planet: The Painless Remedy for our Energy and Environmental Crises*, 2008. http://www.prescriptionfortheplanet.com.
22. Nuclear Energy Agency, *Uranium 2009: Resources, Production and Demand*, OECD NEA and IAEA, 2010.

AFTERWORD

To finish this book we thought a few general points might be useful.

First, if IFR renewal is to begin how must its implementation proceed? Enough paper studies have been done; indeed, more than enough. Any beginning must deal with real things, tangible things, things you can touch, experiments yielding real measurements. And real design work must begin.

Cost is always the challenge. The design and institutional arrangements, in fact everything, must emphasize the need for reduced cost. Simplicity of design, simplicity in operations, in recycle, in waste management, certainly; but in getting anything real done these days it is the legal framework that needs attention. The first plant, whatever it is called, will really be a "first of a kind," an experimental plant, really part of an experimental program. The aircraft industry gives us an example of how things can be done. In that industry a prototype is built, tested and perfected. Only then production in quantity begins.

U.S. laws now state that electricity-producing nuclear plants in the U.S. must be licensed. This requirement is a significant barrier to successful new reactor development in the U.S. Costs alone will assure that everything will have to be done elsewhere. The impact of the early NRC role in CRBR on cost and schedule is evidence of the negative effect of licensing too early in development of a new reactor. Licensing should come with the commercial stages and be based on experience and lessons learned from the prototype, not on exhaustive paper studies before it is ever built. Once the prototype is perfected, licensing based on its proven qualities makes sense.

A primary purpose of the prototype is straightforward. It must demonstrate that fast reactors can be built on time, on schedule and on budget.

Its design constraints are straightforward too. It should have all the necessary features of the IFR as developed in EBR-II. The pool, the metal fuel, the fuel handling system, and on and on down the list; everything that succeeded brilliantly should be provided to this first IFR. EBR-II was the most successful experimental reactor in the world. It needed a few years of break-in initially. A break-in period to perfect the design of the first IFR should be expected, making alterations if necessary, and then operate, and operate, and operate. Demonstration beyond question of the ease of operation of properly designed fast reactors is a goal of this plant and it is crucial.

Supporting the reactor design and construction there must be a closely connected Laboratory devoted to this technology. Who would suggest that our technology of today, or more truly that of fifteen years ago, is the best that can be done? Development must be continued. Rote procedures should be de-emphasized; few for development work, more, but applied with discrimination, for operations. Experimental reactor operation by trained and skilled operators should have procedures recognized as appropriate by the operators themselves.

IFR technology depends on economic recycle. The electrochemical process development is crucial. The R&D work most needed is resumption of an all-out development effort to perfect the electrorefining processes. This has two parts.

First is picking up the R&D on plutonium cathode development and second is a prototype processing facility for treatment of oxide LWR spent fuel. The first should be easy, requiring only the will and the charter to do it. It can be done. This is an area where the work can just be picked up as the old Argonne resources for this are still largely in place. The hot cells are there at INL. The chemical engineering crew is there, still fairly young and very knowledgeable, and they need the charter to do this work.

The second is to move ahead with the 100 ton pyroprocessing plant focused on processing spent fuel from the present reactors as described in the last chapter. Some such facility is on the critical path as spent fuel from present plants is a driver for, or a barrier against, all nuclear construction, LWR or IFR.

A full-scale IFR power plant scaled down to prototype in a fairly complete design is required. Design work led by Chang was done at Argonne in the eighties by people who had worked on the design of EBR-II, and also by MacDonald and associates at Atomics International, a group very experienced in sodium work. But the best choice needs care and - like it or not - we have the time.

Any significant next steps require real changes in governmental attitudes toward a realistic energy future. Righting the mistakes in practice and in legalities that is our legacy today may seem impossible but effort to get this done is crucial, and in time may not be as difficult as it seems today.

Finally, in concluding our work two quotes from the final pages of Richard Rhodes' short book, "Nuclear Renewal: Common Sense About Energy" (1993) seem apt. In the first he quotes David Lilienthal, Chairman both of the TVA, 1941-46, and of the Atomic Energy Commission, 1947-50, from Lilienthal's book, "Atomic Energy: A New Start" (1980):

> "We rely heavily on nuclear power to keep our economy going....For the near- and long-term future, the energy we now have and can count on, from all sources, is not

> enough. Except for temporary periods, it has never been enough, and it never will be enough for the kind of developing country we are, with our population steadily increasing and our desires and incomes expanding without long term letup. I have listened for years to assertions that we don't need more energy; they have always and everywhere been wrong, and they are just as wrong today as they have been throughout the history of energy and industrialized economies. Energy is part of a historic process, a substitute for the labor of human beings. As human aspirations develop, so does the demand for and the use of energy develop and grow. This is the basic lesson of history."

Rhodes then goes on,

> "Satisfying human aspirations is what our species invents technology to do. Some Americans, secure in comfortable affluence, may dream of a simpler and smaller world. However noble such a dream appears to be, its hidden agenda is elitist, selfish and violent. Millions of children die every year for lack of adequate resources—clean water, food, medical care—and the development of those resources is directly dependent on energy supplies. The real world of real human beings needs more energy. With nuclear power, that energy can be generated cleanly and without destructive global warming."

The passage of time has diminished neither the power nor the urgency of these words.

ACKNOWLEDGEMENTS

In beginning this book we were thinking of a volume on fast reactor technology in general to be done in a manner suited to the more technically inclined of the general public. There had been advances in this technology that had not been adequately covered in the literature of the time, we didn't think, and we felt that a book on this area of nuclear technology could play a useful role. However, at about this time the enthusiastic advocacy of the IFR in the writings of Tom Blees, Steve Kirsch, Terry Robinson, Joe Shuster, Barry Brook and Jim Hansen began to appear. In books and articles they outlined the merits of the Integral Fast Reactor and advocated its urgent deployment. Written by these highly technically literate non-specialists in the technology, they provided a general understanding of the IFR and what its implications for energy supplies would be for the future. And they did this admirably, describing accurately and vividly the capabilities of the IFR and the reasons for urgency in its deployment. They could only touch on the technology underlying it, however, and the why and how of the technology that caused it to work as it did, and the influence of the history of its development on the development itself, were obvious to us as being very important too. These things then became the focus of our efforts in this book.

We wish to acknowledge the debt we owe to these men therefore for their forthright and effective advocacy and we want to underline the importance of the encouragement that their efforts gave us in undertaking and completing this book.

There is a particular and very special acknowledgement that we would like to make, one that is in quite another class from the others. The importance of Allan Schriesheim and his late wife Beatrice to all this can be stated very simply: without Al's unwavering support of the IFR development over the whole of that politically difficult IFR decade and Bea's unquestioning faith in its worth, there would have been no IFR technology and there would have been no book. To them we owe this special acknowledgement.

As we finish this manuscript, we are reminded once again of our good fortune in the outstanding technical and personal qualities of the leaders of the five principal IFR development tasks: Leon Walters on fuels; Les Burris and Jim Battles on pyroprocess; John Marchaterre on safety, John Sackett on safety too, and on all matters relating to EBR-II; Dave Wade on core design; and Mike Lineberry on fuel cycle demonstration. Experts all, the best in the world in their fields, their leadership was critical to all the technical accomplishments of the IFR decade. The IFR development was a team effort above all and many, many people made outstanding, often brilliant, technical discoveries and advancements.

We know that in singling out a few people by name we risk slighting others. We can only say this: we know who you are and what you did, and the importance of it, and so do you, and so do all your colleagues. First rate work is recognized and deeply respected by others, and the importance of it is recognized most of all by us, the authors. We also single out the steadfast support of Ray Hunter throughout the program.

For the book itself we are indebted to our colleagues, Bob Benedict, Walt Deitrich, Art Goldman, Hussein Khalil, Leo LeSage, Mike Lineberry, Harold McFarlane, John Sackett, and Al Sattelberger for their encouragement in the book project and for their valuable comments and feedbacks on the manuscript. Our special thanks go to John Ackerman for his guidance and editing the chapters on principles of electrorefining. Thanks also are due to Barry Brook for a thorough technical editing of the entire manuscript and also for his kind words in the Foreword to our book.

We are also indebted to many individuals who provided technical consultations and helped with information and data: Chris Grandy, Bob Hill, T.K. Kim, Bob Phipps, Chad Pope, Mike Simpson, Temi Taiwo, Dee Vaden, Mark Williamson, Jim Willit, and Won Sik Yang. We thank Brea Grischkat and Linda Legerski for their help in formatting the manuscript.

The courtesy of Argonne National Laboratory for making the historical photographs and technical illustrations available to us is greatly appreciated.

Finally, our book project would not have been completed without the strong support from our families, in particular, of course, our wives Kay and Ok Ja, and our children, Hilary, Chris, Megan, Salinda, Alice, Dennis and Eugene. Their continuous encouragement, patience and generosity with family time are appreciated from the bottom of our hearts.

CET and YIC

APPENDIX A

DETAILED EXPLANATION OF THE BASIS OF THE ELECTROREFINING PROCESS

The electrochemical part of the process, electrorefining—the heart of IFR pyroprocessing—is not well understood by most non-chemists. Those not working in the field often make broad statements about its capabilities without fully understanding what it can do, and more importantly, what it can't do, which have led to much misunderstanding. In this Appendix we will give a step-by-step explanation of how the process works, from the most basic phenomena and understandings through to calculations that predict the product and the experiments and measurements that check those calculations. The treatment given is unique—although the basis for most of the analytical and experimental material given here has been published, there has been no other published explanation of the IFR electrorefining process from beginning to end.

Once again, we are indebted to John Ackerman, Argonne's fine electrochemist, for his generous advice and help and in his editing assistance in much of this exposition. In seeking simplicity there may be some error in detail, but the major phenomena are as they are understood at the present time. Responsibility for the former lies with the authors; for assurance of the latter we are in debt to Dr. Ackerman.

A.1 Introduction

The central element of the IFR pyroprocess is a form of electrorefining. But whereas in industries where metals like aluminum, copper, and zinc are electrorefined a single product is purified from a feed composed of an impure-form product material; in the IFR process the situation is very different. Here, the process must deal with most if not all of the elements found on earth and several that are "manmade." That's what the spent fuel is. IFR spent fuel will have had very high burnup, and its composition, while still uranium in large part, will have quantities of fission products too (as a rule highly radioactive), as well as fissionable actinides like plutonium—in total about a hundred elements of different kinds, each with its own chemical properties.

It is up to this single process to separate the reusable nuclear material from the actual nuclear waste. Everything depends on it. The compactness, the inexpensiveness, the effectiveness of the separations, the entire ability to recycle the actinides and remove the fission products depends on that one piece of equipment. And it, in turn, is based on electrochemistry, a fascinating science right at the intersection of electricity and chemistry, one in which even the expert may find some puzzling aspects even today.

A.2 Electrorefining Is an Electrochemical Process—But What Does That Mean Exactly?

Chemical reactions, or more precisely, a specific class of chemical reactions, give rise to electrical phenomena. They are the basis for all of electrochemistry. The electrical phenomena happen when an electrode (a conductor of electrons) is immersed in an electrolyte (which contains ions—electrically charged—and can conduct electricity), thus forming an "electrochemical cell." The reactions occur naturally, right at the interface between electrode and electrolyte. They convert chemical energy to electrical energy naturally, as a battery does. But if, on the other hand, the object of a process is to have a particular *chemical* reaction occur, the reverse can be done: an imposed voltage can cause the desired chemical reaction to take place. Current flows, electrical energy is converted to chemical energy, and chemical bonds are formed or are broken as appropriate to the process. This is what happens in electroplating, for example—and *this is what happens in the IFR pyroprocess.*

The IFR electrolyte is a molten mixture of lithium chloride and potassium chloride salts (which themselves are ionic compounds). This mixture has a relatively low melting point, lower than either of its elemental constituents. In the IFR process, when electrical current flows, the current in the electrolyte IS the flow of ions of uranium, plutonium, and the higher actinides, from one electrode to another. At least that is the common way of thinking about it. The electrolyte salts themselves are in the form of ions as well, and it is said that the ions of the electrolyte salts themselves do not carry current. As we shall see later, even at this basic level controversy lingers as to how the current is actually carried through the electrolyte. But there is no question about the result. The net current is exactly what it would be if, at the voltages we impose, the current were carried by the uranium and other actinide ions and them alone. When a positive actinide ion is created at the anode, instantaneously one is reduced to neutral metal at the cathode. (The charge balance in this way is maintained.) Thinking of the net current as carried by the actinides that are our product exactly describes the result, and that's good enough for us.

The actinide metals (uranium, plutonium, americium, etc.) interact with the electrolyte and gradually dissolve into it as positively charged ions that are collected on appropriately designed negative electrodes (cathodes). Two quite different cathode designs are used in the IFR process. Each collects its own product, one collecting uranium and the other the higher actinides, principally plutonium. Uranium is collected as a fairly pure metal; plutonium and the higher actinides are collected all together as a metallic mixture. These are the two valuable products.

Spent fuel contains close to a hundred different elements, so the array of possible reactions can make the process difficult to follow. But the necessary result is straightforward. The fission products must be separated from the valuable product (to be treated in a next step and disposed of in a further step), while the actinide product, now relatively free of fission products, must be produced in a form suitable for further treatment to allow their recycle back into the reactor.

The chopped-up fuel pieces themselves are the anode of our electrochemical cell. When they gradually dissolve into the electrolyte, the first and very important separation takes place. The most chemically active (the most driven to react with ions in the electrolyte) of the fission products, which are also responsible for much of the radioactivity, react immediately with the ionic compound, UCl_3. They displace the uranium and form their own chlorides. The displaced uranium remains for the present as metal atoms in the electrolyte. At the voltages for actinide refining, the chlorides of the active fission products are stable. They remain in the salt until they are removed as waste in a later operation. Positively charged ions diffuse through the electrolyte toward the cathode. The uranium and higher actinides preferentially deposit on the cathodes *because the higher stability of both the chlorides of the electrolyte materials and the chlorides of the dissolved active metal chlorides prevents them from also "reducing" to metals and depositing on the cathode.* The actinides react electrochemically because *their* chloride stabilities match the voltage range chosen specifically to overcome their stability.

The ionic chlorides are always essentially completely dissociated (think of a sea of ions) and the current or voltage that is applied does nothing to affect that. The individual positive ions are not associated with any single individual negative ion. There are no UCl_3 molecules for example, except as statistical fluctuations. A U^{+++} ion does not have three individual Cl^- ions associated with it. Instead, the electrolyte containing the metal chlorides can be thought of as a sea of chloride ions, Cl^-, with the positive ions wandering about in it. There is a general drift of positive ions from the anode where they are created to the cathode where they are collected. They drift over under the combined influence of the imposed voltage and the voltage generated by chemical reactivity in the electrolyte itself.

The electrons needed to neutralize the actinide ions for collection are removed from the actinides and the active metals right at the anode surface by electrochemical reaction, and once the electrons are dislodged they go through the rest of the electrical circuit outside the cell, via a power supply and wiring, to the cathode. Simultaneously, by a matching electrochemical reaction at the cathode, the electrons, negatively charged, are consumed by the positively charged actinide ions, converting them back to metals again. Then they deposit as metals on the cathode. The two electrochemical reactions are the mechanism for transferring electrons to and from the metal ions across the boundaries formed by the electrode surfaces. In this way the electrical circuit is completed.

Typically the current is controlled and made as high as possible, because current is a direct measure of the product stream. But it is subject to a cutoff to stop the voltage from rising high enough to start to transport other elements as well.

The spent fuel constituents, once dissolved, distribute throughout the electrorefiner. Some stay in the anode basket; some deposit on or in cathodes, depending on the cathode design; some stay fixed in the electrolyte itself; and some go to the layer of liquid cadmium in place below the electrolyte. Where, and in what concentrations, things go in the electrorefiner is an important matter. The ability to calculate this, even approximately, is important to understanding and optimizing the process. It is done by making a simplifying assumption: that the elements reach an equilibrium distribution in the electro-refiner. The process can then be analyzed and further optimized on this basis.

For the equilibrium assumption to hold (approximately), it can be seen that the reactions of the metals from the anode with the electrolyte salt, ionizing them, must be rapid enough to dominate over the processing rates themselves. In development, therefore, it was postulated that the various elements do distribute in concentrations approximately those of equilibrium, and subsequent measurements have shown this to be so, at least to the precision of the measurement techniques. The concentrations of elements in the various parts of the electrorefiner are now calculated with useful precision.

A.3 Principles of Electrorefining: What Are the Basic Phenomena Here? What Is Fundamental?

Atoms, ions, electrons, and molecules are the fundamental particles—but the bases of our process go no deeper than the molecular level. Chemical reactions are in fact just these very small particles interacting (colliding) with each other. *Tiny energy changes occur in these interactions, and the energy changes determine what happens in the process.* Quantum theory is needed to accurately describe properties of particles as infinitesimally small as these, but we need deal with such theory only in the most cursory way for an understanding perfectly adequate to our purposes.

As noted above, the interactions right at the interface between the electrodes and the electrolyte determine much of the important behavior. The voltage drops are not gradual across our electrolytic cell; the principal drop is set up just a molecular layer or so from the interfaces. After this, mainly transport phenomena take over. Ions are transported through the electrolyte—positive ions to the cathode, negative to the anode. A further small voltage drop develops, but with the usual electrode spacing, the principal voltage differences remain right at the electrode interfaces. There ions are no longer transported; they must diffuse through a static layer of electrolyte that clings to the electrode surface.

To move from the microscopic molecular level to the macroscopic, engineering, everyday, useful level, we must deal in the very large amounts of molecules involved in the process. This requires some of the relationships, constants, and units that chemists use. For sizes practical and measurable in the everyday world, amounts of these particles are defined in terms of a *mole* of material. A mole is a unit of amount, giving quantities of particles on the order of 10^{23}. That is a huge amount of these infinitesimally small particles, of course, but these amounts are necessary to bring us to sizes and amounts that are at the practical level and can be measured in grams, kilograms and the like. Moles do this for us. (To give an idea of the immensity of the number of molecules in a mole—if molecules were grains of sand, a mole of grains would be a cube some tens of miles high.)

We want to see, first, what reactions can take place; second, which ones actually do; and third, when they do not, why they do not. And after that, very importantly, what voltage will force the reactions to occur that we want (and not too many reactions that we don't want). We want to put numbers to all this. Magnitudes, quantities—how do we get those?

The forces driving the reactions are the tiny energy changes that occur at the molecular level. Classical thermodynamics, which deals precisely with the unalterable realities of energy relationships (e.g. energy cannot be created or destroyed), gives us exactly the tools we need to determine what is and isn't possible. Chemical reactions are possible only if the energy changes that would result from the reaction "are favorable." What this means is that the energy content of the products of the reaction must be less than the sum of the energy contents of the reacting materials. If energy changes are not favorable in this way, a spontaneous reaction is not possible. We can, however, force a reaction to occur by changing the energy relationships with a voltage imposed between the two electrodes.

The thermodynamic relationships give us energy changes in terms whose sizes are measurable—just what we need. In fact, as a rule, many of the numbers needed for calculations of what is possible for processes have been obtained from other calculations and measurements and tabulated in the literature.

Thermodynamic understandings, relationships and calculations tell us what reaction *can* occur. But they are silent on what *will* occur. Not every reaction that can occur will occur. There is a barrier in energy at the molecular level that must be surmounted for reactions to proceed, even those that have the energy necessary to continue once the barrier is surmounted. Once it is surmounted, the reaction will proceed. To overcome energy barriers, additional "activation energy" is required. (The "ignition" of an ordinary match is an everyday example.) We will need to look briefly at these "activation energies" as well.

Finally, a principal part of the process does not involve electrotransport at all. When we think of the process solely as electrorefining, we implicitly leave out phenomena that are very important to our process. Electrorefining allows the actinides to be collected. But more generally, it is separations we want: separations of the actinides, but not individually, from all the rest. And "the rest" are most of the elements in the periodic table. The first big separation is into a few groups of elements—uranium and the higher actinides separated from two groups of fission products, one stable in the electrolyte as chlorides, others as metals remaining in the anode or collected in the cadmium below. Each of the fission products groups is then to be recovered and put into a final waste form for disposal.

In outline, these are the bases for the IFR process. We will go further into the details of each of these areas in the next sections. But to summarize briefly, quantum mechanics give us an understanding of the tiny energy changes and relationships between the molecule and the electron, the fundamental quantities. These explain how and why electrode-electrolyte relationships exist, and their basic electrical nature. Definitions of chemical quantities, constants and relationships, bring in the mole as the unit of measurement, which lifts the amounts from the molecular to everyday practical levels. Then the simple relationships of classical thermodynamics tell us what reactions are made possible by their energy changes, and give us the tools to calculate these energy relationships. Finally, an energy barrier, called the "activation energy" must be overcome for practical reactions actually to occur, and we will look at that briefly.

Moving onward, the principal concepts chemists use to understand the basis for the chemistry and thermodynamics underlying the behavior are three—activity coefficients, free energies, and equilibrium coefficients.

1. Activity coefficients specify the degree to which a substance actually takes part in a chemical reaction. Ideally, the relevant quantity is simply its concentration, but where it departs from this, the activity coefficient provides the necessary correction.
2. Free energy changes drive the chemical reactions, and the magnitude of the free energy change in a possible reaction determines the driving force as well as its direction, forward or reverse.

3. Equilibrium coefficients give the degree to which reactions go to completion, and thus give the concentrations both of the original substances and of the product formed after everything settles into equilibrium.
4. Equilibrium coefficients are closely related to free energies. The greater the equilibrium coefficient, the more complete the reaction. The relationship is $\Delta G = -RT\,ln K$, for the mathematically inclined. In this expression ΔG is the free energy change driving the reaction, the equilibrium coefficient is K (= $\exp(-\Delta G/RT)$). Reactions have an exponential dependence on the free energy change. The relative amounts of reactant and product are very sensitive to the magnitude of ΔG, and even at moderate values of ΔG, reactions will go on to completion for all practical purposes.

Most of what we need to know of equilibria in our process can be derived from these concepts. Basically we want to know "what goes where" in the electrorefiner and the chemical form of the "what" (is it in a compound, or is it a metal?). And we want to know how these things change as we tinker with the process to give the products we want and to do it in the most effective possible manner.

A.4 "Redox Reaction" Is the Basis of All Electrochemical Phenomena

Redox reactions are chemical reactions of a very particular kind. They involve electrons in the outermost shell only, of atoms, ions, or molecules. Specifically, an electron is transferred to or from that outer shell of electrons. The importance of the redox reaction can be stated simply: it is the basis of all of electrochemistry. When an ion accepts an electron, it is said to have been reduced (in electrical charge), and when an atom gives up an electron, it is said to have been oxidized (not as obvious a usage, but in any case, removal of the electron leaves an ion that is positive in charge). These reactions happen simultaneously in the electrorefiner: an oxidizing reaction at the positive electrode, balanced by a reducing reaction at the negative electrode. Overall charge neutrality is maintained. The paired reduction and oxidation reactions are called redox reactions for short.

The electrons flowing through the external circuit to the cathode were stripped from atoms of the anode material (our spent fuel). Positively charged ions of anode material are thus left at the electrode-electrolyte surface when the electrons depart, and they go on into the electrolyte. For uranium and the other actinide atoms, three electrons are stripped, so the ions created and now in the electrolyte are triply charged. Because electrons are electrically charged, the electrical energy necessary to transfer them to or from electrodes can be supplied by an imposed voltage. Whether a particular element at the electrode surface will actually transfer electrons depends on the magnitude of voltage applied.

The underlying reason for these phenomena lies in the quantum mechanical behavior of electrons in the atom. In an accurate picture of the atom, the electron energy levels (the only energies possible for the atom's electrons) are sharply defined. Electrons can have only certain energies. Whether electrons are actually present at an allowed energy level depends on whether the electrons available have the energy to do so. The highest energy level populated contains the outermost electrons, those of highest energy, and it is these electrons that are involved in all the reactions of chemistry.

Metals, though (in an atomic lattice, for instance), do not have the sharply defined, separated energy levels of a single atom. In metals, atoms are closely packed. Their electron energy levels strongly overlap one another, so much so that continuums of energy levels are created in a solid metal, with electrons present from the lowest energy on up. The levels become so numerous and so close together that they form bands of energy, where electrons can move freely, really from one level to another, but in effect at the same energy as so many levels have almost the same energy. These continuums of energy levels, individual bands of energy, often separated significantly from each other in energy, are the electrical "conduction bands."

The energy of the highest populated level can be changed by supplying electrical energy. When the highest populated energy level of an anode material (uranium, say) has a lower energy than the lowest energy level of the conduction band of the material of the connecting wire, its energy does not allow it to move from the anode material to the wire, and nothing will happen. However, when voltage is applied so that the energy of the populated level rises above the lowest level of the conduction band, the energies become favorable for electron transfer from the atom to the conduction band of the conducting metal. The result is a stripping (oxidation) of these electrons from and consequent dissolution of the anode material. The stripped electrons then constitute the electrical current through the external circuit to the cathode.

Whether the reaction *will* occur, in any practical sense, depends upon the *rate* of the electron transfer reaction. It may be large enough to be practical, or it may be small, or zero, depending on the kinetics (the molecular energies) of the electron transfer—a subject we will go into below. But this is the mechanism of electron transfer that allows electrorefining to be done.

A.5 Other Phenomena Play a Part

A moment's reflection will suggest that there have to be transport phenomena at work here. The actinide ions have to make their way from the anode through the electrolyte to the cathode. This doesn't happen instantly when an actinide positive ion is created at the anode. Bulk transfer across the electrolyte, from one electrode to the other, is necessary, and it takes place at speeds the eye could probably track if ions could be seen. Individual ions take a meandering path, driven toward the cathode by the electrical field of the voltage applied. Such phenomena must be accounted for, but in the main it is sufficient to note that rapid collection generally has the electrodes as close together as other practical considerations allow, minimizing the distance the ions must travel.

Other transport phenomena are involved as well, somewhat more complicated, but which may have implications for further optimization of the process and so should be mentioned. Close to the electrodes is a quiescent thin layer of electrolyte where there is no bulk movement. Ions must traverse this by diffusing through it. Such processes can be calculated using commonly understood diffusion theory. The "diffusion layer" adds an increment to the necessary voltage to drive the process. The effect may be strong enough to set up a significant difference between the composition of the bulk electrolyte and the composition of electrolyte right at the electrode surface where the reduction reactions lower ionic concentrations. So, for example, it is possible, but not certain, that this effect under the right conditions might be strong enough to alter the electrolyte composition at the surface sufficiently to improve higher actinide collection. The electrolyte composition right at the surface is the relevant composition for collection of product, and if the current is high enough, perhaps the composition at the surface could be altered enough that plutonium and the higher actinides, which otherwise would not, could deposit without need for the liquid cadmium cathode that is necessary at present. Eliminating cadmium to the extent possible would be an advance, as it's messy to work with. Some work has been underway investigating these possibilities.

A.6 Thermodynamics Enter in this Way

The driving force for the redox reaction is energy. But what is energy, really? It's actually a rather subtle concept, whose ambiguities are usually swept aside by saying it is "the ability to do work," or to "make changes." That is certainly its practical effect, and that's probably all we need to know. As a practical matter, energy has two forms only, and the two are easily identified and pictured in our minds: *kinetic energy*, the energy of motion—motion of objects, of waves, and in our case, of atoms and of electrons; and *potential energy*, stored energy—energy stored by virtue of position, as in gravitational energy, or, as in our case, in chemical (and, for fission, nuclear) bonds.

Electrical energy is kinetic energy; it's the movement of electrons—in an ordered direction—along a wire, say. And in our case, it causes chemical bonds to be broken so molecules can reform in different configurations, often as different substances (compounds or crystals) in fact. Chemical energy, on the other hand, is stored energy: energy stored in the bonds of atoms and molecules as potential energy. Reactions result when there is a reduction in the stored chemical energy from the reactants to the product—like a weight losing height under the force of gravity. The driving force is the energy transfer.

Energy transfer is the very subject of thermodynamics. The name in fact means movement of heat or energy. There are two parts to the thermodynamics of this. First, and most straightforwardly, a reaction can occur if energy is released by the reaction (felt as heat, a decrease in "enthalpy", another name for total energy). Second, if the energy of the product is less concentrated, more spread out, less useful, after the reaction, there is an increase in *"entropy,"* a somewhat elusive concept, but it too is an energy change—an energy loss, in fact. The two, enthalpy change and entropy change, may not act in the same direction, but if the net of the two overall is a decrease in energy content, the reaction can take place.

The net change is called the *free energy*: that is, it is the energy free to drive the reaction. It is the maximum energy available from a reaction for conversion to other forms of energy. It can be looked at as a "chemical potential," or as a voltage, in fact. For it is a potential energy, and may be thought of as the energy actually available to "flow downhill" and do useful work. As noted previously, there are energy barriers to the reaction proceeding spontaneously, but when those are overcome (we used ignition as an example), reactions are free to proceed.

The classic thermodynamic relationships, elegant and simple, apply. Using them, precise calculations can be made of the energy exchanges. These in turn define what redox reaction can occur and under what conditions. Thermodynamics gives us the quantitative information we need.

For those with some mathematical background, these considerations are summarized in the expression

$$G = U + TS. \qquad (1)$$

The free energy is denoted by G (for Gibbs, the originator of the concept). Enthalpy, U, is a measure of the energy intrinsic to the compound, its "internal energy." Entropy, S, is that portion of that energy unavailable for anything useful. It's a loss of energy that comes with the rearrangement of atoms and molecules in the reaction. T is the absolute temperature in Kelvin, showing that the entropy effect rises as temperature (and in turn, molecular motion) increases.

At constant temperature the change in G, ΔG, in forming a compound is given by equation (2) below. If ΔG is negative (think of this as a well, with the depth given by the magnitude of ΔG), the reaction forming the compound will tend to proceed spontaneously. If the magnitude of the negative ΔG is large (the well is deep), the tendency to spontaneity is great.

$$\Delta G = \Delta U + T\Delta S, \text{ at constant temperature.} \quad (2)$$

In keeping with our analogy of a well, the more negative the free energy of formation (of the compound) the deeper the well, and the more stable the resulting compound is.

Now recall that in our IFR process, the electrolyte is made up of two chloride compounds, salts of lithium and potassium, with a melting point of about 350°C, and the process operates at about 500°C. In the electrolyte the elements that form the most stable of the compounds—that is, really (chemically) active metals like sodium, cesium, and strontium, the ones with the most negative free energies of chloride compound formation, form their compounds by reacting with the chlorides with less negative free energies of formation. In particular, uranium and plutonium are displaced from their chloride compounds by the formation of these more stable compounds. To maintain the desired concentrations of the uranium and plutonium (and other transuranic elements) so that electrorefining can proceed at a useful rate, a small amount of cadmium chloride that has a much less negative free energy of chloride formation is added to the electrolyte. The free energy difference between the uranium and the cadmium in chloride formation causes the desired reaction to occur: The uranium and plutonium that had been displaced as metals, but not in collectable form on the cathode, are oxidized by displacing the cadmium in the cadmium chloride, bringing them back to uranium and plutonium chlorides in the desired amounts. The small amount of cadmium metal so formed (at the anode) may cling or it may drip into the pool below.

It should be noted that cadmium chloride ($CdCl_2$) is by no means the only compound useful for this; other oxidants, such as iron chloride ($FeCl_2$) or uranium chloride (UCl_3) itself, might be considered. They might be advantageous in that they do not form liquids.

The information necessary to predict the reactions that will and won't occur is given in the table of free energies of chloride formation. The most relevant elements are given in Table A-1. [1]

The table shows the three groups of chlorides, separated in free energies of formation. This separation into groups is the basis for the electrochemical separations the process provides. The first group is the active metals that form the most stable chlorides. They are also the most radioactive of the fission products that go to the waste, and they stay in the electrolyte until they are stripped out in later waste processing. The second are the uranium and transuranics which electro-transport, the only elements that are actually "electrorefined." They are the product, and the very reason for the process. They become fuel once again—to be recycled back into the reactor. If left in the waste, however, they would be the principal contributors to the long-lived toxicity of nuclear waste. The third are the metals with still less stable chlorides, iron and the noble metals particularly, which do not form stable chlorides in the presence of more active elements; they merely collect as metals in the cadmium pool below the electrolyte, or remain as hulls in the anode basket.

Table A-1. Free energies of chloride formation at 500°C, - kcal/g-eq*

Elements that remain in salt (very stable chlorides)		Elements efficiently electro transported		Elements that remain as metals (less stable chlorides)	
$BaCl_2$	87.9	$CmCl_3$	64.0	$ZrCl_2$	46.6
CsCl	87.8	$PuCl_3$	62.4	$CdCl_2$	32.3
RbCl	87.0	$AmCl_3$	62.1	$FeCl_2$	29.2
KCl	86.7	$NpCl_3$	58.1	$NbCl_5$	26.7
$SrCl_2$	84.7	UCl_3	55.2	$MoCl_4$	16.8
LiCl	82.5			$TcCl_4$	11.0
NaCl	81.2			$RhCl_3$	10.0
$CaCl_2$	80.7			$PdCl_2$	9.0
$LaCl_3$	70.2			$RuCl_4$	6.0
$PrCl_3$	69.0				
$CeCl_3$	68.6				
$NdCl_3$	67.9				
YCl_3	65.1				

*The terminology kcal/g-eq is to be read as kilocalories per mass in grams of material interacting with one mole of electrons. The sign of these numbers is understood to be negative.

Note that the fission products, which are the great majority of elements in the electrorefiner, whether above or below the actinides in free energy, are not touched at all by the refining process. Their chemistry isolates them and enables them to be recovered and processed later as waste. The chloride compound electrolyte was selected specifically for this reason: It provided the distinct separations necessary to the process.

The chlorides with the greatest free energy of formation (that is, the highest negative numbers in the table) are the alkali metals—lithium, sodium, and, significant because of its radioactivity, cesium; the alkali earths, beryllium and so on; and significant also for its radioactivity, strontium. Strontium and cesium are the most troublesome of the fission products. Their radioactive isotopes are created in quantity, they have penetrating gamma rays, and their half-life is a few decades, assuring they will be around for quite a while—a few hundred years, in fact—and radioactive in quantities high enough to be concerned about. But these fission product chlorides are highly stable; they go to the electrolyte and they stay there. Later in our process they are processed as waste.

The least stable, with the lowest free energies, are the metals themselves: cadmium, steel cladding, the alloying zirconium, and the transition metals normally used as structural materials, characterized by high strengths, high melting points and high boiling points. They are never in the salt. Low in stability as a chloride, these metals remain metals throughout. They remain in the anode basket or in the liquid cadmium layer below the salt.

In the middle group are the fuel isotopes (zirconium to some degree may be picked up too) and a few rare earths. They exist as chlorides in the electrolyte, the elements to be electro refined.

Uranium is deposited on a steel cathode as a dendritic (with tentacles) deposit, quite pure, with some adhering salt. Plutonium and the other transuranics, in the presence of uranium chloride, will not deposit that way. Their stability in chloride form is greater than that of uranium chloride—that is, their free energies are more negative than uranium. So instead of depositing as metal at the cathode, they immediately react with the uranium chloride and form their more stable higher-actinide chlorides once again. Thus, when reduced, they just exchange right back into the electrolyte as chlorides. In the presence of ample uranium chloride, as is normally the case when electrorefining the bulk of the uranium, they cannot be collected this way.

If they are to be collected, something must be done to change the free energy relationships. Or, possibly (and there is some evidence for this), the higher actinides might be collected on a metallic cathode by allowing the voltage to increase to pick up the higher actinides, after reducing the uranium chloride concentrations very far below the concentrations of higher actinides in the electrolyte. Uranium, if present at all, would still appear in a substantial amount in the product, and the higher actinides would come as a group.

The IFR process must use both effects to collect plutonium and the higher actinides—that is, both altering the free energies and reducing uranium chloride concentrations. Free energy relationships are altered by use of a liquid cadmium cathode instead of solid steel. The higher actinides as metals form compounds with the metal cadmium, "intermetallic compounds," whose effect is to lower the free energies of formation of chlorides by the amount of free energy used in forming the intermetallic compound. The free energies of the intermetallic compounds then almost match that of uranium itself; they do remain higher, but only slightly. The intermetallic compounds stabilize the higher actinide elements in the cadmium. In this way, by using the two different cathode types, it is possible to separate transuranics from uranium, not perfectly by any means, but adequately to provide IFR fuel of appropriate fissile content.

Because the intermetallics, once formed, still have slightly higher free energies of chloride formation than uranium, the uranium chloride concentrations in the electrolyte must be reduced well below those of the higher-actinide chlorides. In effect, the plutonium chloride concentration must be increased very substantially over the uranium chloride concentration to make up for the slightly higher free energy of chloride formation of the plutonium in the plutonium-cadmium intermetallic. Even with this, substantial uranium is still collected in the cadmium cathode. The proportion of uranium in the product depends on the ratio of plutonium chloride to uranium chloride in the electrolyte. As we shall see, for a plutonium/uranium product high enough in plutonium to be useful for IFR fuel fabrication, the uranium chloride concentration can only be a fraction of the plutonium chloride concentration in the salt.

This, then, is how thermodynamics, through free energies, plays its part in the overall process—and a very important part it is. We now turn to the next key phenomenon governing the important reactions in the electrorefiner—the reaction kinetics.

A.7 Kinetics and Activation Energies

The originator of the principal concept we now turn to was the great Swedish chemist, Arrhenius. A century ago he argued that for reactants to transform into products, they must first acquire a minimum amount of energy, an activation energy E_a, and at an absolute temperature T the fraction of molecules with a kinetic energy greater than E_a can be calculated from a Maxwell-Boltzmann distribution of molecular energies.

The Maxwell-Boltzmann distribution is similar in kind to the usual bell-shaped distribution in the "normal distribution" seen in all kinds of mathematical descriptions of phenomena, error distributions about a mean, for example. But the Maxwell-Boltzmann distribution has a lopsided (left-skewed) bell shape, with a high energy "tail" trailing on out from the maximum. (The implication of this is that in reactions where the rates are tiny—back reactions, say, where a tiny amount of the product once formed reforms the original reactants—there will always be some reaction, even if very small. And as we shall see this phenomenon is important to our process).

Molecular energies increase with temperature. The distribution of energies broadens out as temperature is increased, and thus the maximum moves to higher energies. The fraction of molecules with energies above the activation energy thus increases with increasing temperature, proportional to an exponential of the form exp(-Ea/RT). Ea is the activation energy (as it increases, the fraction above the barrier decreases and the reaction lessens), RT is the energy corresponding to the temperature T, and R is a constant (the gas constant, converting temperature to energy units).

The simple (and as it turns out, remarkably accurate) Arrhenius equation given below shows the dependence of the rate constant, k, of chemical reactions on temperature and on activation energy, E_a,

The rate constant, $k = A\exp(-Ea/RT)$. (3)

As we shall see, the rate constant calculated in this way allows prediction of the degree to which reactions go to completion and is fundamental to understanding of our process.

The exponential shows the importance of temperature to the rate of a reaction. But we don't need mathematics; everyday experience tells us the same thing—soap and water, for example, are more effective hot than cold. The exponential tells us why and in what way the temperature affects the rate of reactions. This simple exponential is very important. It appears again and again in the kinetics of chemical reactions. It will take a central place in distributions of the substances in the electrorefiner, which we will turn to now.

A.8 Understanding Important Basic Behavior: The Power of Equilibria

Nearly every element in the periodic table is present in spent fuel at some concentration—either as products of fission or as products of neutron capture, particularly in uranium. All the elements below uranium in weight are present, some highly radioactive, but most with relatively short radioactive lives. Present also are actinide elements, newly created, heavier than uranium, and radioactive (but for the most plentiful of them, not particularly so). Some have very long-lasting radioactive lives. A little less than a hundred different elements are present, each distributed according to the chemical principles relevant to it. Understanding what goes where and in what form it exists is possible from the simple principles of equilibrium, applied to the rates of the reactions of each of the elements. Such calculations satisfactorily determine the distribution of each element in the electro refiner.

The principles of equilibrium are very simple, very general, and very powerful. They apply to all sorts of processes and phenomena, all the way from chemistry on a microscopic scale to astronomy on the greatest scale. Very simply, the principle is this: In any reaction, the forward rate must eventually equal the backward rate. That's it.

Eventually the amount of product will build up enough that the backward rate (and there will always be a backward rate) equals the forward rate of product formation. The backward rate—the rate at which the product of a reaction dissociates into the original components of the reaction—may be small, as we have noted. In fact it may be microscopically small, and present only due to highly unlikely statistical fluctuations. But equality of forward and backward rates comes when the concentrations of reactants have so decreased and the amount of product has so increased that the two rates overall are equal. In those reactions where the backward rate is extremely small and present only because of statistical fluctuations, reactions go to almost perfect completion. And where clean separations are important, that's what is wanted.

A fission product can go entirely (well, almost) into the electrolyte and stay there because this is the equilibrium result. And importantly, we can calculate its distribution—where it is and in what quantity. Equating the forward and backward rates, and taking the ratio of the two gives the "equilibrium constant." From the equilibrium constant we can predict the degree of separation of each element, and that is exactly what the IFR chemists do, in a computer program that handles the multitude of elements.

To the present time, the rates of electrorefining do not appear to affect the assumption of equilibrium, at least to the accuracy the distributions could be measured. Running the process faster may bring in issues that lessen the accuracy of the equilibrium assumption. Today the process can be run fast enough to be practical, and equilibrium distributions seem to hold but the rates at present may not be optimal. Eventually there may be economic pressure to run the process faster, perhaps to a point where the kinetics of the process do become important.

The things that matter to the rate of a reaction are not surprising: They are the concentrations of the reactants, or rather the concentrations that actually contribute effectively to the reaction, along with the factor that accounts for the fraction of molecules actually having high enough energies to surmount the activation barrier. Concentrations are relevant because reactions depend on collisions taking place, and the more tightly packed the reactants are—the higher the concentrations—the more collisions there will be. And the factor that accounts for the necessity of molecular energies to equal or exceed the activation energy is just the familiar Arrhenius exponential, $\exp(-Ea/RT)$.

The rate of a reaction then is given by the relationship $A\exp(-Ea/RT)$ multiplied by the concentrations contributing to the reaction. In this expression A is a constant (the pre-exponential constant) which cancels out in the important concept that we have been coming to—that of the *equilibrium constant.*

Equilibrium in a chemical system is a dynamic thing. The concentrations at equilibrium are unchanging because at equilibrium the "forward" and "backward" reactions are equal. Both rates are given by the rate constant, k (as defined in Eq. 3 above), multiplied by the concentrations of reactants relevant to the reaction, forward or backward.

That is, rate = Aexp(-E/RT) x C_1 x C_2, where C_1 and C_2 are concentrations of reactants. The forward rate is Aexp(-E_f /RT) x C_{1f} x C_{2f} , and the backward rate is Aexp(E_b/RT) x C_{1b} x C_{2b}.

The subscripts f and b refer to the forward and backward concentrations respectively. Equal at equilibrium, the two rates can be equated. The constant A cancels. The ratio of the two exponentials gives the equilibrium constant, and the concentrations adjust to give their forward to backward ratio that at value at equilibrium.

So the equilibrium constant is two things. It is the ratio of the reactant and product equilibrium concentration ratios, and it is the ratio of the forward and backward rate constants expressed as the exponential, exp(- (E_f-E_b)/RT).

E_f and E_b are the free energies of formation of the forward and backward reactions in the redox "exchange reactions" of interest, and calculable from the relevant free energies, so we have the means for calculation of the ratio of equilibrium concentrations. This is of central interest to our process. We want to know what the concentrations of the constituents of our products are. We very much want to know how the product concentrations are affected by changes in the concentrations of the elements in the electrolyte. And we want particularly to know how the plutonium to uranium ratio in the product changes with the concentration of plutonium chloride and uranium chloride in the electrolyte.

However, we have one more step to take. As we have implied, often the reactant contributes to a reaction differently than its concentration would suggest. To take this into account, "chemical activities" are defined in place of concentrations per se. Denoted by "a," they replace concentrations in the expression for the equilibrium constant. In an ideal case, concentrations are the correct quantity exactly contributing to reactions. The use of "activities" rather than concentrations corrects for non-ideal behavior. Activities in turn are defined to be the concentration multiplied by an "activity coefficient." The latter is simply the necessary correction to account for the degree to which concentration actually contributes to a reaction, which can differ from the actual concentrations due to interactions between molecules not important in the ideal case. In short, it's a fudge factor. But where the activity coefficients are constant with concentration over the concentrations of interest, concentrations can be used directly, and this appears to be the case for much of the IFR process. [2]

Constant activity coefficients are a good enough assumption as long as the reactants are present in concentrations where they remain dissolved—that is, for unsaturated conditions. This comes up in a case important to the cadmium cathode. Calculations are simplified by the assumption of constant activity coefficients, as activities are then directly proportional to concentration, from zero to its value at saturation in cadmium where precipitation of the intermetallic compound begins.

An example of great interest to our process is the degree of enrichment of the Pu to U ratio in the cathode possible from a given ratio of $PuCl_3/UCl_3$ in the electrolyte. The measured ratio in the electrolyte to the ratio in the cadmium cathode is given as 1.88 for Pu/U, for example, from the Ackerman and Johnson results. [3] This says that at equilibrium, the ratio of plutonium to uranium in our cathode will be only 1/1.88 of the ratio in the electrolyte. The corollary is that for reasonable enrichments of the material in the cathode—our input fuel material—the ratio of uranium to plutonium in the electrolyte must be far lower than the ratio commonly used for uranium deposition.

As we will discuss in some detail in a later section, effects of actinide saturation of the liquid cadmium cathode have importance in determining its composition at the end of the run. After saturation of the plutonium in the cadmium, metallic $PuCd_6$ can be deposited for a time and will act to increase the Pu/U product ratio. But increasing the $PuCl_3/UCl_3$ ratio from the ratio used for uranium deposition before using the cadmium cathode at all is an absolute necessity. The maximum attainable $PuCl_3/UCl_3$ ratio is determined by the solid cathode process—we only run the solid cathode until plutonium impurity in the solid cathode uranium product becomes excessive. The minimum $PuCl_3/UCl_3$ ratio that is usable comes when more U than is acceptable begins to deposit in the Cd cathode. The $PuCl_3/UCl_3$ ratio must be high enough to make the product ratio adequate for its subsequent use in fuel enrichment.

We want to establish as well as we can the systematics of the ratios of plutonium (and other actinide) concentrations to the uranium concentration in the cadmium cathode product as a function of their ratio as chlorides in the electrolyte. Our final product can be no more plutonium rich than whatever the ratio of plutonium to uranium in the cadmium turns out to be. After that, we can further dilute the product with uranium where we need to in fuel fabrication, but there are no more steps in the process that could enrich it further.

The equilibrium distributions of plutonium and uranium between the electrolyte and the cadmium cathode can be calculated from Equation 4 below. This expression looks complicated, but all it amounts to is equating the two different ways of writing the equilibrium coefficient, one in terms of in terms of free energy change (the exponential) and the other in terms of activities (concentrations multiplied by their appropriate activity coefficients).

$$\text{Exp}(-\Delta G_o/RT) = [\gamma_U/\gamma_{Pu}]\,[\chi_U/\chi_{Pu}]\,[\gamma_{PuCl3}/\gamma_{UCl3}]\,[\chi_{PuCl3}/\chi_{UCl3}], \text{ where} \quad (4)$$

Pu and U are the two elements we are interested in,
γ = activity coefficient,
χ = concentration expressed as mol fraction,
ΔG_o is the difference between the standard free energies of chloride formation of $PuCl_3$ and UCl_3,
R is the gas constant, 1.987 cal/K, and
T is temperature, ^oK

The exponential is calculable from the difference in the free energies of chloride formation of the two metals. The activity coefficients have been measured and are tabulated in the chemical literature, and the ratio of the concentrations of uranium and plutonium chlorides in the salt is measurable, so the ratio of plutonium to uranium in the cadmium to be expected—the number we want—can be calculated.

Also very important is how well the fission products are separated from the actinide product. The degree of separation of the waste from the product quantifies the success of this portion of the process. Separation factors measuring the degree of separation of many of the important fission products from uranium product have been measured. [3] Such separation factors are defined first in terms of a distribution coefficient, D, for the individual element. Using the important fission product cesium as an example, the distribution coefficient is the ratio of the cesium in the electrolyte (as cesium chloride, of course) to cesium in the metallic product. (A successful separation will have a very small cesium concentration in the metallic product, for example.) Then the ratio of its distribution to the distribution coefficient of uranium, defined similarly, gives the separation coefficient for cesium with respect to uranium separation. (Strictly, corrections for valences may be necessary in some cases. That need not concern us here, where we need only the principles involved.)

D_{cs} = concentration of cesium chloride/ concentration of cesium metal, and with a similar definition for other elements, notably uranium, for which we get:

The separation factor of cesium with respect to uranium: $SF_{cs/u} = D_{cs}/D_u$.

The separation factors tell us how clean the separations will be. Their basis, once again, is the difference in free energies of formation of the chlorides of the various elements relative to that of uranium. Some measured separation factors are given in Table A-2, taken from Ackerman and Johnson. [3]

The values vary from separation factors of 43.1 to $1.6\text{x}10^9$ for the listed fission products, and from 1.88 to 3.52 for the higher actinides. These are perfectly adequate fission product separations for IFR fuel, whereas for the actinides, small separations between plutonium and both uranium and the actinide above plutonium assure the desirable (for safeguarding purposes) imperfect actinide separations.

Table A-2. Separation Factors Relative to Uranium in LiCl –KCl Salt at 775K

U	1
Pu	1.88
Np	2.12
Am	3.08
Cm	3.52
Pr	43.1
Nd	44.0
Ce	49
La	130
Gd	150
Dy	500
Y	6000
Sm, Eu, Li, Ba, Sr	$>10^{10}$

A.9 Actinide Saturation in Liquid Cadmium: A Key to Enhanced Plutonium Depositions

Saturation of the liquid cadmium can play a role in the effective operation of a liquid cadmium cathode. With it comes the ability to control, within limits, the possible composition and amount of the actinide product. [4]

A.9.1 Solution, Solubility and Saturation

A "solution" is defined very precisely to mean a "solute" (in our case uranium or the plutonium-cadmium compound $PuCd_6$) dissolved in a "solvent" (in our case, liquid cadmium). A solution is one uniform homogeneous phase; the solute is dissolved, and there is no solid in solution. (Think of sugar in water; up to the solubility limit, sugar dissolves and the water is still clear: only one phase is there. Beyond the limit, the solution is no longer clear; there is another phase, sugar sludge probably, in it.) If a solution is saturated, no more solute can go "into solution." Saturated means exactly that, no more and no less: no more solute "in solution." More can deposit, but the amount that is "in solution" is fixed and cannot increase any more. More U added to a cathode saturated in U will result in the formation of another phase; "pure" metallic U will deposit and it won't be dissolved "in solution." More Pu added to a cadmium solution already saturated in plutonium will result in formation of a solid, metallic $PuCd_6$.

The reason for these careful distinctions is that the relative concentrations of plutonium and uranium going into the cathode change markedly from the state when neither of the elements is saturated to the state when one element, but not the other, is saturated. The third state, when both are saturated, has still different characteristics, but as we shall see, we will not choose to operate in that regime at all.

The electrorefining process brings the actinides to the liquid cadmium cathode. At the start they go into solution. But only up to a certain point, as each has a very well defined solubility limit. At concentrations in the range of 2 percent of the cadmium amount, their solubility limits are reached. The mode of deposition changes dramatically, and it affects both the ratio of plutonium to uranium and the product amount.

To be clear, species in solution are present at the atomic and molecular level, *e.g.* "$PuCd_6$ molecules in cadmium solution." The solution is homogeneous; the molecules are uniformly distributed in it. Any new phase will be separate from the solution phase; there'll be an identifiable boundary between the two. Ice and liquid water are separate phases. A solution is precisely one phase. Anything that is not dissolved and uniformly distributed on an atomic scale throughout the liquid—particulates, say—is in some other phase. Add a small amount of plutonium to cadmium. Initially there will be one phase—a liquid of composition that varies as plutonium increases and the proportion of pure liquid cadmium decreases slightly as PuCd6 is formed in solution. At saturation, another phase, solid $PuCd_6$, appears, but the composition *of the liquid phase* will not change any more.

A.9.2 Effect of Saturation on Chemical Activity

And now we come to the key point. *The activity of a solute present in a solution is independent of the amount of solute once the solution is saturated in that solute.* Its value is fixed. This is basic. It is vital to understanding liquid cadmium cathode operation. It is a basis for gathering useful amounts of plutonium in the cathode.

The free energy of the reaction of plutonium in cadmium (or $PuCd_6$ in cadmium) with the uranium chloride in the electrolyte still favors exchanging plutonium for uranium in uranium chloride. Not all the difference in free energies of formation of plutonium and uranium chloride is eliminated by the formation of plutonium intermetallic. This is ninety percent of the driving force for the exchange reaction: plutonium for uranium in uranium chloride has been removed, but some still remains in the direction of removal of the plutonium metal by exchange with uranium in uranium chloride. (As opposed to the "back reaction" of $PuCl_3$ in the presence of U breaking up to form Pu metal and UCl_3.) Something more must be done than simply providing the liquid cadmium.

Two things, in fact, are done. The uranium chloride concentration in the electrolyte is drawn down so the ratio of plutonium chloride to uranium chloride in the electrolyte is brought up to the point where the back reaction becomes significant. But importantly too, the cathode should be operated according to principles recognizing the effects of saturation, which have not been explicit in the work to the present time. Our task is to thoroughly examine these principles and their effects. They are at the center of this important segment of the process. [4]

A.9.3 Effect of Inter-metallic Compound Formation

The intermetallic compound $PuCd_6$, a compound of two metals, makes it possible to deposit plutonium in the presence of (a very limited concentration of) uranium chloride. In the compound $PuCd_6$ in solution, the plutonium activity is tiny. At saturation it is only $4x10^{-6}$. This value is determined by the free energy change in the formation of $PuCd_6$, and this is how it works:

Pu + 6Cd <--> $PuCd_6$ is the reaction between the Pu metal and the Cd liquid metal.

$K_{eq} = \exp(-\Delta G^o{}_f/RT) = \exp(19.1/1.536) = 2.52x10^5$;

$\Delta G^o{}_f$ for the reaction forming $PuCd_6$ is -19.1 kcal/mol, as listed in tables of free energies of formation.

$K_{eq} = 2.52x10^5 = a_{PuCd6}/a_{pu} * a_{cd6}$

In the saturated solution the cadmium activity is unity to a good approximation and the activity of $PuCd_6$ (in $PuCd_6$), of course, is also one.

So $K_{eq} = 1/(1*a_{pu})$

$a_{pu} = 1/2.52x10^5 = 4x10^{-6}$.

Whether we think of the plutonium species as present as $PuCd_6$ molecules in the unsaturated Cd solution or as Pu atoms isn't important—the mol fractions of plutonium come out the same, as the effect depends only on concentration. The form of the plutonium can be thought of either way. In calculations, either picture, atom or molecule, is fine, as long as consistency is maintained—that is, the appropriate values are used for the activities—for plutonium in the first case, or for $PuCd_6$ in the other. We will see this demonstrated in some calculations that follow later.

A.9.4 The First Cathode Regime: Uranium and Plutonium Unsaturated

We will now outline the calculation of the actinide composition in solution in the liquid cadmium cathode at any point in their depositions up to the point when the concentration of one or the other, uranium or plutonium, is saturated. Our goal is to understand what takes place, and to be able to predict the ratio of plutonium to uranium concentrations in this regime.

The first step is to establish how we are to calculate the activities of the uranium and plutonium in solution. Their activities change, related as they are to concentrations, as electrorefining continues and concentrations of the actinides increase. In an unsaturated solution of actinides in liquid cadmium, the actinide contents are low enough that "ideal" behavior is a reasonable assumption: strong chemical interactions that would "almost" lead to the formation of other phases aren't significant at the dilute concentrations in this regime. Their activity coefficients are therefore taken as constant and the activities of uranium and plutonium are assumed to change linearly with their concentrations. That is, they're taken to be directly dependent on concentration. This "ideal solution assumption" is a tool for simplifying calculations, and can be expected to always be at least a little bit "wrong," but again, in these circumstances not significantly so.

The reaction we are getting at is plutonium chloride when reduced to metal at the cathode then reacting with uranium chloride in the salt. When this happens, the plutonium disappears from the cathode and dissolves back into the electrolyte. We must minimize this, and to assess the extent of this reaction, we need to calculate its equilibrium coefficient. The reaction is shown in the balanced equation below. The equations giving the equilibrium coefficient follow from it. (As we shall see later, it's possible to write the reaction taking $PuCd_6$ and Cd into account explicitly. As long as we are consistent in our definition of the appropriate activities, the equilibrium constant we calculate will be the same.) But the simplest, still accurate, form is the reaction as written:

$$Pu + UCl_3 \leftarrow\!\!\rightarrow U + PuCl_3.$$

Our calculations always are for near-equilibrium electrotransport, a satisfactory assumption at the rates of electrorefining we use. Simply put, for purposes of calculation we assume we are at equilibrium. The equilibrium constant, implicitly taking the plutonium-cadmium intermetallic into account, is given by the sum of the free energy changes. The relevant change of free energy is the difference between the free energies of chloride formation of Pu (-187.2 kcal/mol) and U (-165.6

kcal/mol), which is -21.5 kcal/mol, and $PuCd_6$ (-19.1 kcal/mol). The equilibrium constant is then:

$$K_{eq} = \exp(\Delta G^o_f / RT) = 4.69,$$

$$\text{as } \Delta G^o_f = \Delta G^o_{f\ Pu} - \Delta G^o_{f\ U} - \Delta G^o_{f\ PuCd6}$$

$$= (-21.5) - (-19.1) = -2.4 \text{ kcal/mol, and } RT = 1.536 \text{ kcal/mol.}$$

This number, 4.69, is the value of the equilibrium constant for this reaction. It is the important number, as it gives us the basis for calculation of the ratio of the concentrations of plutonium to uranium in the cathode for any concentration up to saturation of one or the other.

In the notation of Eq. 4, the equilibrium distribution of plutonium and uranium between the electrolyte and the cadmium cathode given by the ratios of the activities is:

$$(a_U * a_{PuCl3})/(a_{Pu} * a_{UCl3}) = [\gamma_U / \gamma_{Pu}]\, [\chi_U / \chi_{Pu}]\, [\gamma_{PuCl3}/\gamma_{UCl3}]\, [\chi_{PuCl3}/\chi_{UCl3}] = 4.69 \qquad (5)$$

γ = activity coefficient,
χ = concentration expressed as mol fraction,
ΔG^o_f is the difference between relevant standard free energies of chloride formation of $PuCl_3$ and UCl_3
R is the gas constant, 1.987 cal/K, and
T is temperature, °K.

The activity at saturation is a fixed and known number, available for uranium or plutonium from tables of such quantities. The activity at a concentration less than saturation is directly proportional to its fraction of the saturation concentration. The ratio of $PuCl_3/UCl_3$ in the salt at the start of an electrorefining run is measurable, and is therefore a known quantity.

Then, in accordance with the equilibrium equation above, a ratio of actinide metal activities in the cadmium will be established from the onset of deposition. The ratios of cathode actinide metal and electrolyte actinide chloride activities will satisfy the equilibrium constant expression above *and will continue to do so as deposition progresses.* This goes on smoothly, with no dramatic changes, until the solution becomes saturated with either uranium or plutonium.

A.9.5 The Second Cathode Regime: One Saturated, Plutonium Saturation Is the Important Choice

No *Solid Phase* $PuCd_6$ will form until the Pu concentration reaches its solubility limit (nor will uranium metal form until its solubility limit is reached). When the solution becomes saturated, either with uranium or with plutonium, no more of the saturating metal can enter solution; it has to deposit as a separate phase. *But its activity never changes from the saturation value.* Activity of a *pure* solid or liquid is always unity, so a pure uranium phase has uranium activity of one and $PuCd_6$ has an activity of one. But the activity of the *plutonium itself in* $PuCd_6$ follows from the free energy of formation of the $PuCd_6$ and has a value of $4x10^{-6}$.

The meaning of activities may have remained somewhat puzzling at this point, so let us digress briefly to clarify the basis of chemical "activity." Activity is dependent upon, and its magnitude is defined by, the free energy that drives the tendency to react. It is proportional to the exponential, $\exp(-\Delta G^o{}_f /RT)$, where ΔG is the free energy driving the reaction that the chemical activity quantifies. The exponential is a measure of the effect of the free energy change, the degree to which the reaction will proceed. Thus the free energy change, ΔG, is proportional to the natural logarithm of the activity (ln a), and if the free energy change is zero, as ln(1) is zero the activity is exactly one. *An activity of one means no reaction at all.* That, of course, is to be expected when a pure material simply adds to the pure material already there. But the point is that an activity of one is no arbitrary normalization; it very specifically means that the species has no tendency to react.

When the plutonium saturates, its activity *does not change from then on*; it is constant. K_{eq} is constant, a_{Pu} is constant at saturation, so the quantity $K_{eq}\ a_{Pu}$ is thus a constant.

$$(K_{eq}).(a_{Pu}) = (a_U).(a_{PuCl3}/a_{UCl3}) \qquad (6)$$

What does this simple little expression say? It says that now when actinides deposit, because uranium has not yet saturated, its concentration, and therefore its activity, a_U, will try to increase. *But* it can only increase to the degree that there is a corresponding decrease in the ratio a_{PuCl3}/a_{UCl3}. The product of a_U and a_{PuCl3}/a_{UCl3} is a constant. Therefore, as more and more actinides are reduced, only some small amount of U can go into the Cd, and a_U will increase slightly because its concentration goes up a little, but a_{PuCl3}/a_{UCl3} must then decrease to preserve the equality. The a_{UCl3} will decrease slightly because some UCl_3 is removed to deposit the uranium in solution. Thus a_{PuCl3} must decrease slightly more than a_{UCl3} does. But the only way the $PuCl_3$ activity can be reduced is to remove it from the electrolyte, and the only way to do this is to deposit Pu as $PuCd_6$ metal. To change a_U (which is proportional to the ratio of concentration at this point to the concentration at saturation) there must be a corresponding change in the ratio a_{PuCl3}/a_{UCl3} in the electrolyte.

In a typical practical electrorefiner configuration, the amount of $PuCl_3$ is considerably greater than the amount of uranium metal in the cadmium, so a lot of plutonium must deposit to move a_U much. Where a_{PuCl3} is considerably larger than a_{UCl3}, (the UCl_3 drawdown has increased the $PuCl_3/UCl_3$ ratio before the run starts) more plutonium tends to be deposited than uranium in any case. These effects tend to deposit plutonium preferentially. Plutonium metal deposition does not affect the plutonium activity; it stays constant at the saturation value, so the plutonium deposition goes on, in theory, until either the cadmium or actinide chlorides are used up, or the cadmium becomes saturated with U. While it goes on, the amount of cadmium decreases as $PuCd_6$ is formed, increasing the U metal *concentration* in the cadmium, which will act to bring the uranium to saturation and take us out of this regime.

Looking at the details of the deposit, the concentration of plutonium that is in solution *does not* continue to go up. When more deposits, mixing more with a saturated solution simply causes another plutonium-containing phase to form—the solution itself does not change in concentration at all. At equilibrium (which is all we're talking about) the activity of any species is the same in all its phases. In particular, the activity of the species in a "new" phase is the same as that dissolved in the saturated solution. New plutonium depositing does not go into solution because the solution has all the plutonium it can dissolve. It goes into a new phase: solid, metallic $PuCd_6$. The plutonium in the $PuCd_6$ phase reacts *exactly* as plutonium dissolved in cadmium does at saturation, because the activities are the same. Incidentally, if plutonium is removed from the saturated cadmium so it is no longer saturated, some $PuCd_6$ metal will disassociate and add enough back to the solution to saturate it again.

Saturation causes preferential deposition of the saturating element as a new phase whose activity does not change from its value at saturation. And that is the important point. Richer actinide cathode contents can result, if practical "engineering" phenomena do not arise to stop the resulting deposition of further plutonium.

A.9.6 The Third and Final Regime

With both U and $PuCd_6$ saturated and their activities unity, they accumulate at the same ratio as the ratio of plutonium chloride to uranium chloride activities (a_{PuCl3}/a_{UCl3}) in the electrolyte, and the ratio does not change until all the cadmium or actinide chloride is used up. (The ratio of the concentrations does differ by the amount of the activity coefficient ratio of $PuCl_3/UCl_3$, a difference amounting to several percent.) This regime is not very important, as inconvenient things happen in the real world. Uranium dendrites form out of the cadmium and grow toward the anode. $PuCd_6$ piles up and falls out of the cathode. These effects may be mitigated by not using up all the cadmium.

A.9.7 Operational Implications

For Pu/U ratios in the cathode enhanced in plutonium by this method, the amount of cadmium should be less than the amount of actinide chlorides in the salt—by as much as is practical. The solid cathode should be run until the plutonium contamination of the uranium product becomes excessive so as to increase the initial a_{PuCl3}/a_{UCl3} sufficiently for cadmium cathode operations to begin. The cadmium cathode in turn should be run until uranium metal deposition begins or other effects intervene. The composition and amount of the anode (driver fuel or blanket material) will "buffer" the salt composition; the cathodes alone cannot determine the composition. For example, many blanket batches have to be run to build up enough Pu to run the cadmium cathode to good effect. A blanket batch in the anode will keep the $PuCl_3/UCl_3$ ratio down during that electrorefining run. Even if only uranium is removed, by a solid cathode only, the plutonium concentration cannot increase much in the salt in any one run. The amount of actinide chloride relative to the amount of actinides in both of the electrodes affects how rapidly the salt composition will change as the electrode compositions change. And these in turn affect how the cathode composition itself changes.

At saturation for both the $PuCd_6$ and uranium, with the activities unity, there is one, and one only, equilibrium ratio of the actinide chloride activities. The actinide chloride activity coefficients are known, so the one $PuCl_3/UCl_3$ concentration ratio for equilibrium can be identified. Its numerical value is of real interest, as it gives us the standard to judge how far and in what direction any given $PuCl_3/UCl_3$ ratio in the salt departs from saturated equilibrium. And this in turn tells us whether plutonium rich or uranium rich deposits can be expected as electrorefining proceeds with a new phase forming. One species, uranium or plutonium, will preferentially deposit to move the ratio toward its equilibrium value.

A.10 Calculation of the Important Criteria

From these considerations we can see the manner in which the actinide chloride ratio, $PuCl_3/UCl_3$, affects plutonium deposition in the cadmium cathode. We will now put numbers to these considerations, calculate the numerical value of this ratio, and note the numerical values for saturation of the actinides in the cadmium solution.

The relevant input numbers are:
UCl_3: $\Delta G^o_f = -162.3$ kcal/mole, activity coefficient = 5.79×10^{-2}, [5]

$PuCl_3$: ΔG^o_f = -183.8 kcal/mole, activity coefficient = 6.62×10^{-2}, [6]
U activity coefficient in solution of any concentration = 89,
U mole fraction at saturation = 1.12×10^{-2},
(activity at saturation = $1.12 \times 10^{-2} \times 89 = 1$),
Pu activity coefficient in solution of any concentration = 2.3×10^{-4},
Pu mole fraction at saturation = 1.81×10^{-2} (activity at saturation = 4×10^{-6}),
$PuCd_6$: ΔG^o_f = -19.1 kcal/mole, and
RT = 1.987 x 773 = 1.536 kcal.

Illustrating the point that the reaction of plutonium in cadmium with uranium chloride in the salt as a reaction of molecular $PuCd_6$ with UCl_3, or as a reaction of atoms of Pu reacting with UCl_3, we will carry out the calculations of the ratio of the $PuCd_6$ to UCl_3 in the salt for equilibrium with plutonium in cadmium in contact with the salt both ways.

A.10.1 The Reaction as Plutonium Chloride Molecules

In this case the reaction is written as $PuCd_6 + UCl_3 \leftrightarrow PuCl_3 + U + 6Cd$.

$PuCd_6$ is reacting with UCl_3 in the salt, and the free energy change in forming $PuCl_3$ from $PuCd_6$ is the free energy of $PuCl_3$ formation minus the free energies of UCl_3 and $PuCd_6$ formations.

$$\Delta G^o_f = \Delta G^o_f(PuCl_3) - \Delta G^o_f(UCl_3) - \Delta G^0_f(PuCd_6) = -183.8 + 162.3 + 19.1$$
$$= -2.4 \text{ kcal/mole.}$$

The equilibrium constant K_{eq} for this reaction is

$$\exp(-\Delta G^o_f/RT) = \exp(2.4/1.536) = \exp(1.56) = 4.7, \text{ and}$$

$$K_{eq} = a_{PuCl3} * a_U * a_{Cd6} / a_{UCl3} * a_{PuCd6}.$$

At saturation (of $PuCd_6$ in cadmium) the $PuCd_6$ activity is unity, the uranium activity is unity (neither reacts with the UCl_3), and the Cd activity, although not strictly unity because it contains some actinides in solution, to a good approximation is unity also.

The equilibrium constant, therefore, reduces to a_{PuCl3} / a_{UCl3} (and this ratio of the activities of the actinides chlorides is equal to $\exp(-\Delta G^o_f/RT) = 4.7$). The ratio of the activity coefficients of the actinide chlorides, $\gamma_{PuCl3}/\gamma_{UCl3}$, is 6.62/5.79 or 1.14; thus the equilibrium ratio of the $PuCl_3/UCl_3$ concentrations in the salt is 4.7/ 1.14 = 4.1.

This is the significant number. The concentration ratio of $PuCl_3$ to UCl_3 for equilibrium when both the uranium and plutonium cadmium intermetallic are saturated in the cadmium and in contact with the salt is 4.1. And this number is the criterion for assessing the effect on the ratio of plutonium to uranium in the product of any particular ratio of plutonium chloride to uranium chloride in the salt.

A.10.2 The Reaction as Plutonium Atoms

In this case, the reaction is written as $Pu + UCl_3 \langle\text{--}\rangle U + PuCl_3$.

ΔG^o_f becomes $\Delta G^o_f (PuCl_3) - \Delta G^o_f (UCl_3) = 21.5$ kcal/mole, and we know from previous sections that the equilibrium constant K_{eq} for the free energy difference of this magnitude becomes very large, 1.2×10^6 in fact.

This is the equilibrium constant for the case without $PuCd_6$ formation, as it is for a solid-rod cathode, for example. From this it is obvious why metallic plutonium, as such, can't exist at any reasonable $PuCl_3/UCl_3$ ratio—in fact, not until the ratio reaches a million or so.

For the liquid cadmium cathode, the plutonium activity corresponding to the plutonium in $PuCd_6$, 4×10^{-6}, is the appropriate value. The uranium activity is unity and the ratio of the activity coefficients of the actinide chlorides is 1.14, as before, so the ratio of the concentrations of the actinide chlorides at equilibrium is,

$$\chi_{PuCl3}/\chi_{UCl3} = (1.2 \times 10^6) \times (4 \times 10^{-6})/1.14 = 4.8/1.14 = 4.2.$$

Thus, within rounding error, the two formulations are identical. The very large equilibrium constant, 1.2×10^6, which leads to an extremely high $PuCl_3$ to UCl_3 equilibrium ratio in the salt, is counterbalanced by the tiny activity of plutonium atoms in the intermetallic $PuCd_6$, once again giving the reasonable ratio of concentrations of $PuCl_3$ to UCl_3 of about four for equilibrium.

At equilibrium, with U and Pu both saturated in cadmium and in contact with a salt containing actinide chlorides, the $PuCl_3/UCl_3$ ratio in the salt is 4.1.

Whether plutonium is regarded as present as $PuCd_6$ molecules in the unsaturated cadmium solution or as plutonium atoms does not matter, if consistency in calculation in maintained, as we have done. Two calculations, same result. $PuCd_6$ with activity 1 and Pu with activity 4×10^{-6} are equivalent if we base our calculation on the correct chemical equation and corresponding equilibrium constant.

So we now know that the equilibrium ratio of $PuCl_3$ to UCl_3 when both U and $PuCd_6$ are at saturation concentrations is about four. At saturation, a_U/a_{Pu} is

0.25×10^6. The activity coefficients of uranium and plutonium are listed above, so substituting, we get:

$$\chi_U/\chi_{Pu} = (a_U/a_{Pu}) x (g_{Pu}/g_U) = (0.25 \times 10^6) x (2.3 \times 10^{-4}/89)$$
$$= 0.646, \text{ or inverting, about 1.55 to 1, plutonium to uranium.}$$

This is the composition when both uranium and plutonium are saturated, with the actinide chloride ratio at its equilibrium value. It is not the cathode composition, unless by happy coincidence. It is the composition of the actinide product at the precise point of saturation of both uranium and plutonium before any solid phases have formed. As such, it gives a practical feel for the cathode behavior to be expected. But it surely isn't all we want to know. The total *amount* in all phases of U, Pu, and Cd in the cathode is the totality of our product. It is the total amounts of each that are important to us.

At equilibrium when the concentration ratio of $PuCl_3$ to UCl_3 is 4.1, the actinide ratio Pu/U in the cadmium is about 1.55 in the liquid cadmium phase. But if the ratio of actinide chlorides in the salt differs substantially from 4.1, the ratio of the metals in cadmium may differ greatly from this. With one actinide saturated—plutonium, say—the plutonium to uranium ratio will be greater in the cadmium at a concentration ratio of $PuCl_3$ to UCl_3 greater than 4.1, and less at a lesser ratio. If both are saturated, the $PuCl_3/UCl_3$ equilibrium ratio can only be 4.1. The increase or decrease of the Pu/U ratio in the cathode toward equilibrium value is slow if both are unsaturated, rapid if just one is saturated. But again, by designing to saturate only plutonium, the ratio of the amounts of plutonium to uranium can be increased to get adequate plutonium enrichments at reasonable actinide chloride ratios in the electrolyte.

The principle involved in all of this is that Gibbs Free Energy change is still in the opposite direction to the reaction we would like, but by loading up the ratio of $PuCl_3/UCl_3$ heavily in favor of $PuCl_3$ an equilibrium is achieved that allows us a useful ratio of Pu/U in the cathode. The exact value of the equilibrium coefficient isn't important for purposes of understanding the process; the fact that it is an important criterion is. However, the value 4.1 is probably adequate. If the numerical value changes due to better measurements, well and good; here we seek insight.

The aggregate cathode composition is the sum of the amounts of Cd + U + Pu in the saturated solution, in the U metallic phase (if present) and in the $PuCd_6$ metallic phase (if present). *It does not depend on the history of the compositions—how they got to their present values—at all.* Regimes 1 and 2 only count if the final Cd solution is unsaturated (regime 1), or saturated with only one actinide (regime 2). Regime 2 is our aim with plutonium: the actinide that saturates the cadmium phase.

For those who are curious, the electrorefiner will come to equilibrium at any $PuCl_3/UCl_3$ ratio in the salt. Left alone, it will transfer uranium and plutonium around until the salt and everything in contact with the salt are in equilibrium, and it will do this at reasonable rates. It may be that equilibrium is reached only with the outer layers of the anode—on the order of millimeters, not nanometers—because of slow diffusion through the solid fuel feedstock, but the changes, once the fuel has been in the electrorefiner for a few hours, or so, will be small as electro-transport progresses. Picture a situation where uranium and plutonium are saturated in the cadmium cathode, and the $PuCl_3/UCl_3$ ratio is two (for a short while). The electrorefiner is far from equilibrium. The plutonium in the cathode will react immediately with the UCl_3 in the salt to form $PuCl_3$ and U until the system is once again in equilibrium. The salt would become more plutonium rich just sitting there with no current passing. (And the product will be degraded in plutonium.) Depending on the amounts of actinides in the salt, and in the cathode, and in the anode as well, if it has actinides in it, the cathode would no longer be plutonium saturated. However, in this case it would probably remain uranium saturated.

A consistent way of thinking about all this is provided by the equilibrium expression.

1. At an actinide chloride ratio of precisely 4.1, the uranium and the $PuCd_6$ activities will maintain the same ratio right from the beginning, to the extent that their activity coefficients are given by the "ideal approximation" —the saturation value multiplied by the concentration divided by the saturation concentration. Both will saturate at the same time. All are nicely in equilibrium.

2. At an actinide chloride ratio of two, the activity ratio is half the saturated ratio; the uranium will saturate first while the $PuCd_6$ is only half way to saturation. The UCl_3 concentration has to be decreased to get to the equilibrium value of 4.1, so uranium will preferentially deposit trying to increase the actinide chloride ratio toward 4.1. In this regime, the cathode is acting pretty much like a solid cathode, and the resulting overall deposit will be largely U.

3. At the other extreme—at a ratio of eight, say—the opposite effects take place; the activity ratio is twice the saturated equilibrium ratio, the $PuCd_6$ will saturate. and $PuCd_6$ will deposit as the system evolves to decrease the $PuCl_3$ concentration and move the actinide chloride ratio toward 4.1. In this case the deposit will be plutonium rich, above the 1.55 Pu/U value for the both saturated case 1, above.

Off equilibrium in actinide chloride ratios for saturation, the system tries to achieve saturation and get back to the saturated cadmium equilibrium ratio of actinide chlorides.

Practical considerations limit what can be done. Criticality considerations are important. The amount of fuel that the anode baskets can contain has limits. The cathode, too, is almost certainly limited in capacity. The actinide content of the salt may be high relative to the amounts in the anode and cathode, making changes in its composition tedious. For plutonium deposition, the uranium chloride concentration can be decreased by saturating the cadmium with U and preferentially removing U to the cathode. But for this the cadmium cathode is not ideal. The better choice practically, and what is actually done, is that at higher $PuCl_3/UCl_3$ ratios the solid cathode only is used to remove "pure" uranium. In this way the $PuCl_3/UCl_3$ ratios necessary for adequate plutonium depositions are in place before the cadmium cathode is introduced. But even then the composition of the anode feed may make higher $PuCl_3/UCl_3$ ratios difficult to achieve—still possible, but perhaps only after several batches have been processed. And what is possible in practice may be limited by effects such as deposits growing out of the cathode.

Finally, it is of interest to note that with everything at equilibrium the ratio of plutonium to uranium in the product of 1.55 is a factor of 2.65 less than the 4.1 ratio of plutonium chloride to uranium chloride in the electrolyte. This gives an indication of the difficulty of a pure product—whatever the $PuCl_3/UCl_3$ ratios are in the salt the Pu/U product ratio will be degraded from this by a considerable factor.

A.11 Adding to Understanding of the Process by a Brief Description of its Development

The IFR work began in the mid-eighties in a small electrorefiner with a few hundred grams of material to transport. Initially a liquid cadmium *anode* was felt to be necessary for good electrical contact for dissolution of the fuel. It was found that uranium transported and adhered to the cathode well, but plutonium did not—not an expected result. Equilibrium calculations that followed made clear what was happening. As plutonium chloride was reduced at the cathode, the more plentiful uranium chloride was acting as a sink for the plutonium metal by the exchange reaction described above. The plutonium remained in the salt. Experimentation then turned to liquid cadmium for the cathode, and the process became one of depositing the uranium first on the solid cathode and then the plutonium on the liquid cadmium. The expected effect of the *liquid* cadmium was to sharply reduce the free energy of chloride formation difference between plutonium and uranium, and indeed that was what was found: Plutonium then deposited in the presence of uranium chloride. But it was also found that there was a very strict limit to the ratio of concentration of uranium chloride to that of plutonium chloride that can be present and still deposit plutonium. The initial equilibrium calculations suggested a ratio of two or so, plutonium chloride to uranium chloride, and further experiments showed that the higher the ratio the greater the plutonium to uranium ratio in the product. *But if any uranium chloride at all is present, uranium deposits as well. The process will not deposit pure plutonium from a plutonium- uranium mixture.* Nor will it deposit pure plutonium in the presence of the higher actinides.

A.12 Electrorefining Results: Measurements and Experimental Observations

The thermodynamic, kinetic, and transport phenomena we described in earlier sections are sufficient for an understanding of the main features of the IFR processes. The experiments used to develop and define the processes were in large measure small-scale, a few hundred grams perhaps, although there was a large electro refiner at Argonne in Illinois that opened the way for the larger electro refiners installed in the Fuel Conditioning Facility (FCF) in Idaho. Today we do not have to rely entirely on the early measurements, for even though the IFR program was terminated in 1994, we have a considerable history of full-scale electrorefining runs going back to that year. The treatment of EBR-II spent fuel in the fifteen years since has given us some very significant data.

The results of the larger scale electrorefining runs carried out in the FCF provide proof of the process at a practical scale. The status of development is of great interest, and the fact that these large-scale runs were not done for developmental purposes, but rather as more or less routine operation, is also important. Indeed, they continue today, to safely treat spent core and blanket fuel rods from EBR-II for long-term storage, reuse in reactors, or disposal in a repository. The runs are routine operations done for that specific purpose. But they are important also as pilot plant demonstrations of the IFR processes.

A.12.1 The Plutonium Recovery Experiments

The termination of the IFR reactor program in 1994 closed off the scale up of the plutonium and higher actinide recovery process. Nothing more was done through the Clinton administration. But development and then routine operation of the spent fuel processes for disposal of the EBR-II fuel and blanket as waste continued. This main process recovered uranium only, which was to be as free of plutonium as possible; isolated the fission product waste for safe disposal; and recovered a limited amount of plutonium and higher actinides for storage. The quantities that have been processed are significant. In particular, processing over three tons of uranium from spent fuel and blanket elements provides a comfortable basis for routine operation of that important part of the process.

In processing spent fuel and blanket pins from EBR-II for uranium recovery, a substantial amount of plutonium builds up in the electrorefiner. The buildup is slow because the blanket elements contain only a percent or two of plutonium. However, the amount of plutonium buildup that can be tolerated must be kept within limits, for criticality reasons if for no other. When the Bush administration came in, some work on plutonium recovery was picked up again to facilitate drawdown of the plutonium chloride in the electrolyte.

Four kilogram-scale extractions of plutonium from the electrolyte to a cadmium cathode have now been carried out. Three have been reported in the literature [7], and a fourth done later [8-9] is included here as well. They are analyzed below. The important ratio in the electrolyte of $PuCl_3/UCl_3$ varied from run to run, but all were in a range that from the considerations presented in the previous section would be predicted to give adequate plutonium deposition. (Two were only just in that range.) In all four, a deposit of a kilogram or more was sought and two of the four deposits were over a kilogram.

As expected, the lower the ratio of plutonium chloride to uranium chloride in the electrolyte, the less plutonium in the product, but from the considerations described above, also as expected, all had considerable uranium deposited along with the plutonium. The highest plutonium to uranium ratio had an initial plutonium chloride concentration ten times and a final concentration almost six times the uranium chloride concentration. The product still contained 30 percent uranium. We will look at these very important results in some detail below. Their importance is clear. They are the only engineering-scale plutonium deposition results that we are aware of, to the present time.

The equilibrium calculations demonstrating the effects of plutonium to uranium concentration ratios in the electrolyte on plutonium deposition in the cadmium cathode will be presented below with sufficient information for the calculations to be reproduced by those mathematically inclined. We will show two important calculations.

1. The first calculation demonstrates how closely the measured results of the Idaho actinide runs agree with our calculations and models and give us a measure of the degree to which the methodology can be trusted. We have calculated the equilibrium coefficient for the reaction of plutonium metal with uranium chloride to be 1.2×10^6 from the free energies in the Arrhenius exponential. But an equilibrium coefficient can also be derived from the measured results of each run. What kind of agreement will we find?

2. The second calculation uses the methodology we have described in the previous sections to calculate the most important of the measured results—the plutonium to uranium ratio in the cadmium cathode. The degree of

agreement with the measured ratio will allow us an estimate of how nearly we will be able to predict this important ratio as a function of the $PuCl_3/UCl_3$ ratio for use in the future.

A.12.2 The Experimental Conditions and Measured Results

The input parameters are the experimental conditions for the three actinide runs listed in Table A-3. The composition of the fuel in the anode is given in Table A-4. The measured results are given in Table A-5, all taken from Vaden et. al. [7]

Table A-3. Transuranic Recovery Experiments: Experimental Conditions

Input Parameters	Run 1	Run 2	Run 3
Initial U concentration in salt, wt%	0.27	0.53	0.69
Initial Pu concentration in salt, wt%	2.89	2.66	2.54
Initial Pu to U ratio	10.80	5.06	3.67
Final Pu to U ratio	5.70	3.32	2.87
Initial heavy metal in salt, wt%	3.16	3.19	3.23
Initial heavy metal in anode, kg	2.00	2.00	4.10
Integrated current, 10^6 coulombs	1.81	2.13	2.19
Average current, amperes	20	20	20

Table A-4. Composition of EBR-II Blanket Fuel

Nuclide	Average	Maximum	Minimum
U-238, %	98.94	99.63	97.86
Pu-239, %	0.80	1.57	0.14
U-235, %	0.20	0.22	0.18
Nd, ppm	92	246	8
Ce, ppm	53	142	4
La, ppm	29	77	2
Pr, ppm	27	73	2
Np-237, ppm	11	22	1
Am-241, ppb	776	8300	2

The first thing we wish to know is the state of saturation of the cadmium. In these experiments the cathode contained 26 kg of cadmium. The number of moles of cadmium were 26000/112.4 = 232 moles. Uranium saturates at a concentration of 1.12%, 0.0112 x 232 = 2.60 moles U, or 2.6 x 238 = 618.8 grams U. Plutonium saturates at a concentration of 1.81% = 0.0181 x 232 = 4.20 moles Pu, or 4.2 x 239 =1003.8 grams Pu. *Thus the saturated amounts of the metals in the cadmium cathode are 618.8 grams U and 1003.8 grams of Pu.*

The anode in all three runs was EBR-II blanket material, substantially uranium, only 1 percent or so was plutonium. The effect of this anode composition is to increase the uranium content in the electrolyte as cadmium cathode electrorefining proceeds. Uranium largely replaces plutonium in the electrolyte reduced at the cathode. The measured results are given in Table A-5.

Table A-5. Results of Transuranic Recovery Experiments

	Run 1	Run 2	Run 3
Heavy metal recovered, g	1365	1739	1313
Pu in casting furnace ingot, g	1024	1080	492
Np-237 in casting furnace ingot, g	4.5	1.2	0.6
U in salt at end of test, wt%	0.47	0.74	0.83
Pu in salt at end of test, wt%	2.67	2.46	2.38
Recovered metal theoretical, 10^6 coulomb	1.65	2.11	1.59
Integrated current, 10^6 coulomb	1.81	2.13	2.09
Transport efficiency, %	91.5	99.2	76.1

A.12.3 Equilibrium Constants Derived From the Measured Quantities of Each Run

Our reaction is $Pu + UCl_3 \text{<-->} U + PuCl_3$.

Our calculated value for the equilibrium constant for the reaction is fixed at 1.2×10^6, unchanged from run to run. We will compare this to the values derived from the measured results of the three electrorefining runs. We want to see how well calculations based on the assumptions we use, principally that of equilibrium, are able to predict results.

The required input data is shown in Table A-6 below. We are indebted to Dee Vaden of INL for explanation of the details of the experiments, published in reference 7. The important calculations given in Table A-7 below were done by John Ackerman [4] and we are indebted to him for their use.

Table A-6. The Input Data Used

Atomic weight of plutonium-239	239
Atomic weight of uranium-238	238
Atomic weight of chlorine naturally occurring	35.5
Atomic weight of cadmium naturally occurring	112.4
Activity coefficient ratio of $PuCl_3/UCl_3$	1.14
Activity coefficient of uranium in cadmium	89
Activity coefficient of plutonium in cadmium	2.3×10^{-4}

Tracing through the calculation, it starts with the state of saturation of the actinide elements in the cadmium. All calculations are based on molar amounts, not weights, so we calculate the total molar amounts, and then the molar amounts actually in solution. The cathode contained 26 kg of cadmium, as shown in line 1, and the other experimentally determined quantities follow in lines 2 to 7. From these numbers, the activity ratio of the chlorides in the salt and the moles of cadmium, uranium, and plutonium in the cathode are calculated and shown in lines 8 to 12. Knowing that uranium saturates at a concentration of 1.12% and plutonium at a concentration of 1.81% gives the maximum possible moles of the two actinides in solution; this is shown in lines 13 and 14.

The *maximum* saturation amounts are 2.59 moles or 616.4 grams of U and 4.19 moles or 1001.4 grams of Pu. (The amounts can be smaller due to the loss of cadmium to $PuCd_6$ formation, as shown in lines 15 to 21. Line 15 was obtained by iteration, changing line 15 until the sum of line 21 and line 15 is equal to line 12. The corrections however are small.)

The amounts in solution and in the metal phase phases are summarized in lines 22 to 25. The final mol fractions in solution are given by lines 19/(17 +18+19) for uranium, and line 17/(17+18+19) for plutonium. Multiplication by the appropriate activity coefficients, 89 and 2.3×10^{-4} respectively, gives the relevant activities, and their ratio divided by the same ratio of the actinide chlorides gives the calculated equilibrium constant.

The calculated Keq for each run is given in line 34, the Keq given by the free energy exponential is given in line 35, and the difference is given in line 36, defined for this purpose as the difference between the two divided by the average of the two, to imply that neither is a standard by which the other result should be judged.

Table A-7. Basic Data and Calculations for the Idaho Actinide Runs
(m) indicates the data from Reference 7

#			RUN 1	RUN 2	RUN 3
1	(m)	Measured kg Cd in cathode,initial	26.00	26.00	26.00
2		Assumed kg Cd in cathode, final	26.00	26.00	26.00
3	(m)	Measured kg U+Pu in cathode(heavy metal recovered)	1.37	1.74	1.31
4	(m)	Measured kg Pu in cathode, final	1.02	1.08	0.49
5		Assumed kg U in cathode, final	0.34	0.66	0.82
6	(m)	Final $PuCl_3$ concentration, wt %	2.67	2.46	2.38
7	(m)	Final UCl_3 concentration, wt%	0.47	0.74	0.83
8		Final Concentration ratio,$PuCl_3/UCl_3$	5.66	3.31	2.86
9		Final activity ratio, $PuCl_3/UCl_3$	6.48	3.79	3.27
10		Mols of Cd in cathode	231.32	231.32	231.32
11		Mols of U in cathode	1.43	2.77	3.45
12		Mols of Pu in cathode	4.28	4.52	2.06
13		Calculated Max mol Pu in solution	4.19	4.19	4.19
14		Calculated Max mol U in solution	2.59	2.59	2.59
Correction for loss of Cd to $PuCd_6$					
15		Mol Pu in $PuCd_6$ phase	0.11	0.37	0.00
16		Mol Cd in $PuCd_6$ Phase	0.66	2.23	0.00
17		Mol Pu in solution phase	4.17	4.15	2.06
18		Mol Cd in solution phase	230.66	229.08	231.32
19		Mol U in solution phase	1.43	2.57	2.59
20		Mol U in U metal phase	0.00	0.20	0.86
21		Calc mol Pu in Cd solution	4.17	4.15	4.19
Summary: amounts in metal phase and amounts in solution					
22		g Pu in $PuCd_6$ phase	26.2	89.0	0.0
23		g U in U metal phase	0.0	48.4	204.4
24		g Pu in Cd solution	997.8	991.0	492.0
25		g U in Cd solution	341.0	610.6	616.6
26		Saturation g Pu in Cd solution	1001.4	1001.4	1001.4
27		Saturation g U in Cd solution	616.4	616.4	616.4
28		Pu % saturation	100%	100%	49%
29		U % saturation	55%	100%	100%
Calculated Keq from solution numbers					
30		Final mol fraction U in Cd	0.00606	0.01088	0.01097
31		Final mol fraction Pu in Cd	0.01767	0.01758	0.00872
32		Final activity U in Cd	0.5397	0.9684	0.9772
33		Final activity Pu in Cd	4.06E-6	4.04E-6	2.01E-6
34		Calculated Keq: Pu+UCl_3 <-> U + $PuCl_3$	0.86E+6	0.91E+6	1.59E+6
35		K_{eq} calculated from exp(-ΔG°_f/RT)	1.20E+6	1.20E+6	1.20E+6
36		% difference calculated K_{eq} vs K_{eq} from exp(-ΔG°_f/RT)	-33%	-28%	28%
37		Average Difference		-11%	

The calculated K_{eq} for each of the three runs is 0.86×10^6, 0.91×10^6, and 1.59×10^6, which are to be compared to the 1.2×10^6 value given by the free energies. The degree of agreement is gratifying, given the many assumptions, the sensitivity of the exponential to small changes, and the uncertainties in the measured numbers as well. The numbers were obtained from runs done to reduce plutonium amounts in the electrolyte, not as carefully controlled experiments.

It is also appropriate to note here that considerable understanding of the process can be gained by study of the step-by-step calculation in the table above.

A.12.4 Calculation of the Important Ratio: The Pu/U ratio in the Product

Our next calculation gives the expected richness of the plutonium deposit. How well is this important ratio estimated from these simple considerations? We know from the calculations in Table A-7 the degree of saturation of the two actinides when each run was terminated. The runs were to go until a kilogram of plutonium was deposited. How much uranium can we expect in each deposit as we change the $PuCl_3/UCl_3$ ratio?

The actinide ratios in the salt at the end of the runs are given in line 8 of Table A-7. They ranged downward from 5.66 to 3.31 to 2.86, which are to be compared to 4.1, the calculated $PuCl_3/UCl_3$ ratio at equilibrium with both plutonium and uranium saturated in the cadmium. In Run 1, plutonium was saturated, uranium was not; in Run 2, both were saturated; and in Run 3, uranium was saturated, plutonium was not. Where amounts greater than saturation amounts were present in the product, the amounts in grams of the excess are listed in lines 22 and 23 of the table.

The table lists calculations based on the actual measured mol fractions, and from that, a deduced equilibrium constant. We would therefore expect perfect agreement between the Pu/U ratios from the equilibrium constant derived in this way. In other words, if we were to calculate the Pu/U ratios from the K_{eq} based on measured quantities, it would merely serve as a consistency check of the calculation and add no new information. The more interesting number is the ratio calculated with the K_{eq} of $1.2x10^6$, which will then give the Pu/U ratio that would be predicted prior to the run.

With K_{eq} $1.2x10^6$, the appropriate value for the plutonium activity is that of Pu in $PuCd_6$—that is, $4x10^{-6}$.

The reaction is $Pu + UCl_3 \text{<-->} PuCl_3 + U$.

The equilibrium constant is $K_{eq} = (a_{PuCl3}/a_{UCl3})/(a_{Pu}/a_U)$.

The uranium activity is $a_U = K_{eq}*a_{Pu}/(a_{PuCl3}/a_{UCl3})$.

Our calculation is of the amount of each of the actinides in-solution.
The runs in every case went slightly beyond the saturation of at least one of the

actinides. We therefore calculate the in-solution amounts and make relatively minor corrections for the additional amounts accumulating as metal phase where saturation is exceeded. The numerical values for each of the variables are taken from Table A-7.

Run1: Plutonium is saturated, uranium isn't

The $PuCl_3/UCl_3$ ratio in the salt at the end of the run was 5.7, so the actinide chloride ratio was richer in $PuCl_3$ than the equilibrium ratio of 4.1 for U and Pu both saturated. The 1024 grams of Pu was just over plutonium saturation (1001.4 g); 341 grams of U is nowhere near uranium saturation (616.4 g).

$a_U = (1.2x10^6 \text{ x } 4.0x10^{-6})/6.48 = 0.74$ (of saturation),
Uranium in solution, calculated = 0.74x616.4 = 456 g (341.0 g measured),
Plutonium in solution, calculated = 1000.4g (compare to 997.8 g, the small difference due to cadmium loss in $PuCd_6$ solid),
Pu/U ratio in solution, calculated = 2.2,
Pu as $PuCd_6$ metal = 26 g,
Total Pu in cathode, calculated = 1027 g, and
Calculated Pu/U ratio, total amounts = 2.25.

As the $PuCl_3/UCl_3$ ratio of 5.7 is above the equilibrium ratio of 4.1, the system will try to equilibrate by removing $PuCl_3$, or by increasing the UCl_3. Less uranium would be expected in the product than is calculated based on the $PuCl_3/UCl_3$ ratio of 5.7.

Run2: Plutonium and Uranium both are saturated

With 1080 grams of Pu and 659 grams of U in the product, once again the plutonium is saturated, but now the uranium is as well. Both concentrations are slightly in excess of saturation. The $PuCl_3/UCl_3$ ratio at the end of the run was 3.32, less than the equilibrium ratio of 4.1, and some plutonium would be expected to migrate back into the salt as the system tries to equilibrate. The expected plutonium to uranium ratio in the cadmium with both saturated and the system in equilibrium is 1.55 (as calculated in Section A.10.2). The small amounts of uranium (48g) as metal and plutonium (89g as $PuCd_6$ metal) won't affect the calculated ratio much.

Saturated amounts are 616.4g of uranium, 1001.4g of plutonium,
Pu/U ratio in solution, calculated = 1.6,
U as metal = 48 g,
Pu as $PuCd_6$ metal = 89 g,
Total U in cathode = 659 g,
Total Pu in cathode = 1090 g, and
Calculated Pu/U ratio, total amounts = 1.6.

Run3: Plutonium isn't, Uranium is saturated

With 492 grams of plutonium (saturation 1001.6 g) and 821 grams of uranium (saturation 616.4 g), the U is saturated, the Pu is not. The difference between the 821 g found for uranium and the saturation value of 616.4 we can assume is uranium "run-up" on saturation. The actinide ratio in the salt at completion was 2.87, well below the equilibrium value of 4.1 for both saturated, so once again we would expect plutonium to tend to migrate from the product.

$a_{Pu} = (a_{PuCl3}/a_{UCl3})* (a_U)/1.2x10^6 = 3.27/1.2x10^6 = 2.72x10^{-6}$,
Pu fraction of saturation = 2.72/4.0 = 0.68,
Plutonium in solution = 0.68*1001.4 = 681 g,
Uranium in solution = 616.4 g,
Pu/U ratio in-solution = 681/616.4 = 1.1,
Uranium as metal = 204 g,
Total U in cathode = 820 g,
Total Pu in cathode = 616 g, and
Calculated Pu/U ratio, total amounts = 0.75.

The amount of plutonium in the product is calculated to 681 grams, the amount found 492 grams. Again we can expect less plutonium in the cadmium as the system tries to equilibrate by means of some plutonium migrating back into the electrolyte.

A.12.5 Discussion of the Calculated Results

The calculated equilibrium constant for the reaction is *fixed*, as it depends only on the free energy difference of the reaction. Each ratio of $PuCl_3/UCl_3$ will give the activity ratio of uranium and plutonium in the cadmium appropriate for it up to the point of saturation of one of the actinides. Some alteration of the ratio can then be expected due to the saturation effects described, but none of the runs went far past saturation of either plutonium or uranium. Further, equilibration effects can be expected to somewhat alter the ratios found as well.

The experimental conditions also add some uncertainty to the measured product numbers. There was uncertainty in some small amount of product remaining as dross in the cathode processor crucible, and in the formation of dendrites external to the cathode in run number three. We are indebted to Dee Vaden of INL for explanation of these details of the experiments, published in reference 7.

A.12.6 The Bottom Line: Comparison of the Calculated Results with Measurement

The three kilogram-scale extractions of plutonium from the electrolyte to a cadmium cathode whose results we have been calculating in this section were reported in the literature in 2008. [7] A fourth run has been done, which the Idaho National Laboratory has kindly given approval to reference. [8] As shown in Table A-8, Run 4 is very similar to Run 3. *It is a valuable check on the reproducibility of the measured data.*

Table A-8. Consistency of Data Run to Run

	Run 3	Run 4
Initial $PuCl_3/UCl_3$ ratio (meas.)	3.7	3.45
Final $PuCl_3/UCl_3$ ratio (meas.)	2.86	2.85
Final $PuCl_3$ concentration, wt % (meas.)	2.38	2.39
Final UCl_3 concentration, wt% (meas.)	0.83	0.84
Pu metal in cathode, kg (meas.)	0.49	0.55
U metal in cathode, kg (meas.)	0.82	0.77
Final activity ratio, $PuCl_3/UCl_3$	3.26	3.27
Pu in $PuCd_6$ phase, g	0.0	0.0
U in U metal phase, g	204.4	171
Pu in Cd solution, g	492	553
U in Cd solution,g	616.6	616.6
Pu % of saturation	49	56
U % of saturation	100	100

The first six entries are measured values; the next seven give the meaning of the measurements, converting concentrations to activities, the amounts of plutonium or uranium in the metal phase (not in cadmium solution), the amounts actually in solution, and where they are not saturated, the degree of saturation. The degree of agreement between runs is excellent, lending confidence to all four results.

We now calculate the expected result of each run with its final $PuCl_3/UCl_3$ ratio and compare the calculated results to the measured values from Table A-8.

The important practical result is the richness of the plutonium deposit—the ratio of plutonium to uranium in the deposit. How well can this important ratio be estimated from our simple considerations? We know from Table A-7 the actual degree of saturation of the two actinides when each run was terminated. The runs were to go until a kilogram of plutonium was deposited. How much uranium can we expect in each deposit as the $PuCl_3/UCl_3$ ratio varies in the three runs?

The final actinide chloride ratios in the salt ranged downward from 5.66 to 3.31 to 2.86 in the three runs, which are to be compared to 4.1, the calculated ratio $PuCl_3/UCl_3$ at equilibrium with both plutonium and uranium saturated in the cadmium. In Run 1, plutonium was saturated, uranium was not; in Run 2, both were saturated, and in Run 3, plutonium was not saturated, uranium was. Where amounts greater than saturation amounts were present in the product, the amounts in grams of the metallic phases are listed in Table A-7.

The measured results are compared to calculation by our simple equilibrium expressions in Table A-9 below.

Table A-9. Calculated vs. Measured Pu/U Product Ratio

	Initial Actinide Chloride Ratio	Final Actinide Chloride Ratio	Pu/U Product ratio		
			Measured	Calculated in-solution	Calculated total
Run 1	10.2	5.7	3.0	2.2	2.25
Run 2	5.1	3.3	1.6	1.6	1.6
Run 3	3.7	2.9	0.6	1.1	0.75
Run 4	3.5	2.9	0.7	1.1	0.78

Each ratio of actinide chlorides will give the concentration ratio of uranium and plutonium in the cadmium relevant to it up to the point of saturation of one of the actinides. For a $PuCl_3/UCl_3$ concentration ratio of 4.1 in the salt, everything is in equilibrium and the ratio of plutonium to uranium in the cathode will be 1.55. For actinide chloride concentration ratios above and below 4.1, the U/Pu ratio in the cadmium will follow the relationships given by the equilibrium coefficient expressions. The degree of agreement between the measurements and our calculations based on equilibrium is gratifying.

Run 1

The $PuCl_3/UCl_3$ ratio of 5.7 is above the equilibrium ratio of 4.1, and the system will try to equilibrate by subtracting uranium from the cathode to increase the UCl_3 in the electrolyte and bring the $PuCl_3/UCl_3$ ratio down. Less uranium than calculated based on the $PuCl_3/UCl_3$ ratio of 5.7 was found in the product.

Run 2

The $PuCl_3/UCl_3$ ratio of 3.32 is somewhat below the equilibrium ratio of 4.1. The amounts are both are slightly in excess of saturation. The actinide ratio in the salt now is less than the equilibrium ratio of 4.1, and some of the deposited plutonium would be expected to migrate back into the salt as the system tries to equilibrate.

However, the ratio of plutonium to uranium in the deposit does agree with that calculated based on the $PuCl_3/UCl_3$ ratio of 3.32.

Run 3

The $PuCl_3/UCl_3$ ratio of 2.86 is further below the equilibrium ratio. The system will try to add plutonium from the cathode to add $PuCl_3$ to the electrolyte, reducing the amount of plutonium in the cathode. Here, less plutonium than calculated based on the $PuCl_3/UCl_3$ ratio 2.87 was found in the product.

Run 4

The $PuCl_3/UCl_3$ ratio of 2.85 is essentially the same as in Run 3. The results are the same as well, to the accuracy of calculation and measurement. The run provides assurance of the consistency of results but no new information.

All four measurements show that even with high plutonium to uranium ratios in the salt, the plutonium to uranium ratio in the product will be much lower, by a factor varying but in the range of the 2.65 previously calculated for lowering of the product Pu/U ratio from the equilibrium $PuCl_3/UCl_3$ ratio of 4.1 in the salt. In other words, getting close to a pure plutonium product without accompanying uranium is extremely difficult, and the presence of actinides above plutonium makes a pure plutonium product impossible.

The measurements show that a perfectly adequate plutonium enrichment of the product for reactor fuel recycle is possible with reasonable $PuCl_3/UCl_3$ ratios in the salt. The calculations show that the equilibrium assumption allows prediction of the results of these very significant experiments surprisingly well.

In summary, we now have the means to estimate the ratio of Pu/U in the product as a function of the $PuCl_3/UCl_3$ ratio in the electrolyte salt. At equilibrium with both uranium and plutonium saturated in the cadmium, the ratio of Pu/U product is 1.55. The $PuCl_3/UCl_3$ ratio of Run 2 (3.32) is nearest to the calculated equilibrium value of 4.1. It gave both a measured and a calculated value of the Pu/U ratio of 1.6, agreement that is certainly fortuitous, but is gratifying in any case. The conclusion to be drawn overall is that the important Pu/U ratio can be predicted with useful accuracy from these simple considerations and very simple expressions.

Neither calculation nor the experiment can be taken to be highly accurate. Many approximations were made in these simple calculations. The assumption of linearity in activities prior to saturation, the assumption of equilibrium or near- equilibrium, the rough corrections for saturation effects; all these and more are relevant. The values of several of the constants used may change with higher accuracy measurements in the future. There were uncertainties in various elements of the experimental results, where uncertainties entered in the amount of product remaining as dross in the cathode processor crucible in Run 1 and in the formation of dendrites external to the cathode in Run 3. But the systematic tracing of the experimental results displayed by the calculations, as summarized in the table above, does strongly suggest that the equilibrium calculations predict surprisingly well the results found in these very significant experiments.

We conclude that the important plutonium-uranium ratio in the product can be predicted with useful precision by these rather simple concepts and the techniques of the calculations based upon them. More, we conclude that they provide us with a very adequate understanding of the important phenomena in IFR electrorefining.

References

1. W.H. Hannum, Ed., "The Technology of the Integral Fast Reactor and its Associated Fuel Cycle," *Progress in Nuclear Energy*, 31, nos. 1/2, Special Issue, 1997.
2. J. P. Ackerman and T.R. Johnson, "Partition of Actinides and Fission Products between Metal and Molten Salt Phases: Theory, Measurement and Application to Pyroprocess Development." Actinides-93 International Conference, Santa Fe, New Mexico, September 19-24, 1993.
3. J. P. Ackerman and T.R. Johnson, "New High-Level Waste Management Technology for IFR Pyroprocessing Wastes," Global '93 International Conference on Future Nuclear Systems, Emerging Fuel Cycles and Waste Disposal Options, Seattle, Washington, September 12-17, 1993.
4. J. P. Ackerman, private communication. The authors are indebted to Dr. Ackerman first for pointing out the effects of cathode saturation and his follow-on detailed discussions with us of such effects.
5. L. Yang, et al, *Physical Chemistry of Process Metallurgy,* 2:925-943, G. R. St. Pierce, Ed., Interscience Publishers, New York, 1961.
6. G. M. Campbell and J.A. Leary, "Thermodynamic Properties of Plutonium Compounds from EMF Measurements," LA-3399, Los Alamos National Laboratory, March 1966.
7. D. Vaden, S. X. Li, B. R. Westphal, and K. B. Davies. "Engineering-Scale Liquid Cadmium Cathode Experiments," *Nuclear Technology*, 162, May 2008.
8. D. Vaden, private communication. We are indebted to Dee Vaden for helping us with our understanding of the experimental data, as well as for pointing out some uncertainties in the experimental results not mentioned in Reference 7, above.
9. C. Pope, private communication. We are indebted to Chad Pope for his sustained efforts to provide us the data for Run #4, done after the work published in Ref 7, above, and for his efforts in obtaining the necessary approvals to release it.

ACRONYMS

AEC	Atomic Energy Commission
AECL	Atomic Energy of Canada Limited
ANL	Argonne National Laboratory
ANL-E	Argonne National Laboratory-East
ANL-W	Argonne National Laboratory-West
ANS	American Nuclear Society
APDA	Atomic Power Development Associates
ASPO	Association for the Study of Peak Oil and Gas
ATWS	Anticipated transient without scram
BR	Breeding ratio
BWR	Boiling water reactor
CANDU	Canada Deuterium-Uranium Reactor
CBI	Chicago Bridge & Iron Company
CEFR	China Experimental Fast Reactor
CEO	Chief executive officer
CDFR	Commercial Demonstration Fast Reactor
CFR	Code of federal regulation
CGE	Canadian General Electric
CIAE	China Institute of Atomic Energy
CNNC	China National Nuclear Corporation
CP	Chicago Pile
CPFR	China Prototype Fast Reactor
CRBR	Clinch River Breeder Reactor
CRIEPI	Central Research Institute of Electric Power Industry
DFR	Dounreay Fast Reactor
DOE	Department of Energy
EBWR	Experimental Boiling Water reactor
EEDB	Energy economics data base
EFL	Experimental Fuel Laboratory
EPA	Environmental Protection Agency
FARET	Fast Reactor Test Facility
FBTR	Fat Breeder Test Reactor
FCF	Fuel Cycle Facility or Fuel Conditioning Facility
FFTF	High Flux Test Facility
FMF	Fuel Manufacturing Facility
GE	General Electric
GEM	Gas expansion module
GNEC	General Nuclear Engineering Company
GW	Giga watt

HCDA	Hypothetical core disruptive accident
HFEF	Hot Fuel Examination Facility
HM	Heavy metal
HWR	Heavy water reactor
HTR	High Temperature Gas-cooled Reactor
IAEA	International Atomic Energy Agency
ICRP	International Commission on Radiological Protection
IEA	International Energy Agency
IFR	Integral Fast Reactor
IHX	Intermediate heat exchanger
INFCE	International Nuclear Fuel Cycle Evaluation
INL	Idaho National Laboratory
JAEA	Japan Atomic Energy Agency
JAPC	Japan Atomic Power Company
JSFR	JAEA Sodium-cooled Fast Reactor
KAERI	Korea Atomic Energy Research Institute
LAMPRE	Los Alamos Molten Plutonium Reactor Experiment
LANL	Los Alamos National Laboratory
LMFBR	Liquid metal fast breeder reactor
LNG	Liquefied natural gas
LSPB	Large Scale Prototype Breeder
LWR	Light water reactor
MIT	Massachusetts Institute of Technology
MOX	Mixed oxide
MW	Megawatt
MWD/T	Megawatt-days per tone
NAC	Nuclear Assurance Corporation
NAS	National Academy of Sciences
NASAP	Nonproliferation Alternative Systems Assessment program
NPD	Nuclear Power Demonstration
NRC	Nuclear Regulatory Commission
NRTS	National Reactor Testing Station
NWPA	National Waste Policy Act
OCS	Operator control station
PFBR	Prototype Fast Breeder Reactor
PFR	Prototype Fast Reactor
PNC	Power Reactor and Nuclear Fuel Cycle Development Corporation
PRISM	Power Reactor Innovative Small Module
PUREX	Plutonium Uranium Extraction
QA	Quality assurance
QC	Quality control
R&D	Research and development
RBCB	Run beyond cladding breach
RDT	Reactor Development and Technology, a Division of AEC

RSRC	Reactor Safety review Committee
SAF	Semi-Automated Fabrication
SAFR	Sodium Advanced Fast Reactor
SEFOR	Southwest Experimental Fast Oxide Reactor
SFR	Sodium-cooled fast reactor
SRE	Sodium Reactor Experiment
SS	Stainless steel
STR	Submarine Thermal Reactor
THORP	Thermal Oxide Reprocessing Plant
TMI	Three Mile Island
TOR	*Traitement des Oxydes Rapides*
TREAT	Transient Reactor Test Facility
UOX	Uranium oxide
ZPPR	Zero Power Physics Reactor

ABOUT THE AUTHORS

Dr. CHARLES E. TILL received his Ph.D. in nuclear engineering from the Imperial College, University of London, in 1960. Early in his career he worked on a variety of reactor concepts, including the U.K. gas-cooled reactor, the Canadian heavy water reactor and the U.S. light water reactor upon joining Argonne National Laboratory in 1963. There, after a year or two, Dr. Till became been deeply involved in the development of the fast breeder reactor. From 1980 onward, as Associate Laboratory Director for Engineering Research, Till led the large Argonne reactor development program for seventeen of its most innovative and productive years. He created the Integral Fast Reactor concept and spearheaded the development of its underlying technologies. An advanced reactor technology with revolutionary improvements in safety, nuclear waste disposal, and resource usage, this was a major effort involving a thousand to two thousand engineers and supporting staff and carried out over the ten year period from 1984 to 1994 at Argonne's two sites, its main laboratory in Illinois, and its big reactor facilities on the desert in Idaho. A Fellow of American Nuclear Society and recipient of its Walker Cisler Medal for distinguished contributions to fast reactor development, he was elected to the National Academy of Engineering in 1989.

Dr. YOON IL CHANG received his Ph.D. in nuclear science from the University of Michigan in 1971. After a short time at Nuclear Assurance Corporation working on nuclear fuel cycle services, he joined Argonne National Laboratory in 1974, hired initially by Till as a reactor analyst. With the initiation of the Integral Fast Reactor program in 1984, as Till's deputy and as the program's General Manager, he managed the program through its ten years of development. Bringing all the many parts on IFR program together in a coherent and focused program, it was Chang who saw to its progress day by day, month by month. So directed and nurtured, the program saw success after success in the many areas necessary to complete the entire Integral Fast Reactor system—the physics and reactor core design, the electrochemical processes, the fuel design and fabrication, the reactor safety, reactor engineering and reactor operations, are principal examples. Upon Till's retirement in 1998, Dr. Chang succeeded him as Associate Laboratory Director for Engineering Research, and also served as Interim Laboratory Director. The recipient of outstanding alumni awards from the University of Michigan and Seoul National University, a Fellow of American Nuclear Society and recipient of its Walker Cisler Medal, he received the Department of Energy's Ernest Orlando Lawrence Award in 1994 for his technical leadership role in the IFR development.

Printed in Great Britain
by Amazon.co.uk, Ltd.,
Marston Gate.